Table of Contents

Bioinformatics Data Skills

Vince Buffalo

Beijing · Boston · Farnham · Sebastopol · Tokyo

Bioinformatics Data Skills

by Vince Buffalo

Printed in the United States of America.

Published by O'Reilly Media, Inc., 1005 Gravenstein Highway North, Sebastopol, CA 95472.

O'Reilly books may be purchased for educational, business, or sales promotional use. Online editions are also available for most titles (*http://safaribooksonline.com*). For more information, contact our corporate/institutional sales department: 800-998-9938 or corporate@oreilly.com.

Editors: Courtney Nash and Amy Jollymore	**Indexer:** Ellen Troutman
Production Editor: Nicole Shelby	**Interior Designer:** David Futato
Copyeditor: Jasmine Kwityn	**Cover Designer:** Ellie Volckhausen
Proofreader: Kim Cofer	**Illustrator:** Rebecca Demarest

June 2015:   First Edition

Revision History for the First Edition
2015-06-30:   First Release

See *http://oreilly.com/catalog/errata.csp?isbn=9781449367374* for release details.

978-1-449-36737-4

[LSI]

To my (rather large) family for their continued support: Mom, Dad, Anne, Lisa, Lauren, Violet, and Dalilah; the Buffalos, the Kihns, and the Lambs.

And my earliest mentors for inspiring me to be who I am today: Randy Siverson and Duncan Temple Lang.

Part III.   Practice: Bioinformatics Data Skills

Preface

This book is the answer to a question I asked myself two years ago: "What book would I want to read *first* when getting started in bioinformatics?" When I began working in this field, I had programming experience in Python and R but little else. I had hunted around for a terrific introductory text on bioinformatics, and while I found some good books, most were not targeted to the daily work I did as a bioinformatician. A few of the texts I found approached bioinformatics from a theoretical and algorithmic perspective, covering topics like Smith-Waterman alignment, phylogeny reconstruction, motif finding, and the like. Although they were fascinating to read (and I do recommend that you explore this material), I had no need to implement bioinformatics algorithms from scratch in my daily bioinformatics work—numerous terrific, highly optimized, well-tested implementations of these algorithms already existed. Other bioinformatics texts took a more practical approach, guiding readers unfamiliar with computing through each step of tasks like running an aligner or downloading sequences from a database. While these were more applicable to my work, much of those books' material was outdated.

As you might guess, I couldn't find that best "first" bioinformatics book. *Bioinformatics Data Skills* is my version of the book I was seeking. This book is targeted toward readers who are unsure how to bridge the giant gap between knowing a scripting language and practicing bioinformatics to answer scientific questions in a robust and reproducible way. To bridge this gap, one must learn data skills—an approach that uses a core set of tools to manipulate and explore any data you'll encounter during a bioinformatics project.

Data skills are the best way to learn bioinformatics because these skills utilize time-tested, open source tools that continue to be the best way to manipulate and explore changing data. This approach has stood the test of time: the advent of high-throughput sequencing rapidly changed the field of bioinformatics, yet skilled bioinformaticians adapted to this new data using these same tools and skills. Next-generation data was, after all, just data (different data, and *more* of it), and master bioinformaticians had the essential skills to solve problems by applying their tools to this new data. *Bioinformatics Data Skills* is written to provide you with training in these core tools and help you develop these same skills.

The Approach of This Book

Many biologists starting out in bioinformatics tend to equate "learning bioinformatics" with "learning how to run bioinformatics software." This is an unfortunate and misinformed idea of what bioinformaticians actually do. This is analogous to thinking "learning molecular biology" is just "learning pipetting." Other than a few simple examples used to generate data in Chapter 11, this book doesn't cover running bioinformatics software like aligners, assemblers, or variant callers. Running bioinformatics software isn't all that difficult, doesn't take much skill, and it doesn't embody any of the significant challenges of bioinformatics. I don't teach how to run these types of bioinformatics applications in *Bioinformatics Data Skills* for the following reasons:

- It's easy enough to figure out on your own
- The material would go rapidly out of date as new versions of software or entirely new programs are used in bioinformatics
- The original manuals for this software will always be the best, most up-to-date resource on how to run a program

Instead, the approach of this book is to focus on the skills bioinformaticians use to explore and extract meaning from complex, large bioinformatics datasets. Exploring and extracting information from these datasets is the fun part of bioinformatics research. The goal of *Bioinformatics Data Skills* is to teach you the computational tools and data skills you need to explore these large datasets as you please. These data skills give you freedom; you'll be able to look at any bioinformatics data—in any format, and files of any size—and begin exploring data to extract biological meaning.

Throughout *Bioinformatics Data Skills*, I emphasize working in a robust and reproducible manner. I believe these two qualities—reproducibility and robustness—are too often overlooked in modern computational work. By *robust*, I mean that your work is resilient against silent errors, confounders, software bugs, and messy or noisy data. In contrast, a fragile approach is one that does not decrease the odds of some type of error adversely affecting your results. By *reproducible*, I mean that your work can be repeated by other researchers and they can arrive at the same results. For this to be the case, your work must be well documented, and your methods, code, and data all need to be available so that other researchers have the materials to reproduce everything. Reproducibility also relies on your work being robust—if a workflow run on a different machine yields a different outcome, it is neither robust nor fully reproducible. I introduce these concepts in more depth in Chapter 2, and these are themes that reappear throughout the book.

Why This Book Focuses on Sequencing Data

Bioinformatics is a broad discipline, and spans subfields like proteomics, metabolomics, structure bioinformatics, comparative genomics, machine learning, and image processing. *Bioinformatics Data Skills* focuses primarily on handling sequencing data for a few reasons.

First, sequencing data is abundant. Currently, no other "omics" data is as abundant as high-throughput sequencing data. Sequencing data has broad applications across biology: variant detection and genotyping, transcriptome sequencing for gene expression studies, protein-DNA interaction assays like ChIP-seq, and bisulfite sequencing for methylation studies just to name a few examples. The ways in which sequencing data can be used to answer biological questions will only continue to increase.

Second, sequencing data is terrific for honing your data skills. Even if your goal is to analyze other types of data in the future, sequencing data serves as great example data to learn with. Developing the text-processing skills necessary to work with sequencing data will be applicable to working with many other data types.

Third, other subfields of bioinformatics are much more domain specific. The wide availability and declining costs of sequencing have allowed scientists from all disciplines to use genomics data to answer questions in their systems. In contrast, bioinformatics subdisciplines like proteomics or high-throughput image processing are much more specialized and less widespread. Still, if you're interested in these fields, *Bioinformatics Data Skills* will teach you useful computational and data skills that will be helpful in your research.

Audience

In my experience teaching bioinformatics to friends, colleagues, and students of an intensive week-long course taught at UC Davis, most people wishing to learn bioinformatics are either biologists, or computer scientists/programmers. Biologists wish to develop the computational skills necessary to analyze their own data, while the programmers and computer scientists wish to apply their computational skills to biological problems. Although these two groups differ considerably in biological knowledge and computational experience, *Bioinformatics Data Skills* covers material that should be helpful to both.

If you're a biologist, *Bioinformatics Data Skills* will teach you the core data skills you need to work with bioinformatics data. It's important to note that *Bioinformatics Data Skills* is *not* a how-to bioinformatics book; such a book on bioinformatics would quickly go out of date or be too narrow in focus to help the majority of biologists. You will need to supplement this book with knowledge of your specific research and system, as well as the modern statistical and bioinformatics methods that your subfield uses. For example, if your project involves aligning sequencing reads to a reference genome, this book won't tell you the newest and best alignment software for your particular system. But regardless of which aligner you use, you will need to have a thorough understanding of alignment formats and how to manipulate alignment data —a topic covered in Chapter 11. Throughout this book, these general computational and data skills are meant to be a solid, widely applicable foundation on which the majority of biologists can build.

If you're a computer scientist or programmer, you are likely already familiar with some of the computational tools I teach in this book. While the material presented in *Bioinformatics Data Skills* may overlap knowledge you already have, you will still learn about the specific formats, tools, and approaches bioinformaticians use in their work. Also, working through the examples in this book will give you good practice in applying your computational skills to genomics data.

The Difficulty Level of *Bioinformatics Data Skills*

Bioinformatics Data Skills is designed to be a thorough—and in parts, dense—book. When I started writing this book, I decided the greatest misdeed I could do would be to treat bioinformatics as a subject that's easier than it truly is. Working as a professional bioinformatician, I routinely saw how very subtle issues could crop up and adversely change the outcome of the analysis had they not been caught. I don't want your bioinformatics work to be incorrect because I've made a topic artificially simple. The depth at which I cover topics in *Bioinformatics Data Skills* is meant to prepare you to catch similar issues in your own work so your results are robust.

The result is that sections of this book are quite advanced and will be difficult for some readers. Don't feel discouraged! Like most of science, this material *is hard*, and may take a few reads before it fully sinks in. Throughout the book, I try to indicate when certain sections are especially advanced so that you can skip over these and return to them later.

Lastly, I often use technical jargon throughout the book. I don't like using jargon, but it's necessary to communicate technical concepts in computing. Primarily it will help you search for additional resources and help. It's much easier to Google successfully for "left outer join" than "data merge where null records are included in one table."

Assumptions This Book Makes

Bioinformatics Data Skills is meant to be an intermediate book on bioinformatics. To make sure everyone starts out on the same foot, the book begins with a few simple chapters. In Chapter 2, I cover the basics of setting up a bioinformatics project, and in Chapter 3 I teach some remedial Unix topics meant to ensure that you have a solid grasp of Unix (because Unix is a large component in later chapters). Still, as an intermediate book, I make a few assumptions about you:

You know a scripting language

This is the biggest assumption of the book. Except for a few Python programs and the R material (R is introduced in Chapter 8), this book doesn't directly rely on using lots of scripting. However, in learning a scripting language, you've already encountered many important computing concepts such as working with a text editor, running and executing programs on the command line, and basic programming. If you do not know a scripting language, I would recommend learning Python while reading this book. Books like *Bioinformatics Programming Using Python* by Mitchell L. Model (O'Reilly, 2009), *Learning Python, 5th Edition*, by Mark Lutz (O'Reilly, 2013), and *Python in a Nutshell, 2nd*, by Alex Martelli (O'Reilly, 2006) are great to get started. If you know a scripting language other than Python (e.g., Perl or Ruby), you'll be prepared to follow along with most examples (though you will need to translate some examples to your scripting language of choice).

You know how to use a text editor

It's essential that you know your way around a text editor (e.g., Emacs, Vim, Text-Mate2, or Sublime Text). Using a word processor (e.g., Microsoft Word) will not work, and I would discourage using text editors such as Notepad or OS X's TextEdit, as they lack syntax highlighting support for common programming languages.

You have basic Unix command-line skills

For example, I assume you know the difference between a terminal and a shell, understand how to enter commands, what command-line options/flags and arguments are, and how to use the up arrow to retrieve your last entered command. You should also have a basic understanding of the Unix file hierarchy (including concepts like your home directory, relative versus absolute directories, and root directories). You should also be able to move about and manipulate the directories and files in Unix with commands like cd, ls, pwd, mv, rm, rmdir, and mkdir. Finally, you should have a basic grasp of Unix file ownership and permissions, and changing these with chown and chmod. If these concepts are unclear, I would recommend you play around in the Unix command line first (carefully!) and consult a good beginner-level book such as *Practical Computing for Biologists* by Steven Haddock and Casey Dunn (Sinauer, 2010) or *UNIX and Perl to the Rescue* by Keith Bradnam and Ian Korf (Cambridge University Press, 2012).

You have a basic understanding of biology

Bioinformatics Data Skills is a BYOB book—*bring your own biology*. The examples don't require a lot of background in biology beyond what DNA, RNA, proteins, and genes are, and the central dogma of molecular biology. You should also be familiar with some very basic genetics and genomic concepts (e.g., single nucleotide polymorphisms, genotypes, GC content, etc.). All biological examples in the book are designed to be quite simple; if you're unfamiliar with any topic, you should be able to quickly skim a Wikipedia article and proceed with the example.

You have a basic understanding of regular expressions

Occasionally, I'll make use of regular expressions in this book. In most cases, I try to quickly step through the basics of how a regular expression works so that you can get the general idea. If you've encountered regular expressions while learning a scripting language, you're ready to go. If not, I recommend you learn the basics —not because regular expressions are used heavily throughout the book, but because mastering regular expressions is an important skill in bioinformatics. *Introducing Regular Expressions* by Michael Fitzgerald (O'Reilly) is a great introduction. Nowadays, writing, testing, and debugging regular expressions is easier than ever thanks to online tools like *http://regex101.com* and *http://www.debuggex.com*. I recommend using these tools in your own work and when stepping through my regular expression examples.

You know how to get help and read documentation

Throughout this book, I try to minimize teaching information that can be found in manual pages, help documentation, or online. This is for three reasons:

- I want to save space and focus on presenting material in a way you can't find elsewhere

- Manual pages and documentation will always be the best resource for this information

- The ability to quickly find answers in documentation is one of the most important skills you can develop when learning computing

This last point is especially important; you don't need to remember all arguments of a command or R function—you just need to know where to find this information. Programmers consult documentation *constantly* in their work, which is why documentation tools like man (in Unix) and help() (in R) exist.

You can manage your computer system (or have a system administrator)

This book does not teach you system administration skills like setting up a bioinformatics server or cluster, managing user accounts, network security, managing disks and disk space, RAID configurations, data backup, and high-performance computing concepts. There simply isn't the space to adequately cover these important topics. However, these are all very, very important—if you don't have a system administrator and need to fill that role for your lab or research group, it's essential for you to master these skills, too. Frankly, system administration skills take years to master and good sysadmins have incredible patience and experience in handling issues that would make most scientists go insane. If you can employ a full-time system administrator shared across labs or groups or utilize a university cluster with a sysadmin, I would do this. Lastly, this shouldn't need to be said, but just in case: constantly back up your data and work. It's easy when learning Unix to execute a command that destroys files—your best protection from losing everything is continual backups.

Supplementary Material on GitHub

The supplementary material needed for this book's examples is available in the GitHub repository (*http://github.com/vsbuffalo/bds-files*). You can download material from this repository as you need it (the repository is organized by chapter), or you can download everything using the Download Zip link. Once you learn Git in Chapter 5, I would recommend cloning the repository so that you can restore any example files should you accidentally overwrite them.

Try navigating to this repository now and poking around so you're familiar with the layout. Look in the Preface's directory and you'll find the *README.md* file (*http:// bit.ly/bds-readme*), which includes additional information about many of the topics I've discussed. In addition to the supplementary files needed for all examples in the book, this repository contains:

- Documentation on how all supplementary files were produced or how they were acquired. In some cases, I've used makefiles or scripts (both of these topics are covered in Chapter 12) to create example data, and all of these resources are available in each chapter's GitHub directory. I've included these materials not only for reproducible purposes, but also to serve as additional learning material.

- Additional information readers may find interesting for each chapter. This information is in each chapter's *README.md* file. I've also included other resources like lists of recommended books for further learning.

- Errata, and any necessary updates if material becomes outdated for some reason.

I chose to host the supplementary files for *Bioinformatics Data Skills* on GitHub so that I could keep everything up to date and address any issues readers may have. Feel free to create a new issue on GitHub (*http://github.com/vsbuffalo/bds-files/issues*) should you find any problem with the book or its supplementary material.

Computing Resources and Setup

I've written this entire book on my laptop, a 15-inch MacBook Pro with 16 GB of RAM. Although this is a powerful laptop, it is much smaller than the servers common in bioinformatics computing. All examples are designed and tested to run a machine this size. Nearly every example should run on a machine with 8 GB of memory.

All examples in this book work on Mac OS X and Linux—other operating systems are not supported (mostly because modern bioinformatics relies on Unix-based operating systems). All software required throughout the book is freely available and is easily installable; I provide some basic instructions in each section as software installation is needed. In general, you should use your operating system's package management system (e.g., `apt-get` on Ubuntu/Debian). If you're using a Mac, I highly recommend Homebrew, a terrific package manager for OS X that allows you to easily install software from the command line. You can find detailed instructions on Homebrew's website (*http://brew.sh*), Most important, Homebrew maintains a collection of scientific software packages called homebrew-science (*http://github.com/Homebrew/homebrew-science*), including the bioinformatics software we use throughout this book. Follow the directions in homebrew-science's *README.md* to learn how to install these scientific programs.

Organization of This Book

This book is composed of three parts: Part I, containing one chapter on ideology; Part II, which covers the basics of getting started with a bioinformatics project; and Part III, which covers bioinformatics data skills. Although chapters were written to be read sequentially, if you're comfortable with Unix and R, you may find that you can skip around without problems.

In Chapter 1, I introduce why learning bioinformatics by developing data skills is the best approach. I also introduce the ideology of this book, and describe reproducible and robust bioinformatics and some recommendations to apply in your own work.

Part II of *Bioinformatics Data Skills* introduces the basic skills needed to start a bioinformatics project. First, we'll look at how to set up and manage a project directory in Chapter 2. This may seem like trivial topic, but complex bioinformatics projects demand we think about project management. In the frenzy of research, there will be files *everywhere*. Starting out with a carefully organized project can prevent a lot of hassle in the future. We'll also learn about documentation with Markdown, a useful format for plain-text project documentation.

In Chapter 3, we explore intermediate Unix in bioinformatics. This is to make sure that you have a solid grasp of essential concepts (e.g., pipes, redirection, standard input and output, etc.). Understanding these prerequisite topics will allow you to focus on analyzing data in later chapters, not struggling to understand Unix basics.

Most bioinformatics tasks require more computing power than we have on our personal workstations, meaning we have to work with remote servers and clusters. Chapter 4 covers some tips and tricks to increase your productivity when working with remote machines.

In Chapter 5, we learn Git, which is a version control system that makes managing versions of projects easy. Bioinformatics projects are filled with lots of code and data that should be managed using the same modern tools as collaboratively developed software. Git is a large, powerful piece of software, so this is a long chapter. However, this chapter was written so that you could skip the section on branching and return to it later.

Chapter 6 looks at data in bioinformatics projects: how to download large amounts of data, use data compression, validate data integrity, and reproducibly download data for a project.

In Part III, our attention turns to developing the essential data skills all bioinformaticians need to tackle problems in their daily work. Chapter 7 focuses on Unix data tools, which allow you to quickly write powerful stream-processing Unix pipelines to process bioinformatics data. This approach is a cornerstone of modern bioinformatics, and is an absolutely essential data skill to have.

In Chapter 8, I introduce the R language through learning exploratory data analysis techniques. This chapter prepares you to use R to explore your own data using techniques like visualization and data summaries.

Genomic range data is ubiquitous in bioinformatics, so we look at range data and range operations in Chapter 9. We'll first step through the different ways to represent genomic ranges, and work through range operations using Bioconductor's IRanges package to bolster our range-thinking intuition. Then, we'll work with genomic data using GenomicRanges. Finally, we'll look at the BEDTools Suite of tools for working with range data on the command line.

In Chapter 10, we learn about sequence data, a mainstay of bioinformatics data. We'll look at the FASTA and FASTQ formats (and their limitations) and work through trimming low-quality bases off of sequences and seeing how this affects the distribution of quality scores. We'll also look at FASTA and FASTQ parsing.

Chapter 11 focuses on the alignment data formats SAM and BAM. Understanding and manipulating files in these formats is an integral bioinformatics skill in working with high-throughput sequencing data. We'll see how to use Samtools to manipulate these files and visualize the data, and step through a detailed example that illustrates some of the intricacies of variant calling. Finally, we'll learn how to use Pysam to parse SAM/BAM files so you can write your own scripts that work with these specialized data formats.

Most daily bioinformatics work involves writing data-processing scripts and pipelines. In Chapter 12, we look at how to write such data-processing pipelines in a robust and reproducible way. We'll look specifically at Bash scripting, manipulating files using Unix powertools like find and xargs, and finally take a quick look at how you can write pipelines using Make and makefiles.

In bioinformatics, our data is often too large to fit in our computer's memory. In Chapter 7, we saw how streaming with Unix pipes can help to solve this problem, but Chapter 13 looks at a different method: out-of-memory approaches. First, we'll look at Tabix, a fast way to access information in indexed tab-delimited files. Then, we'll look at the basics of SQL through analyzing some GWAS data using SQLite.

Finally, in Chapter 14, I discuss where you should head next to further develop your bioinformatics skills.

Code Conventions

Most bioinformatics data has one thing in common: it's large. In code examples, I often need to truncate the output to have it fit into the width of a page. To indicate that output has been truncated, I will always use [...] in the output. Also, in code examples I often use variable names that are short to save space. I encourage you to use more descriptive names than those I've used throughout this book in your own personal work.

Conventions Used in This Book

The following typographical conventions are used in this book:

Italic
> Indicates new terms, URLs, email addresses, filenames, and file extensions.

`Constant width`
> Used for program listings, as well as within paragraphs to refer to program elements such as variable or function names, databases, data types, environment variables, statements, and keywords.

`Constant width bold`
> Shows commands or other text that should be typed literally by the user.

`Constant width italic`
> Shows text that should be replaced with user-supplied values or by values determined by context.

 This element signifies a tip or suggestion.

This element signifies a general note.

 This element indicates a warning or caution.

Using Code Examples

This book is here to help you get your job done. In general, if example code is offered with this book, you may use it in your programs and documentation. You do not need to contact us for permission unless you're reproducing a significant portion of the code. For example, writing a program that uses several chunks of code from this book does not require permission. Selling or distributing a CD-ROM of examples from O'Reilly books does require permission. Answering a question by citing this book and quoting example code does not require permission. Incorporating a significant amount of example code from this book into your product's documentation does require permission.

We appreciate, but do not require, attribution. An attribution usually includes the title, author, publisher, and ISBN. For example: "*Bioinformatics Data Skills* by Vince Buffalo (O'Reilly). Copyright 2015 Vince Buffalo, 978-1-449-36737-4."

If you feel your use of code examples falls outside fair use or the permission given above, feel free to contact us at *permissions@oreilly.com*.

Safari® Books Online

 *Safari Books Online* is an on-demand digital library that delivers expert content in both book and video form from the world's leading authors in technology and business.

Technology professionals, software developers, web designers, and business and creative professionals use Safari Books Online as their primary resource for research, problem solving, learning, and certification training.

Safari Books Online offers a range of plans and pricing for enterprise, government, education, and individuals.

Members have access to thousands of books, training videos, and prepublication manuscripts in one fully searchable database from publishers like O'Reilly Media, Prentice Hall Professional, Addison-Wesley Professional, Microsoft Press, Sams, Que, Peachpit Press, Focal Press, Cisco Press, John Wiley & Sons, Syngress, Morgan Kaufmann, IBM Redbooks, Packt, Adobe Press, FT Press, Apress, Manning, New Riders, McGraw-Hill, Jones & Bartlett, Course Technology, and hundreds more. For more information about Safari Books Online, please visit us online.

How to Contact Us

Please address comments and questions concerning this book to the publisher:

O'Reilly Media, Inc.
1005 Gravenstein Highway North
Sebastopol, CA 95472
800-998-9938 (in the United States or Canada)
707-829-0515 (international or local)
707-829-0104 (fax)

We have a web page for this book, where we list errata, examples, and any additional information. You can access this page at *http://bit.ly/Bio-DS*.

To comment or ask technical questions about this book, send email to *bookquestions@oreilly.com*.

For more information about our books, courses, conferences, and news, see our website at *http://www.oreilly.com*.

Find us on Facebook: *http://facebook.com/oreilly*

Follow us on Twitter: *http://twitter.com/oreillymedia*

Watch us on YouTube: *http://www.youtube.com/oreillymedia*

Acknowledgments

Writing a book is a monumental effort—for two years, I've worked on *Bioinformatics Data Skills* during nights and weekends. This is in addition to a demanding career as a professional bioinformatician (and for the last five months of writing, as a PhD student). Balancing work and life is already difficult enough for most scientists; I now know that balancing work, life, and writing a book is nearly impossible. I wouldn't have survived this process without the support of my partner, Helene Hopfer.

I thank Ciera Martinez for continually providing useful feedback and helping me calibrate the tone and target audience of this book. Cody Markelz tirelessly provided feedback and was never afraid to say when I'd missed the mark on a chapter—for this, all readers should be thankful. My friend Al Marks deserves special recognition not only for proving valuable feedback on many chapters, but also for introducing me to computing and programming back in high school. I also thank Jeff Ross-Ibarra for inspiring my passion for population genetics and presenting me with challenging and interesting projects in his lab. I owe a debt of gratitude to the entire UC Davis Bioinformatics Core for the fantastic time I spent working there; thanks especially to Dawei Lin, Joe Fass, Nikhil Joshi, and Monica Britton for sharing their knowledge and granting me freedom to explore bioinformatics. Mike Lewis also deserves a special thanks for teaching me about computing and being a terrific person to nerd out on techie details with. Peter Morrell, his lab, and the "Does[0]compute?" reading group provided lots of useful feedback that I'm quite grateful for. I thank Jorge Dubcovsky—witnessing his tireless pursuit of science has motivated me to do the same. Lastly, I'm indebted to my wonderful advisor, Graham Coop, for his patience in allowing me to finish this book—with this book out of the way, I'm eager to pursue my future directions under his mentorship.

This book was significantly improved by the valuable input of many reviewers, colleagues, and friends. Thank you Peter Cock, Titus Brown, Keith Bradnam, Mike Covington, Richard Smith-Unna, Stephen Turner, Karthik Ram, Gabe Becker, Noam Ross, Chris Hamm, Stephen Pearce, Anke Schennink, Patrik D'haeseleer, Bill Broadley, Kate Crosby, Arun Durvasula, Aaron Quinlan, and David Ruddock. Shaun Jackman deserves recognition for his tireless effort in making bioinformatics software easy to install through the Homebrew and apt-get projects—my readers will greatly benefit from this. I also am grateful for the comments and positive feedback I received from many of the early release readers of this book; the positive reception provided a great motivating push to finish everything. However, as author, I do take full credit for any errors or omissions that have slipped by these devoted reviewers.

Most authors are lucky if they work with one great editor—I got to work with two. Thank you, Courtney Nash and Amy Jollymore, for your continued effort and encouragement throughout this process. Simply put, I wouldn't have been able to do this without you both. I'd also like to thank my production editor Nicole Shelby, copyeditor Jasmine Kwityn, and the rest of the O'Reilly production team for their extremely hard work in editing *Bioinformatics Data Skills*. Finally, thank you, Mike Loukides, for your feedback and for taking an interest in my book when it was just a collection of early, rough ideas—you saw more.

## PART I

Ideology: Data Skills for Robust and Reproducible Bioinformatics

## CHAPTER 1

How to Learn Bioinformatics

Right now, in labs across the world, machines are sequencing the genomes of the life on earth. Even with rapidly decreasing costs and huge technological advancements in genome sequencing, we're only seeing a glimpse of the biological information contained in every cell, tissue, organism, and ecosystem. However, the smidgen of total biological information we're gathering amounts to mountains of data biologists need to work with. At no other point in human history has our ability to understand life's complexities been so dependent on our skills to work with and analyze data.

This book is about learning bioinformatics through developing data skills. In this chapter, we'll see what data skills are, and why learning data skills is the best way to learn bioinformatics. We'll also look at what robust and reproducible research entails.

Why Bioinformatics? Biology's Growing Data

Bioinformaticians are concerned with deriving biological understanding from large amounts of data with specialized skills and tools. Early in biology's history, the datasets were small and manageable. Most biologists could analyze their own data after taking a statistics course, using Microsoft Excel on a personal desktop computer. However, this is all rapidly changing. Large sequencing datasets are widespread, and will only become more common in the future. Analyzing this data takes different tools, new skills, and many computers with large amounts of memory, processing power, and disk space.

In a relatively short period of time, sequencing costs dropped drastically, allowing researchers to utilize sequencing data to help answer important biological questions. Early sequencing was low-throughput and costly. Whole genome sequencing efforts were expensive (the human genome cost around $2.7 billion) and only possible through large collaborative efforts. Since the release of the human genome, sequencing costs have decreased exponentially until about 2008, as shown in Figure 1-1. With the introduction of next-generation sequencing technologies, the cost of sequencing a megabase of DNA dropped even more rapidly. At this crucial point, a technology that was only affordable to large collaborative sequencing efforts (or individual researchers with very deep pockets) became affordable to researchers across all of biology. You're likely reading this book to learn to work with sequencing data that would have been much too expensive to generate less than 10 years ago.

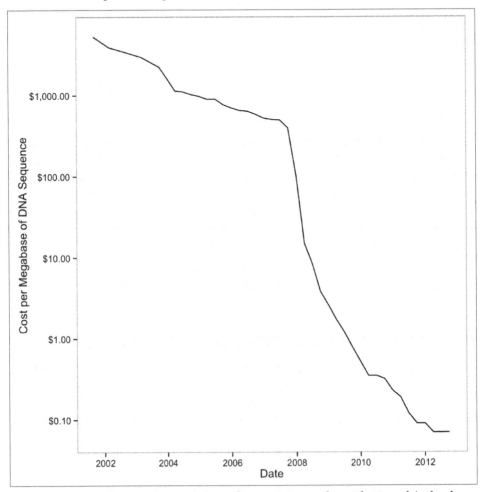

Figure 1-1. Drop of sequencing costs (note the y-axis is on a logarithmic scale); the sharp drop around 2008 was due to the introduction of next-generation sequencing data. (figure reproduced and data downloaded from the NIH (http://genome.gov/sequencing costs))

What was the consequence of this drop in sequencing costs due to these new technologies? As you may have guessed, lots and lots of data. Biological databases have swelled with data after exponential growth. Whereas once small databases shared between collaborators were sufficient, now petabytes of useful data are sitting on servers all over the world. Key insights into biological questions are stored not just in the unanalyzed experimental data sitting on your hard drive, but also spinning around a disk in a data center thousands of miles away.

The growth of biological databases is as astounding as the drop of sequencing costs. As an example, consider the Sequence Read Archive (*http://bit.ly/seq-read*) (previously known as the *Short* Read Archive), a repository of the raw sequencing data from sequencing experiments. Since 2010, it has experienced remarkable growth; see Figure 1-2.

To put this incredible growth of sequencing data into context, consider Moore's Law. Gordon Moore (a cofounder of Intel) observed that the number of transistors in computer chips doubles roughly every two years. More transistors per chip translates to faster speeds in computer processors and more random access memory in computers, which leads to more powerful computers. This extraordinary rate of technological improvement—output doubling every two years—is likely the fastest growth in technology humanity has ever seen. Yet, since 2011, the amount of sequencing data stored in the Short Read Archive has outpaced even this incredible growth, having doubled every year.

To make matters even more complicated, new tools for analyzing biological data are continually being created, and their underlying algorithms are advancing. A 2012 review listed over 70 short-read mappers (Fonseca et al., 2012; see http://bit.ly/hts-mappers). Likewise, our approach to genome assembly has changed considerably in the past five years, as methods to assemble long sequences (such as overlap-layout-consensus algorithms) were abandoned with the emergence of short high-throughput sequencing reads. Now, advances in sequencing chemistry are leading to longer sequencing read lengths and new algorithms are replacing others that were just a few years old.

Unfortunately, this abundance and rapid development of bioinformatics tools has serious downsides. Often, bioinformatics tools are not adequately benchmarked, or if they are, they are only benchmarked in one organism. This makes it difficult for new biologists to find and choose the best tool to analyze their data. To make matters more difficult, some bioinformatics programs are not actively developed so that they lose relevance or carry bugs that could negatively affect results. All of this makes choosing an appropriate bioinformatics program in your own research difficult. More importantly, it's imperative to critically assess the output of bioinformatics programs run on your own data. We'll see examples of how data skills can help us assess program output throughout Part II.

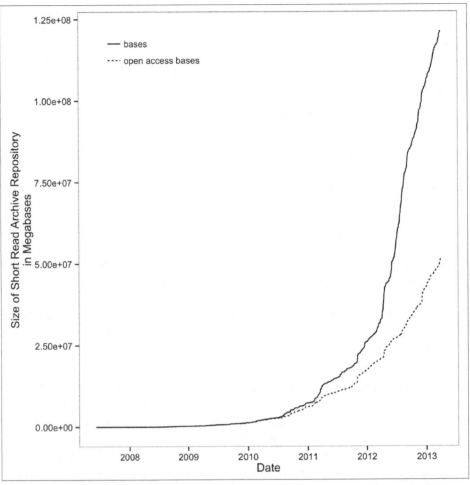

Figure 1-2. Exponential growth of the Short Read Archive; open access bases are SRA submissions available to the public (figure reproduced and data downloaded from the NIH (http://bit.ly/seq-read))

Learning Data Skills to Learn Bioinformatics

With the nature of biological data changing so rapidly, how are you supposed to learn bioinformatics? With all of the tools out there and more continually being created, how is a biologist supposed to know whether a program will work appropriately on her organism's data?

The solution is to approach bioinformatics as a bioinformatician does: try stuff, and assess the results. In this way, bioinformatics is just about having the skills to experiment with data using a computer and understanding your results. The experimental part is easy; this comes naturally to most scientists. The limiting factor for most biologists is having the data skills to freely experiment and work with large data on a computer. The goal of this book is to teach you the bioinformatics data skills necessary to allow you to experiment with data on a computer as easily as you would run experiments in the lab.

Unfortunately, many of the biologist's common computational tools can't scale to the size and complexity of modern biological data. Complex data formats, interfacing numerous programs, and assessing software and data make large bioinformatics datasets difficult to work with. Learning core bioinformatics data skills will give you the foundation to learn, apply, and assess any bioinformatics program or analysis method. In 10 years, bioinformaticians may only be using a few of the bioinformatics software programs around today. But we most certainly will be using data skills and experimentation to assess data and methods of the future.

So what are data skills? They are the set of computational skills that give you the ability to quickly improvise a way of looking at complex datasets, using a well-known set of tools. A good analogy is what jazz musicians refer to as having "chops." A jazz musician with good chops can walk into a nightclub, hear a familiar standard song being played, recognize the chord changes, and begin playing musical ideas over these chords. Likewise, a bioinformatician with good data skills can receive a huge sequencing dataset and immediately start using a set of tools to see what story the data tells.

Like a jazz musician that's mastered his instrument, a bioinformatician with excellent data chops masters a set of tools. Learning one's tools is a necessary, but not sufficient step in developing data skills (similarly, learning an instrument is a necessary, but not sufficient step to playing musical ideas). Throughout the book, we will develop our data skills, from setting up a bioinformatics project and data in Part II, to learning both small and big tools for data analysis in Part III. However, this book can only set you on the right path; real mastery requires learning through repeatedly applying skills to real problems.

New Challenges for Reproducible and Robust Research

Biology's increasing use of large sequencing datasets is changing more than the tools and skills we need: it's also changing how reproducible and robust our scientific findings are. As we utilize new tools and skills to analyze genomics data, it's necessary to ensure that our approaches are still as reproducible and robust as any other experimental approaches. Unfortunately, the size of our data and the complexity of our analysis workflows make these goal especially difficult in genomics.

The requisite of reproducibility is that we share our data and methods. In the pregenomics era, this was much easier. Papers could include detailed method summaries and entire datasets—exactly as Kreitman's 1986 paper did with a 4,713bp Adh gene flanking sequence (it was embedded in the middle of the paper). Now papers have long supplementary methods, code, and data. Sharing data is no longer trivial either, as sequencing projects can include terabytes of accompanying data. Reference genomes and annotation datasets used in analyses are constantly updated, which can make reproducibility tricky. Links to supplemental materials, methods, and data on journal websites break, materials on faculty websites disappear when faculty members move or update their sites, and software projects become stale when developers leave and don't update code. Throughout this book, we'll look at what can be done to improve reproducibility of your project alongside doing the actual analysis, as I believe these are necessarily complementary activities.

Additionally, the complexity of bioinformatics analyses can lead to findings being susceptible to errors and technical confounding. Even fairly routine genomics projects can use dozens of different programs, complicated input parameter combinations, and many sample and annotation datasets; in addition, work may be spread across servers and workstations. All of these computational data-processing steps create results used in higher-level analyses where we draw our biological conclusions. The end result is that research findings may rest on a rickety scaffold of numerous processing steps. To make matters worse, bioinformatics workflows and analyses are usually only run once to produce results for a publication, and then never run or tested again. These analyses may rely on very specific versions of all software used, which can make it difficult to reproduce on a different system. In learning bioinformatics data skills, it's necessary to concurrently learn reproducibility and robust best practices. Let's take a look at both reproducibility and robustness in turn.

Reproducible Research

Reproducing scientific findings is the only way to confirm they're accurate and not the artifact of a single experiment or analysis. Karl Popper, in *The Logic of Scientific Discovery*, famously said: "non-reproducible single occurrences are of no significance to science" (1959). Independent replication of experiments and analysis is the gold standard by which we assess the validity of scientific findings. Unfortunately, most sequencing experiments are too expensive to reproduce from the test tube up, so we increasingly rely on *in silico* reproducibility only. The complexity of bioinformatics projects usually discourages replication, so it's our job as good scientists to facilitate and encourage *in silico* reproducibility by making it easier. As we'll see later, adopting good reproducibility practices can also make your life easier as a researcher.

So what is a reproducible bioinformatics project? At the very least, it's sharing your project's code and data. Most journals and funding agencies require you to share your project's data, and resources like NCBI's Sequence Read Archive exist for this purpose. Now, editors and reviewers will often suggest (or in some cases require) that a project's code also be shared, especially if the code is a significant part of a study's results. However, there's a lot more we can and should do to ensure our projects' reproducibility. By having to reproduce bioinformatics analyses to verify results, I've learned from these sleuthing exercises that the devil is in the details.

For example, colleagues and I once had a difficult time reproducing an RNA-seq differential expression analysis we had done ourselves. We had preliminary results from an analysis on a subset of samples done a few weeks earlier, but to our surprised, our current analysis was producing a drastically smaller set of differentially expressed genes. After rechecking how our past results were created, comparing data versions and file creation times, and looking at differences in the analysis code, we were still stumped—nothing could explain the difference between the results. Finally, we checked the version of our R package and realized that it had been updated on our server. We then reinstalled the old version to confirm this was the source of the difference, and indeed it was. The lesson here is that often replication, by either you in the future or someone else, relies on not just data and code but details like software versions and when data was downloaded and what version it is. This *metadata*, or data about data, is a crucial detail in ensuring reproducibility.

Another motivating case study in bioinformatics reproducibility is the so-called "Duke Saga." Dr. Anil Potti and other researchers at Duke University created a method that used expression data from high-throughput microarrays to detect and predict response to different chemotherapy drugs. These methods were the beginning of a new level of personalized medicine, and were being used to determine the chemotherapy treatments for patients in clinical trials. However, two biostatisticians, Keith Baggerly and Kevin Coombes, found serious flaws in the analysis of this study when trying to reproduce it (Baggerly and Coombes, 2009). Many of these required what Baggerly and Coombes called "forensic bioinformatics"—sleuthing to try to reproduce a study's findings when there isn't sufficient documentation to retrace each step. In total, Baggerly and Coombes found multiple serious errors, including:

- An off-by-one error, as an entire list of gene expression values was shifted down in relation to their correct identifier
- Two outlier genes of biological interest were not on the microarrays used
- There was confounding of treatment with the day the microarray was run
- Sample group names were mixed up

Baggerly and Coombes's work is best summarized by their open access article, "Deriving Chemosensitivity from Cell Lines: Forensic Bioinformatics and Reproducible Research in High-Throughput Biology" (see this chapter's GitHub directory for this article and more information about the Duke Saga). The lesson of Baggerly and Coombes's work is that "common errors are simple, and simple errors are common" and poor documentation can lead to both errors and irreproducibility. Documentation of methods, data versions, and code would have not only facilitated reproducibility, but it likely would have prevented a few of these serious errors in their study. Striving for maximal reproducibility in your project often will make your work more robust, too.

Robust Research and the Golden Rule of Bioinformatics

Since the computer is a sharp enough tool to be really useful, you can cut yourself on it.

— *The Technical Skills of Statistics (1964)* John Tukey

In wetlab biology, when experiments fail, it can be very apparent, but this is not always true in computing. Electrophoresis gels that look like Rorschach blots rather than tidy bands clearly indicate something went wrong. In scientific computing, errors can be *silent*; that is, code and programs may produce output (rather than stop with an error), but this output may be incorrect. This is a very important notion to put in the back of your head as you learn bioinformatics.

Additionally, it's common in scientific computing for code to be run only once, as researchers get their desired output and move on to the next step. In contrast, consider a video game: it's run on thousands (if not millions) of different machines, and is, in effect, constantly being tested by many users. If a bug that deletes a user's score occurs, it's exceptionally likely to be quickly noticed and reported by users. Unfortunately, the same is not true for most bioinformatics projects.

Genomics data also creates its own challenges for robust research. First, most bioinformatics analyses produce intermediate output that is too large and high dimensional to inspect or easily visualize. Most analyses also involve multiple steps, so even if it were feasible to inspect an entire intermediate dataset for problems, it would be difficult to do this for each step (thus, we usually resort to inspecting samples of the data, or looking at data summary statistics). Second, in wetlab biology, it's usually easier to form prior expectations about what the outcome of an experiment might be. For example, a researcher may expect to see a certain mRNA expressed in some tissues in lower abundances than a particular housekeeping gene. With these prior expectations, an aberrant result can be attributed to a failed assay rather than biology. In contrast, the high dimensionality of most genomics results makes it nearly impossible to form strong prior expectations. Forming specific prior expectations on the expression of each of tens of thousands of genes assayed by an RNA-seq experiment is impractical. Unfortunately, without prior expectations, it can be quite difficult to distinguish good results from bad results.

Bioinformaticians also have to be wary that bioinformatics programs, even the large community-vetted tools like aligners and assemblers, may not work well on their particular organism. Organisms are all wonderfully, strangely different, and their genomes are too. Many bioinformatics tools are tested on a few model diploid organisms like human, and less well-tested on the complex genomes from the other parts of the tree of life. Do we really expect that out-of-the-box parameters from a short-read aligner tuned to human data will work on a polyploid genome four times its size? Probably not.

The easy way to ensure everything is working properly is to adopt a cautious attitude, and check everything between computational steps. Furthermore, you should approach biological data (either from an experiment or from a database) with a healthy skepticism that there might be something wrong with it. In the computing community, this is related to the concept of "garbage in, garbage out"—an analysis is only as good as the data going in. In teaching bioinformatics, I often share this idea as the Golden Rule of Bioinformatics:

> ## Never ever trust your tools (or data)

This isn't to make you paranoid that none of bioinformatics can be trusted, or that you must test every available program and parameter on your data. Rather, this is to train you to adopt the same cautious attitude software engineers and bioinformaticians have learned the hard way. Simply checking input data and intermediate results, running quick sanity checks, maintaining proper controls, and testing programs is a great start. This also saves you from encountering bugs later on, when fixing them means redoing large amounts of work. You naturally test whether lab techniques are working and give consistent results; adopting a robust approach to bioinformatics is merely doing the same in bioinformatics analyses.

Adopting Robust and Reproducible Practices Will Make Your Life Easier, Too

Working in sciences has taught many of us some facts of life the hard way. These are like Murphy's law: anything that can go wrong, will. Bioinformatics has its own set of laws like this. Having worked in the field and discussed war stories with other bioinformaticians, I can practically guarantee the following will happen:

- You will almost certainly have to rerun an analysis more than once, possibly with new or changed data. Frequently this happens because you'll find a bug, a collaborator will add or update a file, or you'll want to try something new upstream of a step. In all cases, downstream analyses depend on these earlier results, meaning all steps of an analysis need to be rerun.

- In the future, you (or your collaborators, or advisor) will almost certainly need to revisit part of a project and it will look completely cryptic. Your only defense is to document each step. Without writing down key facts (e.g., where you downloaded data from, when you downloaded it, and what steps you ran), you'll certainly forget them. Documenting your computational work is equivalent to keeping a detailed lab notebook—an absolutely crucial part of science.

Luckily, adopting practices that will make your project reproducible also helps solve these problems. In this sense, good practices in bioinformatics (and scientific computing in general) both make life easier and lead to reproducible projects. The reason for this is simple: if each step of your project is designed to be rerun (possibly with different data) and is well documented, it's already well on its way to being reproducible.

For example, if we automate tasks with a script and keep track of all input data and software versions, analyses can be rerun with a keystroke. Reproducing all steps in this script is much easier, as a well-written script naturally documents a workflow (we'll discuss this more in Chapter 12). This approach also saves you time: if you receive new or updated data, all you need to do is rerun the script with the new input file. This isn't hard to do in practice; scripts aren't difficult to write and computers excel at doing repetitive tasks enumerated in a script.

Recommendations for Robust Research

Robust research is largely about adopting a set of practices that stack the odds in your favor that a silent error won't confound your results. As mentioned above, most of these practices are also beneficial for reasons other than preventing the dreaded silent error—which is all the more reason to include apply the recommendations below in your daily bioinformatics work.

Pay Attention to Experimental Design

Robust research starts with a good experimental design. Unfortunately, no amount of brilliant analysis can save an experiment with a bad design. To quote a brilliant statistician and geneticist:

> To consult the statistician after an experiment is finished is often merely to ask him to conduct a post mortem examination. He can perhaps say what the experiment died of.
>
> —R.A. Fisher

This quote hits close to the heart; I've seen projects land on my desk ready for analysis, after thousands of sequencing dollars were spent, yet they're completely dead on arrival. Good experimental design doesn't have to be difficult, but as it's fundamentally a statistical topic it's outside of the scope of this book. I mention this topic because unfortunately nothing else in this book can save an experiment with a bad design. It's especially necessary to think about experimental design in high-throughput studies, as technical "batch effects" can significantly confound studies (for a perspective on this, see Leek et al., 2010).

Most introductory statistics courses and books cover basic topics in experimental design. Quinn and Keough's *Experimental Design and Data Analysis for Biologists* (Cambridge University Press, 2002) is an excellent book on this topic. Chapter 18 of O'Reilly's *Statistics in a Nutshell, 2nd Edition*, by Sarah Boslaugh covers the basics well, too. Note, though, that experimental design in a genomics experiment is a different beast, and is actively researched and improved. The best way to ensure your multithousand dollar experiment is going to reach its potential is to see what the current best design practices are for your particular project. It's also a good idea to consult your local friendly statistician about any experimental design questions or concerns you may have in planning an experiment.

Write Code for Humans, Write Data for Computers

> Debugging is twice as hard as writing the code in the first place. Therefore, if you write the code as cleverly as possible, you are, by definition, not smart enough to debug it.
>
> —Brian Kernighan

Bioinformatics projects can involve mountains of code, and one of our best defenses against bugs is to write code for humans, not for computers (a point made in the excellent article from Wilson et al., 2012). Humans are the ones doing the debugging, so writing simple, clear code makes debugging easier.

Code should be readable, broken down into small contained components (modular), and reusable (so you're not rewriting code to do the same tasks over and over again). These practices are crucial in the software world, and should be applied in your bioinformatics work as well. Commenting code and adopting a style guide are simple ways to increase code readability. Google has public style guides for many languages (*https://github.com/google/styleguide*), which serve as excellent templates. Why is code readability so important? First, readable code makes projects more reproducible, as others can more easily understand what scripts do and how they work. Second, it's much easier to find and correct software bugs in readable, well-commented code than messy code. Third, revisiting code in the future is always easier when the code is well commented and clearly written. Writing modular and reusable code just takes practice—we'll see some examples of this throughout the book.

In contrast to code, data should be formatted in a way that facilitates computer readability. All too often, we as humans record data in a way that maximizes its readability to us, but takes a considerable amount of cleaning and tidying before it can be processed by a computer. The more data (and metadata) that is computer readable, the more we can leverage our computers to work with this data.

Let Your Computer Do the Work For You

Humans doing rote activities tend to make many mistakes. One of the easiest ways to make your work more robust is to have your computer do as much of this rote work as possible. This approach of automating tasks is more robust because it decreases the odds you'll make a trivial mistake such as accidentally omitting a file or naming output incorrectly.

For example, running a program on 20 different files by individually typing out (or copy and pasting) each command is fragile—the odds of making a careless mistake increase with each file you process. In bioinformatics work, it's good to develop the habit of letting your computer do this sort of repetitive work for you. Instead of pasting the same command 20 times and just changing the input and output files, write a script that does this for you. Not only is this easier and less likely to lead to mistakes, but it also increases reproducibility, as your script serves as a reference of what you did to each of those files.

Leveraging the benefits of automating tasks requires a bit of thought in organizing up your projects, data, and code. Simple habits like naming data files in a consistent way that a computer (and not just humans) can understand can greatly facilitate automating tasks and make work much easier. We'll see examples of this in Chapter 2.

Make Assertions and Be Loud, in Code and in Your Methods

When we write code, we tend to have implicit assumptions about our data. For example, we expect that there are only three DNA strands options (forward, reverse, and unknown), that the start position of a gene is less than the end position, and that we can't have negative positions. These implicit assumptions we make about data impact how we write code; for example, we may not think to handle a certain situation in code if we assume it won't occur. Unfortunately, this can lead to the dreaded silent error: our code or programs receive values outside our expected values, behave improperly, and yet still return output without warning. Our best approach to prevent this type of error is to *explicitly* state and test our assumptions about data in our code using assert statements like Python's assert() and R's stopifnot().

Nearly every programming language has its own version of the assert function. These assert functions operate in a similar way: if the statement evaluated to false, the assert function will stop the program and raise an error. They may be simple, but these assert functions are indispensable in robust research. Early in my career, a mentor motivated me to adopt the habit of using asserts quite liberally—even when it seems like there is absolutely no way the statement could ever be false—and yet I'm continually surprised at how many times these have caught a subtle error. In bioinformatics (and all fields), it's crucial that we do as much as possible to turn the dreaded silent error into loud errors.

Test Code, or Better Yet, Let Code Test Code

Software engineers are a clever bunch, and they take the idea of letting one's computer do the work to new levels. One way they do this is having code test other code, rather than doing it by hand. A common method to test code is called *unit testing*. In unit testing, we break down our code into individual modular units (which also has the side effect of improving readability) and we write additional code that tests this code. In practice, this means if we have a function called add(), we write an additional function (usually in separate file) called test_add(). This test_add() function would call the add() function with certain input, and test that the output is as expected. In Python, this may look like:

```
EPS = 0.00001 # a small number to use when comparing floating-point values

def add(x, y):
    """Add two things together."""
    return x + y

def test_add():
    """Test that the add() function works for a variety of numeric types."""
    assert(add(2, 3) == 5)
    assert(add(-2, 3) == 1)
    assert(add(-1, -1) == -2)
    assert(abs(add(2.4, 0.1) - 2.5) < EPS)
```

The last line of the test_add() function looks more complicated than the others because it's comparing floating-point values. It's difficult to compare floating-point values on a computer, as there are representation and roundoff errors. However, it's a good reminder that we're always limited by what our machine can do, and we must mind these limitations in analysis.

Unit testing is used much less in scientific coding than in the software industry, despite the fact that scientific code is *more* likely to contain bugs (because our code is usually only run once to generate results for a publication, and many errors in scientific code are silent). I refer to this as *the paradox of scientific coding*: the bug-prone nature of scientific coding means we should utilize testing as much or more than the software industry, but we actually do much less testing (if any). This is regrettable, as nowadays many scientific conclusions are the result of mountains of code.

While testing code is the best way to find, fix, and prevent software bugs, testing isn't cheap. Testing code makes our results robust, but it also takes a considerable amount of our time. Unfortunately, it would take too much time for researchers to compose unit tests for every bit of code they write. Science moves quickly, and in the time it would take to write and perform unit tests, your research could become outdated or get scooped. A more sensible strategy is to consider three important variables each time you write a bit of code:

- How many times is this code called by other code?
- If this code were wrong, how detrimental to the final results would it be?
- How noticeable would an error be if one occurred?

How important it is to test a bit of code is proportional to the first two variables, and inversely proportional to the third (i.e., if a bug is very noticeable, there's less reason to write a test for it). We'll employ this strategy throughout the book's examples.

Use Existing Libraries Whenever Possible

There's a turning point in every budding programmer's career when they feel comfortable enough writing code and think, "Hey, why would I use a library for this, I could easily write this myself." It's an empowering feeling, but there are good reasons to use an existing software library instead.

Existing open source libraries have two advantages over libraries you write yourself: a longer history and a wider audience. Both of these advantages translate to fewer bugs. Bugs in software are similar to the proverbial problem of finding a needle in a haystack. If you write your own software library (where a few bugs are bound to be lurking), you're one person looking for a few needles. In contrast, open source software libraries in essence have had many more individuals looking for a much longer time for those needles. Consequently, bugs are more likely to be found, reported, and fixed in these open source libraries than your own home-brewed versions.

A good example of this is a potentially subtle issue that arises when writing a script to translate nucleotides to proteins. Most biologists with some programming experience could easily write a script to do this task. But behind these simple programming exercises lurks hidden complexity you alone may not consider. What if your nucleotide sequences have Ns in them? Or Ys? Or Ws? Ns, Ys, and Ws may not seem like valid bases, but these are International Union of Pure and Applied Chemistry (IUPAC) standard ambiguous nucleotides and are entirely valid in bioinformatics. In many cases, well-vetted software libraries have already found and fixed these sorts of hidden problems.

Treat Data as Read-Only

Many scientists spend a lot of time using Excel, and without batting an eye will change the value in a cell and save the results. I strongly discourage modifying data this way. Instead, a better approach is to treat all data as *read-only* and only allow programs to read data and create new, separate files of results.

Why is treating data as read-only important in bioinformatics? First, modifying data in place can easily lead to corrupted results. For example, suppose you wrote a script that directly modifies a file. Midway through processing a large file, your script encounters an error and crashes. Because you've modified the original file, you can't undo the changes and try again (unless you have a backup)! Essentially, this file is corrupted and can no longer be used.

Second, it's easy to lose track of how we've changed a file when we modify it in place. Unlike a workflow where each step has an input file and an output file, a file modified in place doesn't give us any indication of what we've done to it. If we were to lose track of how we've changed a file and don't have a backup copy of the original data, our changes are essentially irreproducible.

Treating data as read-only may seem counterintuitive to scientists familiar with working extensively in Excel, but it's essential to robust research (and prevents catastrophe, and helps reproducibility). The initial difficulty is well worth it; in addition to safeguarding data from corruption and incorrect changes, it also fosters reproducibility. Additionally, any step of the analysis can easily be redone, as the input data is unchanged by the program.

Spend Time Developing Frequently Used Scripts into Tools

Throughout your development as a highly skilled bioinformatician, you'll end up creating some scripts that you'll use over and over again. These may be scripts that download data from a database, or process a certain type of file, or maybe just generate the same pretty graphs. These scripts may be shared with lab members or even across labs. You should put extra effort and attention into making these high-use or highly shared scripts as robust as possible. In practice, I think of this process as turning one-off scripts into tools.

Tools, in contrast to scripts, are designed to be run over and over again. They are well documented, have explicit versioning, have understandable command-line arguments, and are kept in a shared version control repository. These may seem like minor differences, but robust research is about doing small things that stack the deck in your favor to prevent mistakes. Scripts that you repeatedly apply to numerous datasets by definition impact more results, and deserve to be more developed so they're more robust and user friendly. This is especially the case with scripts you share with other researchers who need to be able to consult documentation and apply your tool safely to their own data. While developing tools is a more labor-intensive process than writing a one-off script, in the long run it can save time and prevent headaches.

Let Data Prove That It's High Quality

When scientists think of analyzing data, they typically think of analyzing experimental data to draw biological conclusions. However, to conduct robust bioinformatics work, we actually need to analyze more than just experimental data. This includes inspecting and analyzing data about your experiment's data quality, intermediate output files from bioinformatics programs, and possibly simulated test data. Doing so ensures our data processing is functioning as we expect, and embodies the golden rule of bioinformatics: don't trust your tools or data.

It's important to never assume a dataset is high quality. Rather, data's quality should be proved through exploratory data analysis (known as EDA). EDA is not complex or time consuming, and will make your research much more robust to lurking surprises in large datasets. We'll learn more about EDA using R in Chapter 8.

Recommendations for Reproducible Research

Adopting reproducible research practices doesn't take much extra effort. And like robust research practices, reproducible methods will ultimately make your life easier as you yourself may need to reproduce your past work long after you've forgotten the details. Below are some basic recommendations to consider when practicing bioinformatics to make your work reproducible.

Release Your Code and Data

For reproducibility, the absolute minimal requirements are that code and data are released. Without available code and data, your research is not reproducible (see Peng, 2001 for a nice discussion of this). We'll discuss how to share code and data a bit later in the book.

Document Everything

The first day a scientist steps into a lab, they're told to keep a lab notebook. Sadly, this good practice is often abandoned by researchers in computing. Releasing code and data is the minimal requirement for reproducibility, but extensive documentation is an important component of reproducibility, too. To fully reproduce a study, each step of analysis must be described in much more detail than can be accomplished in a scholarly article. Thus, additional documentation is essential for reproducibility.

A good practice to adopt is to document each of your analysis steps in plain-text *README* files. Like a detailed lab notebook, this documentation serves as a valuable record of your steps, where files are, where they came from, or what they contain. This documentation can be stored alongside your project's code and data (we'll see more about this in Chapters 2 and 5), which can aid collaborators in figuring out what you've done. Documentation should also include all input parameters for each program executed, these programs' versions, and how they were run. Modern software like R's knitr and iPython Notebooks are powerful tools in documenting research; I've listed some resources to get started with these tools in this chapter's *README* on GitHub.

Make Figures and Statistics the Results of Scripts

Ensuring that a scientific project is reproducible involves more than just replicability of the key statistical tests important for findings—supporting elements of a paper (e.g., figures and tables) should also be reproducible. The best way to ensure these components are reproducible is to have each image or table be the output of a script (or scripts).

Writing scripts to produce images and tables may seem like a more time-consuming process than generating these interactively in Excel or R. However, if you've ever had to regenerate multiple figures by hand after changing an earlier step, you know the merit of this approach. Scripts that generate tables and images can easily be rerun, save you time, and lead your research to be more reproducible. Tools like iPython Notebooks and knitr (mentioned in the previous section) greatly assist in these tasks, too.

Use Code as Documentation

With complex processing pipelines, often the best documentation is well-documented code. Because code is sufficient to tell a computer how to execute a program (and which parameters to use), it's also close to sufficient to tell a human how to replicate your work (additional information like software version and input data is also necessary to be fully reproducible). In many cases, it can be easier to write a script to perform key steps of an analysis than it is to enter commands and then document them elsewhere. Again, code is a wonderful thing, and using code to document each step of an analysis means it's easy to rerun all steps of an analysis if necessary—the script can just simply be rerun.

Continually Improving Your Bioinformatics Data Skills

Keep the basic ideology introduced in this chapter in the back of your head as you work through the rest of the book. What I've introduced here is just enough to get you started in thinking about some core concepts in robust and reproducible bioinformatics. Many of these topics (e.g., reproducibility and software testing) are still actively researched at this time, and I encourage the interested reader to explore these in more depth (I've included some resources in this chapter's *README* on GitHub).

## PART II

Prerequisites: Essential Skills for Getting Started with a Bioinformatics Project

## CHAPTER 2

Setting Up and Managing a Bioinformatics Project

Just as a well-organized laboratory makes a scientist's life easier, a well-organized and well-documented project makes a bioinformatician's life easier. Regardless of the particular project you're working on, your project directory should be laid out in a consistent and understandable fashion. Clear project organization makes it easier for both you and collaborators to figure out exactly where and what everything is. Additionally, it's much easier to automate tasks when files are organized and clearly named. For example, processing 300 gene sequences stored in separate FASTA files with a script is trivial if these files are organized in a single directory and are consistently named.

Every bioinformatics project begins with an empty project directory, so it's fitting that this book begin with a chapter on project organization. In this chapter, we'll look at some best practices in organizing your bioinformatics project directories and how to digitally document your work using plain-text Markdown files. We'll also see why project directory organization isn't just about being tidy, but is essential to the way by which tasks are automated across large numbers of files (which we routinely do in bioinformatics).

Project Directories and Directory Structures

Creating a well-organized directory structure is the foundation of a reproducible bioinformatics project. The actual process is quite simple: laying out a project only entails creating a few directories with mkdir and empty *README* files with touch (commands we'll see in more depth later). But this simple initial planning pays off in the long term. For large projects, researchers could spend years working in this directory structure.

Other researchers have noticed the importance of good project organization, too (Noble 2009). While eventually you'll develop a project organization scheme that works for you, we'll begin in this chapter with a scheme I use in my work (and is similar to Noble's).

All files and directories used in your project should live in a single project directory with a clear name. During the course of a project, you'll have amassed data files, notes, scripts, and so on—if these were scattered all over your hard drive (or worse, across many computers' hard drives), it would be a nightmare to keep track of everything. Even worse, such a disordered project would later make your research nearly impossible to reproduce.

Keeping all of your files in a single directory will greatly simplify things for you and your collaborators, and facilitate reproducibility (we'll discuss how to collaboratively work on a project directory with Git in Chapter 5). Suppose you're working on SNP calling in maize (*Zea mays*). Your first step would be to choose a short, appropriate project name and create some basic directories:

```
$ mkdir zmays-snps
$ cd zmays-snps
$ mkdir data
$ mkdir data/seqs scripts analysis
$ ls -l
total 0
drwxr-xr-x  2 vinceb  staff   68 Apr 15 01:10 analysis
drwxr-xr-x  3 vinceb  staff  102 Apr 15 01:10 data
drwxr-xr-x  2 vinceb  staff   68 Apr 15 01:10 scripts
```

This is a sensible project layout scheme. Here, *data/* contains all raw and intermediate data. As we'll see, data-processing steps are treated as separate subprojects in this *data/* directory. I keep general project-wide scripts in a *scripts/* directory. If scripts contain many files (e.g., multiple Python modules), they should reside in their own subdirectory. Isolating scripts in their own subdirectory also keeps project directories tidy while developing these scripts (when they produce test output files).

Bioinformatics projects contain many smaller analyses—for example, analyzing the quality of your raw sequences, the aligner output, and the final data that will produce figures and tables for a paper. I prefer keeping these in a separate *analysis/* directory, as it allows collaborators to see these high-level analyses without having to dig deeper into subproject directories.

What's in a Name?

Naming files and directories on a computer matters more than you might think. In transitioning from a graphical user interface (GUI) based operating system to the Unix command line, many folks bring the bad habit of using spaces in file and directory names. This isn't appropriate in a Unix-based environment, because spaces are used to separate arguments in commands. For example, suppose that you create a directory named *raw sequences* from a GUI (e.g., through OS X's Finder), and later try to remove it and its contents with the following command:

```
$ rm -rf raw sequences
```

If you're lucky, you'd be warned that there is "No such file or directory" for both `raw` and `sequences`. What's going on here? Spaces matter: your shell is interpreting this `rm` command as "delete both the *raw* and *sequences* files/directories," not "delete a single file or directory called *raw sequences*."

If you're unlucky enough to have a file or directory named either *raw* or *sequences*, this `rm` command would delete it. It's possible to escape this by using quotes (e.g., `rm -r "raw sequences"`), but it's better practice to not use spaces in file or directory names in the first place. It's best to use only letters, numbers, underscores, and dashes in file and directory names.

Although Unix doesn't require file extensions, including extensions in filenames helps indicate the type of each file. For example, a file named *osativa-genes.fasta* makes it clear that this is a file of sequences in FASTA format. In contrast, a file named *osativa-genes* could be a file of gene models, notes on where these *Oryza sativa* genes came from, or sequence data. When in doubt, explicit is always better than implicit when it comes to filenames, documentation, and writing code.

Scripts and analyses often need to refer to other files (such as data) in your project hierarchy. This may require referring to parent directories in your directory's hierarchy (e.g., with `..`). In these cases, it's important to always use *relative paths* (e.g., `../data/stats/qual.txt`) rather than absolute paths (e.g., `/home/vinceb/projects/zmays-snps/data/stats/qual.txt`). As long as your internal project directory structure remains the same, these relative paths will always work. In contrast, absolute paths rely on your particular user account and directory structures details *above* the project directory level (not good). Using absolute paths leaves your work less portable between collaborators and decreases reproducibility.

My project directory scheme here is by no means the only scheme possible. I've worked with bioinformaticians that use entirely different schemes for their projects and do excellent work. However, regardless of the organization scheme, a good bioin-formatician will always document everything extensively and use clear filenames that can be parsed by a computer, two points we'll come to in a bit.

Project Documentation

In addition to having a well-organized directory structure, your bioinformatics project also should be well documented. Poor documentation can lead to irreproduci-bility and serious errors. There's a vast amount of lurking complexity in bioinformat-ics work: complex workflows, multiple files, countless program parameters, and different software versions. The best way to prevent this complexity from causing problems is to document everything extensively. Documentation also makes your life easier when you need to go back and rerun an analysis, write detailed methods about your steps for a paper, or find the origin of some data in a directory. So what exactly should you document? Here are some ideas:

Document your methods and workflows
> This should include full command lines (copied and pasted) that are run through the shell that generate data or intermediate results. Even if you use the default values in software, be sure to write these values down; later versions of the pro-gram may use different default values. Scripts naturally document all steps and parameters (a topic we'll cover in Chapter 12), but be sure to document any command-line options used to run this script. In general, any command that produces results used in your work needs to be documented somewhere.

Document the origin of all data in your project directory
> You need to keep track of where data was downloaded from, who gave it to you, and any other relevant information. "Data" doesn't just refer to your project's experimental data—it's any data that programs use to create output. This includes files your collaborators send you from their separate analyses, gene annotation tracks, reference genomes, and so on. It's critical to record this important data about your data, or *metadata*. For example, if you downloaded a set of genic regions, record the website's URL. This seems like an obvious recommendation, but countless times I've encountered an analysis step that couldn't easily be reproduced because someone forgot to record the data's source.

Document when you downloaded data
> It's important to include when the data was downloaded, as the external data source (such as a website or server) might change in the future. For example, a script that downloads data directly from a database might produce different results if rerun after the external database is updated. Consequently, it's impor-tant to document when data came into your repository.

Record data version information

Many databases have explicit release numbers, version numbers, or names (e.g., TAIR10 version of genome annotation for *Arabidopsis thaliana*, or Wormbase release WS231 for *Caenorhabditis elegans*). It's important to record all version information in your documentation, including minor version numbers.

Describe how you downloaded the data

For example, did you use MySQL to download a set of genes? Or the UCSC Genome Browser? These details can be useful in tracking down issues like when data is different between collaborators.

Document the versions of the software that you ran

This may seem unimportant, but remember the example from "Reproducible Research" on page 6 where my colleagues and I traced disagreeing results down to a single piece of software being updated. These details matter. Good bioinformatics software usually has a command-line option to return the current version. Software managed with a version control system such as Git has explicit identifiers to every version, which can be used to document the precise version you ran (we'll learn more about this in Chapter 5). If no version information is available, a release date, link to the software, and download date will suffice.

All of this information is best stored in plain-text *README* files. Plain text can easily be read, searched, and edited directly from the command line, making it the perfect choice for portable and accessible *README* files. It's also available on all computer systems, meaning you can document your steps when working directly on a server or computer cluster. Plain text also lacks complex formatting, which can create issues when copying and pasting commands from your documentation back into the command line. It's best to avoid formats like Microsoft Word for *README* documentation, as these are less portable to the Unix systems common in bioinformatics.

Where should you keep your *README* files? A good approach is to keep *README* files in each of your project's main directories. These *README* files don't necessarily need to be lengthy, but they should at the very least explain what's in this directory and how it got there. Even this small effort can save someone exploring your project directory a lot of time and prevent confusion. This someone could be your advisor or a collaborator, a colleague trying to reproduce your work after you've moved onto a different lab, or even yourself six months from now when you've completely forgotten what you've done (this happens to everyone!).

For example, a *data/README* file would contain metadata about your data files in the *data/* directory. Even if you think you could remember all relevant information about your data, it's much easier just to throw it in a *README* file (and collaborators won't have to email you to ask what files are or where they are). Let's create some empty *README* files using touch. touch updates the modification time of a file or creates a file if it doesn't already exist. We can use it for this latter purpose to create empty template files to lay out our project structure:

```
$ touch README data/README
```

Following the documentation guidelines just discussed, this *data/README* file would include where you downloaded the data in *data/*, when you downloaded it, and how. When we learn more about data in Chapter 6, we'll see a case study example of how to download and properly document data in a project directory ("Case Study: Reproducibly Downloading Data" on page 120).

By recording this information, we're setting ourselves up to document everything about our experiment and analysis, making it reproducible. Remember, as your project grows and accumulates data files, it also pays off to keep track of this for your own sanity.

Use Directories to Divide Up Your Project into Subprojects

Bioinformatics projects involve many subprojects and subanalyses. For example, the quality of raw experimental data should be assessed and poor quality regions removed before running it through bioinformatics tools like aligners or assemblers (we see an example of this in "Example: Inspecting and Trimming Low-Quality Bases" on page 346). Even before you get to actually analyzing sequences, your project directory can get cluttered with intermediate files.

Creating directories to logically separate subprojects (e.g., sequencing data quality improvement, aligning, analyzing alignment results, etc.) can simplify complex projects and help keep files organized. It also helps reduce the risk of accidentally clobbering a file with a buggy script, as subdirectories help isolate mishaps. Breaking a project down into subprojects and keeping these in separate subdirectories also makes documenting your work easier; each *README* pertains to the directory it resides in. Ultimately, you'll arrive at your own project organization system that works for you; the take-home point is: leverage directories to help stay organized.

Organizing Data to Automate File Processing Tasks

Because automating file processing tasks is an integral part of bioinformatics, organizing our projects to facilitate this is essential. Organizing data into subdirectories and using clear and consistent file naming schemes is imperative—both of these practices allow us to *programmatically* refer to files, the first step to automating a task. Doing something programmatically means doing it through code rather than manually, using a method that can effortlessly scale to multiple objects (e.g., files). Programmatically referring to multiple files is easier and safer than typing them all out (because it's less error prone).

Shell Expansion Tips

Bioinformaticians, software engineers, and system administrators spend *a lot* of time typing in a terminal. It's no surprise these individuals collect tricks to make this process as efficient as possible. As you spend more time in the shell, you'll find investing a little time in learning these tricks can save you a lot of time down the road.

One useful trick is *shell expansion*. Shell expansion is when your shell (e.g., Bash, which is likely the shell you're using) expands text for you so you don't have to type it out. If you've ever typed cd ~ to go to your home directory, you've used shell expansion—it's your shell that expands the tilde character (~) to the full path to your home directory (e.g., /Users/vinceb/). Wildcards like an asterisk (*) are also expanded by your shell to all matching files.

A type of shell expansion called *brace expansion* can be used to quickly create the simple *zmays-snps/* project directory structure with a single command. Brace expansion creates strings by expanding out the comma-separated values inside the braces. This is easier to understand through a trivial example:

```
$ echo dog-{gone,bowl,bark}
dog-gone dog-bowl dog-bark
```

Using this same strategy, we can create the *zmays-snps/* project directory:

```
$ mkdir -p zmays-snps/{data/seqs,scripts,analysis}
```

This produces the same *zmays-snps* layout as we constructed in four separate steps in "Project Directories and Directory Structures" on page 21: *analysis/*, *data/seqs*, and *scripts/*. Because mkdir takes multiple arguments (creating a directory for each), this creates the three subdirectories (and saves you having to type "zmays-snps/" three times). Note that we need to use mkdir's -p flag, which tells mkdir to create any necessary subdirectories it needs (in our case, *data/* to create *data/seqs/*).

We'll step through a toy example to illustrate this point, learning some important shell wildcard tricks along the way. In this example, organizing data files into a single directory with consistent filenames prepares us to iterate over *all* of our data, whether it's the four example files used in this example, or 40,000 files in a real project. Think of it this way: remember when you discovered you could select many files with your mouse cursor? With this trick, you could move 60 files as easily as six files. You could also select certain file types (e.g., photos) and attach them all to an email with one movement. By using consistent file naming and directory organization, you can do the same programmatically using the Unix shell and other programming languages.

We'll see a Unix example of this using shell wildcards to automate tasks across many files. Later in Chapter 12, we'll see more advanced bulk file manipulation strategies.

Let's create some fake empty data files to see how consistent names help with programmatically working with files. Suppose that we have three maize samples, "A," "B," and "C," and paired-end sequencing data for each:

```
$ cd data
$ touch seqs/zmays{A,B,C}_R{1,2}.fastq
$ ls seqs/
zmaysA_R1.fastq zmaysB_R1.fastq zmaysC_R1.fastq
zmaysA_R2.fastq zmaysB_R2.fastq zmaysC_R2.fastq
```

In this file naming scheme, the two variable parts of each filename indicate sample name (zmaysA, zmaysB, zmaysC) and read pair (R1 and R2). Suppose that we wanted to programmatically retrieve all files that have the sample name zmaysB (regardless of the read pair) rather than having to manually specify each file. To do this, we can use the Unix shell wildcard, the asterisk character (*):

```
$ ls seqs/zmaysB*
zmaysB_R1.fastq zmaysB_R2.fastq
```

Wildcards are expanded to all matching file or directory names (this process is known as *globbing*). In the preceding example, your shell expanded the expression zmaysB* to zmaysB_R1.fastq and zmaysB_R2.fastq, as these two files begin with *zmaysB*. If this directory had contained hundreds of *zmaysB* files, all could be easily referred to and handled with shell wildcards.

Wildcards and "Argument list too long"

OS X and Linux systems have a limit to the number of arguments that can be supplied to a command (more technically, the limit is to the total length of the arguments). We sometimes hit this limit when using wildcards that match tens of thousands of files. When this happens, you'll see an "Argument list too long" error message indicating you've hit the limit. Luckily, there's a clever way around this problem (see "Using find and xargs" on page 411 for the solution).

In general, it's best to be as restrictive as possible with wildcards. This protects against accidental matches. For example, if a messy colleague created an Excel file named *zmaysB-interesting-SNPs-found.xls* in this directory, this would accidentally match the wildcard expression zmaysB*. If you needed to process all *zmaysB* FASTQ files, referring to them with zmaysB* would include this Excel file and lead to problems. This is why it's best to be as restrictive as possible when using wildcards. Instead of zmaysB*, use zmaysB*fastq or zmaysB_R?.fastq (the ? only matches a single character).

There are other simple shell wildcards that are quite handy in programmatically accessing files. Suppose a collaborator tells you that the C sample sequences are poor quality, so you'll have to work with just the A and B samples while C is resequenced. You don't want to delete *zmaysC_R1.fastq* and *zmaysC_R2.fastq* until the new samples are received, so in the meantime you want to ignore these files. The folks that invented wildcards foresaw problems like this, so they created shell wildcards that allow you to match specific characters or ranges of characters. For example, we could match the characters U, V, W, X, and Y with either [UVWXY] or [U-Y] (both are equivalent). Back to our example, we could exclude the C sample using either:

```
$ ls zmays[AB]_R1.fastq
zmaysA_R1.fastq zmaysB_R1.fastq
$ ls zmays[A-B]_R1.fastq
zmaysA_R1.fastq zmaysB_R1.fastq
```

Using a range between A and B isn't really necessary, but if we had samples A through I, using a range like zmays[C-I]_R1.fastq would be better than typing out zmays[CDEFGHI]_R1.fastq. There's one very important caveat: ranges operate on *character* ranges, not numeric ranges like 13 through 30. This means that wildcards like snps_[10-13].txt will not match files *snps_10.txt*, *snps_11.txt*, *snps_12.txt*, and *snps_13.txt*.

However, the shell does offer an expansion solution to numeric ranges—through the brace expansion we saw earlier. Before we see this shortcut, note that while wildcard matching and brace expansion may seem to behave similarly, they are slightly different. Wildcards *only expand to existing files that match them*, whereas brace expansions always expand *regardless of whether corresponding files or directories exist or not*. If we knew that files *snps_10.txt* through *snps_13.txt* did exist, we could match them with the brace expansion *sequence expression* like snps_{10..13}.txt. This expands to the integer sequence 10 through 13 (but remember, whether these files exist or not is not checked by brace expansion). Table 2-1 lists the common Unix wildcards.

Table 2-1. Common Unix filename wildcards

Wildcard	What it matches
*	Zero or more characters (but ignores hidden files starting with a period).
?	One character (also ignores hidden files).
[A-Z]	Any character between the supplied alphanumeric range (in this case, any character between A and Z); this works for any alphanumeric character range (e.g., [0-9] matches any character between 0 and 9).

By now, you should begin to see the utility of shell wildcards: they allow us to handle multiple files with ease. Because lots of daily bioinformatics work involves file processing, programmatically accessing files makes our job easier and eliminates mistakes made from mistyping a filename or forgetting a sample. However, our ability to programmatically access files with wildcards (or other methods in R or Python) is only possible when our filenames are consistent. While wildcards are powerful, they're useless if files are inconsistently named. For example, processing a subset of files with names like *zmays sampleA - 1.fastq*, *zmays_sampleA-2.fastq*, *sampleB1.fastq*, *sample-B2.fastq* is needlessly more complex because of the inconsistency of these filenames. Unfortunately, inconsistent naming is widespread across biology, and is the scourge of bioinformaticians everywhere. Collectively, bioinformaticians have probably wasted thousands of hours fighting others' poor naming schemes of files, genes, and in code.

Leading Zeros and Sorting

Another useful trick is to use *leading zeros* (e.g., *file-0021.txt* rather than *file-21.txt*) when naming files. This is useful because lexicographically sorting files (as ls does) leads to the correct ordering. For example, if we had filenames such as *gene-1.txt*, *gene-2.txt*, …, *gene-14.txt*, sorting these lexicographically would yield:

```
$ ls -l
-rw-r--r--  1 vinceb  staff  0 Feb 21 21:24 genes-1.txt
-rw-r--r--  1 vinceb  staff  0 Feb 21 21:24 genes-11.txt
-rw-r--r--  1 vinceb  staff  0 Feb 21 21:24 genes-12.txt
-rw-r--r--  1 vinceb  staff  0 Feb 21 21:24 genes-13.txt
-rw-r--r--  1 vinceb  staff  0 Feb 21 21:24 genes-14.txt
[...]
```

But if we use leading zeros (e.g., *gene-001.txt*, *gene-002.txt*, …, *gene-014.txt*), the files sort in their correct order:

```
$ ls -l
-rw-r--r--  1 vinceb  staff  0 Feb 21 21:23 genes-001.txt
-rw-r--r--  1 vinceb  staff  0 Feb 21 21:23 genes-002.txt
[...]
-rw-r--r--  1 vinceb  staff  0 Feb 21 21:23 genes-013.txt
-rw-r--r--  1 vinceb  staff  0 Feb 21 21:23 genes-014.txt
```

Using leading zeros isn't just useful when naming filenames; this is also the best way to name genes, transcripts, and so on. Projects like Ensembl use this naming scheme in naming their genes (e.g., ENSG00000164256).

In addition to simplifying working with files, consistent file naming is an often overlooked component of robust bioinformatics. Bad sample naming schemes can easily lead to switched samples. Poorly chosen filenames can also cause serious errors when you or collaborators think you're working with the correct data, but it's actually outdated or the wrong file. I guarantee that out of all the papers published in the past decade, at least a few and likely many more contain erroneous results because of a file naming issue.

Markdown for Project Notebooks

It's very important to keep a project notebook containing detailed information about the chronology of your computational work, steps you've taken, information about why you've made decisions, and of course all pertinent information to reproduce your work. Some scientists do this in a handwritten notebook, others in Microsoft Word documents. As with *README* files, bioinformaticians usually like keeping project notebooks in simple plain-text because these can be read, searched, and edited from the command line and across network connections to servers. Plain text is also a future-proof format: plain-text files written in the 1960s are still readable today, whereas files from word processors only 10 years old can be difficult or impossible to open and edit. Additionally, plain-text project notebooks can also be put under version control, which we'll talk about in Chapter 5.

While plain-text is easy to write in your text editor, it can be inconvenient for collaborators unfamiliar with the command line to read. A lightweight markup language called *Markdown* is a plain-text format that is easy to read and painlessly incorporated into typed notes, and can also be rendered to HTML or PDF.

Markdown originates from the simple formatting conventions used in plain-text emails. Long before HTML crept into email, emails were embellished with simple markup for emphasis, lists, and blocks of text. Over time, this became a de facto plain-text email formatting scheme. This scheme is very intuitive: underscores or asterisks that flank text indicate emphasis, and lists are simply lines of text beginning with dashes.

Markdown is just plain-text, which means that it's portable and programs to edit and read it will exist. Anyone who's written notes or papers in old versions of word processors is likely familiar with the hassle of trying to share or update out-of-date proprietary formats. For these reasons, Markdown makes for a simple and elegant notebook format.

Markdown Formatting Basics

Markdown's formatting features match all of the needs of a bioinformatics notebook: text can be broken down into hierarchical sections, there's syntax for both code blocks and inline code, and it's easy to embed links and images. While the Markdown format is very simple, there are a few different variants. We'll use the original Markdown format, invented by John Gruber, in our examples. John Gruber's full markdown syntax specification is available on his website (*http://bit.ly/markdown-doc*). Here is a basic Markdown document illustrating the format:

```
# *Zea Mays* SNP Calling

We sequenced three lines of *zea mays*, using paired-end
sequencing. This sequencing was done by our sequencing core and we
received the data on 2013-05-10. Each variety should have **two**
sequences files, with suffixes `_R1.fastq` and `_R2.fastq`, indicating
which member of the pair it is.

## Sequencing Files

All raw FASTQ sequences are in `data/seqs/`:

    $ find data/seqs -name "*.fastq"
    data/seqs/zmaysA_R1.fastq
    data/seqs/zmaysA_R2.fastq
    data/seqs/zmaysB_R1.fastq
    data/seqs/zmaysB_R2.fastq
    data/seqs/zmaysC_R1.fastq
    data/seqs/zmaysC_R2.fastq

## Quality Control Steps

After the sequencing data was received, our first stage of analysis
was to ensure the sequences were high quality. We ran each of the
three lines' two paired-end FASTQ files through a quality diagnostic
and control pipeline. Our planned pipeline is:

1. Create base quality diagnostic graphs.
2. Check reads for adapter sequences.
3. Trim adapter sequences.
4. Trim poor quality bases.

Recommended trimming programs:

  - Trimmomatic
  - Scythe
```

Figure 2-1 shows this example Markdown notebook rendered in HTML5 by Pandoc. What makes Markdown a great format for lab notebooks is that it's as easy to read in unrendered plain-text as it is in rendered HTML. Next, let's take a look at the Markdown syntax used in this example (see Table 2-2 for a reference).

Zea Mays SNP Calling

We sequenced three lines of *zea mays*, using paired-end sequencing. This sequencing was done by our sequencing core and we received the data on 2013-05-10. Each variety should have **two** sequences files, with suffixes `_R1.fastq` and `_R2.fastq`, indicating which member of the pair it is.

Sequencing Files

All raw FASTQ sequences are in `data/seqs/`:

```
$ find data/seqs -name "*.fastq"
data/seqs/zmaysA_R1.fastq
data/seqs/zmaysA_R2.fastq
data/seqs/zmaysB_R1.fastq
data/seqs/zmaysB_R2.fastq
data/seqs/zmaysC_R1.fastq
data/seqs/zmaysC_R2.fastq
```

Quality Control Steps

After the sequencing data was received, our first stage of analysis was to ensure the sequences were high quality. We ran each of the three lines' two paired-end FASTQ files through a quality diagnostic and control pipeline. Our planned pipeline is:

1. Create base quality diagnostic graphs.
2. Check reads for adapter sequences.
3. Trim adapter sequences.
4. Trim poor quality bases.

Recommended trimming programs:

- Trimmomatic
- Scythe

Figure 2-1. HTML Rendering of the Markdown notebook

Table 2-2. Minimal Markdown inline syntax

Markdown syntax	Result
`*emphasis*`	*emphasis*
`**bold**`	**bold**
`` `inline code` ``	inline code
`<http://website.com/link>`	Hyperlink to *http://website.com/link*
`[link text](http://website.com/link)`	Hyperlink to *http://website.com/link*, with text "link text"

Markdown syntax	Result
`![alt text](path/to/` `figure.png)`	Image, with alternative text "alt text"

Block elements like headers, lists, and code blocks are simple. Headers of different levels can be specified with varying numbers of hashes (#). For example:

```
# Header level 1
## Header level 2
### Header level 3
```

Markdown supports headers up to six levels deep. It's also possible to use an alternative syntax for headers up to two levels:

```
Header level 1
==============

Header level 2
--------------
```

Both ordered and unordered lists are easy too. For unordered lists, use dashes, asterisks, or pluses as list element markers. For example, using dashes:

```
- Francis Crick
- James D. Watson
- Rosalind Franklin
```

To order your list, simply use numbers (i.e., 1., 2., 3., etc.). The ordering doesn't matter, as HTML will increment these automatically for you. However, it's still a good idea to clearly number your bullet points so the plain-text version is readable.

Code blocks are simple too—simply add four spaces or one tab before each code line:

```
I ran the following command:

    $ find seqs/ -name "*.fastq"
```

If you're placing a code block within a list item, make this eight spaces, or two tabs:

```
1. I searched for all FASTQ files using:

        find seqs/ -name "*.fastq"

2. And finally, VCF files with:

        find vcf/ -name "*.vcf"
```

What we've covered here should be more than enough to get you started digitally documenting your bioinformatics work. Additionally, there are extensions to Markdown, such as MultiMarkdown (*http://fletcherpenney.net/multimarkdown/*) and GitHub Flavored Markdown (*http://bit.ly/gh-flavor-md*). These variations add features (e.g., MultiMarkdown adds tables, footnotes, LaTeX math support, etc.) and change some of the default rendering options. There are also many specialized Markdown editing applications if you prefer a GUI-based editor (see this chapter's *README* on GitHub for some suggestions).

Using Pandoc to Render Markdown to HTML

We'll use Pandoc (*http://johnmacfarlane.net/pandoc/*), a popular document converter, to render our Markdown documents to valid HTML. These HTML files can then be shared with collaborators or hosted on a website. See the Pandoc installation page (*http://bit.ly/pan-install*) for instructions on how to install Pandoc on your system.

Pandoc can convert between a variety of different markup and output formats. Using Pandoc is very simple—to convert from Markdown to HTML, use the `--from mark down` and `--to html` options and supply your input file as the last argument:

```
$ pandoc --from markdown --to html notebook.md > output.html
```

By default, Pandoc writes output to standard out, which we can redirect to a file (we'll learn more about standard out and redirection in Chapter 3). We could also specify the output file using `--output output.html`. Finally, note that Pandoc can convert between *many* formats, not just Markdown to HTML. I've included some more examples of this in this chapter's *README* on GitHub, including how to convert from HTML to PDF.

Remedial Unix Shell

The Unix shell is the foundational computing environment for bioinformatics. The shell serves as our interface to large bioinformatics programs, as an interactive console to inspect data and intermediate results, and as the infrastructure for our pipelines and workflows. This chapter will help you develop a proficiency with the necessary Unix shell concepts used extensively throughout the rest of the book. This will allow you to focus on the content of commands in future chapters, rather than be preoccupied with understanding shell syntax.

This book assumes you're familiar with basic topics such as what a terminal is, what the shell is, the Unix filesystem hierarchy, moving about directories, file permissions, executing commands, and working with a text editor. If these topics sound foreign to you, it's best to brush up on using more basic materials (see "Assumptions This Book Makes" on page xvi for some resources). In this chapter, we'll cover remedial concepts that deeply underlie how we use the shell in bioinformatics: streams, redirection, pipes, working with running programs, and command substitution. Understanding these shell topics will prepare you to use the shell to work with data (Chapter 7) and build pipelines and workflows (Chapter 12). In this chapter, we'll also see *why* the Unix shell has such a prominent role in how we do modern bioinformatics. If you feel comfortable with these shell topics already, I suggest reading the first section of this chapter and then skipping to Chapter 4.

Why Do We Use Unix in Bioinformatics? Modularity and the Unix Philosophy

Imagine rather than using the Unix shell as our bioinformatics computing environment, we were to implement our entire project as single large program. We usually don't think of a bioinformatics project as a "program" but it certainly could be—we could write a single complex program that takes raw data as input, and after hours of data processing, outputs publication figures and a final table of results. For a project like variant calling, this program would include steps for raw sequence read processing, read mapping, variant calling, filtering variant calls, and final data analysis. This program's code would be massive—easily thousands of lines long.

While a program like this has the benefit of being customized to a particular variant calling project, it would not be general enough to adapt to others. Given its immense amount of code, this program would be impractical to adjust to each new project. The large codebase would also make finding and fixing bugs difficult. To make matters worse, unless our monolithic program was explicitly programmed to check that data between steps looked error free, a step may go awry (unbeknownst to us) and it would dutifully continue the analysis with incorrect data. While this custom program might be more computationally efficient, this would come at the expense of being fragile, difficult to modify, error prone (because it makes checking intermediate data very difficult), and not generalizable to future projects.

Unix is the foundational computing environment in bioinformatics because its design philosophy is the antithesis of this inflexible and fragile approach. The Unix shell was designed to allow users to easily build complex programs by interfacing smaller modular programs together. This approach is the Unix philosophy:

> This is the Unix philosophy: Write programs that do one thing and do it well. Write programs to work together. Write programs to handle text streams, because that is a universal interface.

> —Doug McIlory

The Unix shell provides a way for these programs to talk to each other (pipes) and write to and read files (redirection). Unix's core programs (which we'll use to analyze data on the command line in Chapter 7) are modular and designed to work well with other programs. The modular approach of the Unix philosophy has many advantages in bioinformatics:

- With modular workflows, it's easier to spot errors and figure out where they're occurring. In a modular workflow, each component is independent, which makes it easier to inspect intermediate results for inconsistencies and isolate problematic steps. In contrast, large nonmodular programs hide potential problems (all you see is its final output data) and make isolating where problems originate difficult.

- Modular workflows allow us to experiment with alternative methods and approaches, as separate components can be easily swapped out with other components. For example, if you suspect a particular aligner is working poorly with your data, it's easy to swap this aligner out with another one. This is possible only with modular workflows, where our alignment program is separate from downstream steps like variant calling or RNA-seq analysis.

- Modular components allow us to choose tools and languages that are appropriate for specific tasks. This is another reason the Unix environment fits bioinformatics well: it allows us to combine command-line tools for interactively exploring data (covered in more depth in Chapter 7), Python for more sophisticated scripting, and R for statistical analysis. When programs are designed to work with other programs, there's no cost to choosing a specialized tool for a specific task—something we quite often do in bioinformatics.

- Modular programs are reusable and applicable to many types of data. Well-written modular programs can be recombined and applied to different problems and datasets, as they are independent pieces. Most important, by remixing modular components, novel problems can be solved with existing tools.

In addition to emphasizing the program modularity and interfacing, McIlroy's quote also mentions *text streams*. We'll address Unix streams in this chapter, but the concept of a *stream* is very important in how we process large data. Definitions of large data may vary, and while a single lane of sequencing data may be big to one lab just getting into sequencing, this is minuscule compared to what larger sequencing centers process each hour. Regardless, a lane of sequencing data is too big to fit in the memory of most standard desktops. If I needed to search for the exact string "GTGAT-TAACTGCGAA" in this data, I couldn't open up a lane of data in Notepad and use the Find feature to pinpoint where it occurs—there simply isn't enough memory to hold all these nucleotides in memory. Instead, tools must rely on streams of data, being read from a source and actively processed. Both general Unix tools and many bioinformatics programs are designed to take input through a stream and pass output through a different stream. It's these text streams that allow us to both couple programs together into workflows and process data without storing huge amounts of data in our computers' memory.

The Many Unix Shells

Throughout this book, I'll refer to the *Unix shell* in general, but there's no single Unix shell. Shells are computer programs, and many programmers have designed and implemented their own versions. These many versions can lead to frustrating problems for new users, as some shells have features incompatible with others.

To avoid this frustration, make sure you're using the Bourne-again shell, or `bash`. Bash is widely available and the default shell on operating systems like Apple's OS X and Ubuntu Linux. You can run `echo $SHELL` to verify you're using bash as your shell (although it's best to also check what `echo $0` says too, because even how you identify your shell differs among shells!). I wouldn't recommend other shells like the C shell (`csh`), its descendant `tcsh`, and the Korn shell (`ksh`), as these are less popular in bioinformatics and may not be compatible with examples in this book. The Bourne shell (`sh`) was the predecessor of the Bourne-again shell (`bash`); but bash is newer and usually preferred.

It's possible to change your shell with the command `chsh`. In my daily bioinformatics work, I use Z shell (`zsh`) and have made this my default shell. Z shell has more advanced features (e.g., better autocomplete) that come in handy in bioinformatics. Everything in this book is compatible between these two shells unless explicitly noted otherwise. If you feel confident with general shell basics, you may want to try Z shell. I've included resources about Z shell in this chapter's *README* file on GitHub.

The last point to stress about the Unix shell is that it's incredibly powerful. With simple features like wildcards, it's trivial to apply a command to hundreds of files. But with this power comes risk: the Unix shell does not care if commands are mistyped or if they will destroy files; the Unix shell is not designed to prevent you from doing unsafe things. A good analogy comes from Gary Bernhardt: Unix is like a chainsaw. Chainsaws are powerful tools, and make many difficult tasks like cutting through thick logs quite easy. Unfortunately, this power comes with danger: chainsaws can cut just as easily through your leg (well, technically more easily). For example, consider:

```
$ rm -rf tmp-data/aligned-reads*  # deletes all old large files
$ # versus
$ rm -rf tmp-data/aligned-reads *  # deletes your entire current directory
rm: tmp-data/aligned-reads: No such file or directory
```

In Unix, a single space could mean the difference between cleaning out some old files and finishing your project behind schedule because you've accidentally deleted everything. This isn't something that should cause alarm—this is a consequence of working with powerful tools. Rather, just adopt a cautious attitude when experimenting or trying a new command (e.g., work in a temporary directory, use fake files or data if you're unsure how a command behaves, and always keep backups). That the Unix shell has the power to allow you to do powerful (possibly unsafe) things is an important part of its design:

> Unix was not designed to stop its users from doing stupid things, as that would also stop them from doing clever things.
>
> —Doug Gwyn

Tackling repetitive large data-processing tasks in clever ways is a large part of being a skillful bioinformatician. Our shell is often the fastest tool for these tasks. In this chapter, we'll focus on some of these Unix shell primitives that allow us to build complex programs from simple parts: streams, redirection, pipes, working with processes, and command substitution. We'll learn more about automating tasks, another important part of the Unix shell, in Chapter 12.

Working with Streams and Redirection

Bioinformatics data is often text—for example, the As, Cs, Ts, and Gs in sequencing read files or reference genomes, or tab-delimited files of gene coordinates. The text data in bioinformatics is often large, too (gigabytes or more that can't fit into your computer's memory at once). This is why Unix's philosophy of handling text streams is useful in bioinformatics: text streams allow us to do processing on a *stream* of data rather than holding it all in memory.

For example, suppose we had two large files full of nucleotide sequences in a FASTA file, a standard text format used to store sequence data (usually of DNA, but occasionally proteins, too). Even the simple task of combining these two large files into a single file becomes tricky once these are a few gigabytes in size. How would this simple task be accomplished without using the Unix shell? You could try to open one file, select and copy all of its contents, and paste it into another file. However, not only would this require loading both files in memory, but you'd also use additional memory making another copy of one file when you select all, copy, and paste. Approaches like this do not scale to the size of data we routinely work with in bioinformatics. Additionally, pasting contents into a file doesn't follow a recommendation from Chapter 1: treat data as read-only. If something went wrong, one of the files (or both!) could easily be corrupted. To make matters worse, copying and pasting large files uses lots of memory, so it's even more likely something could go awry with your computer. Streams offer a scalable, robust solution to these problems.

Redirecting Standard Out to a File

The Unix shell simplifies tasks like combining large files by leveraging streams. Using streams prevents us from unnecessarily loading large files into memory. Instead, we can combine large files by printing their contents to the *standard output* stream and *redirect* this stream from our terminal to the file we wish to save the combined results to. You've probably used the program `cat` to print a file's contents to standard out (which when not redirected is printed to your terminal screen). For example, we can look at the *tb1-protein.fasta* file (available in this chapter's directory on GitHub) by using `cat` to print it to standard out:

```
$ cat tb1-protein.fasta
>teosinte-branched-1 protein
LGVPSVKHMFPFCDSSSPMDLPLYQQLQLSPSSPKTDQSSSFYCYPCSPP
FAAADASFPLSYQIGSAAAADATPPQAVINSPDLPVQALMDHAPAPATEL
GACASGAEGSGASLDRAAAAARKDRHSKICTAGGMRDRRMRLSLDVARKF
FALQDMLGFDKASKTVQWLLNTSKSAIQEIMADDASSECVEDGSSSLSVD
GKHNPAEQLGGGGDQKPKGNCRGEGKKPAKASKAAATPKPPRKSANNAHQ
VPDKETRAKARERARERTKEKHRMRWVKLASAIDVEAAAASVPSDRPSSN
NLSHHSSLSMNMPCAAA
```

`cat` also allows us to print multiple files' contents to the standard output stream, in the order as they appear to command arguments. This essentially concatenates these files, as seen here with the tb1 and tga1 translated sequences:

```
$ cat tb1-protein.fasta tga1-protein.fasta
>teosinte-branched-1 protein
LGVPSVKHMFPFCDSSSPMDLPLYQQLQLSPSSPKTDQSSSFYCYPCSPP
FAAADASFPLSYQIGSAAAADATPPQAVINSPDLPVQALMDHAPAPATEL
GACASGAEGSGASLDRAAAAARKDRHSKICTAGGMRDRRMRLSLDVARKF
FALQDMLGFDKASKTVQWLLNTSKSAIQEIMADDASSECVEDGSSSLSVD
GKHNPAEQLGGGGDQKPKGNCRGEGKKPAKASKAAATPKPPRKSANNAHQ
VPDKETRAKARERARERTKEKHRMRWVKLASAIDVEAAAASVPSDRPSSN
NLSHHSSLSMNMPCAAA
>teosinte-glume-architecture-1 protein
DSDCALSLLSAPANSSGIDVSRMVRPTEHVPMAQQPVVPGLQFGSASWFP
RPQASTGGSFVPSCPAAVEGEQQLNAVLGPNDSEVSMNYGGMFHVGGGSG
GGEGSSDGGT
```

While these files have been concatenated, the results are not saved anywhere—these lines are just printed to your terminal screen. In order to save these concatenated results to a file, you need to *redirect* this standard output stream from your terminal screen to a file. Redirection is an important concept in Unix, and one you'll use frequently in bioinformatics.

We use the operators > or >> to redirect standard output to a file. The operator > redirects standard output to a file and overwrites any existing contents of the file (take note of this and be careful), whereas the latter operator >> appends to the file (keeping the contents and just adding to the end). If there isn't an existing file, both operators will create it before redirecting output to it. To concatenate our two FASTA files, we use `cat` as before, but redirect the output to a file:

```
$ cat tb1-protein.fasta tga1-protein.fasta > zea-proteins.fasta
```

Note that nothing is printed to your terminal screen when you redirect standard output to a file. In our example, the entire standard output stream ends up in the *zea-proteins.fasta* file. Redirection of a standard output stream to a file looks like Figure 3-1 (b).

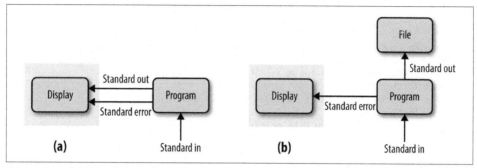

Figure 3-1. (a) Unredirected standard output, standard error, and standard input (the gray box is what is printed to a user's terminal); (b) standard output redirected to a file

We can verify that our redirect worked correctly by checking that the mostly recently created file in this directory is the one we just created (i.e., *zea-proteins.fasta*):

```
ls -lrt
total 24
-rw-r--r--  1 vinceb  staff  353 Jan 20 21:24 tb1-protein.fasta
-rw-r--r--  1 vinceb  staff  152 Jan 20 21:24 tga1-protein.fasta
-rw-r--r--  1 vinceb  staff  505 Jan 20 21:35 zea-proteins.fasta
```

Adding -lrt to the ls lists files in this directory in list format (-l), in reverse (-r) time (-t) order (see man ls for more details). Also, note how these flags have been combined into -lrt; this is a common syntactic shortcut. If you wished to see the newest files at the top, you could omit the r flag.

Redirecting Standard Error

Because many programs use the standard output stream for outputting data, a separate stream is needed for errors, warnings, and messages meant to be read by the user. *Standard error* is a stream just for this purpose (depicted in Figure 3-1). Like standard output, standard error is by default directed to your terminal. In practice, we often want to redirect the standard error stream to a file so messages, errors, and warnings are logged to a file we can check later.

To illustrate how we can redirect both standard output and standard error, we'll use the command ls -l to list both an existing file (*tb1.fasta*) and a file that does not exist (*leafy1.fasta*). The output of ls -l for the existing file *tb1.fasta* will be sent to standard output, while an error message saying *leafy1.fasta* does not exist will be output to standard error. When you don't redirect anything, both streams are output to your terminal:

```
$ ls -l tb1.fasta leafy1.fasta
ls: leafy1.fasta: No such file or directory
-rw-r--r--  1 vinceb  staff  0 Feb 21 21:58 tb1.fasta
```

To redirect each stream to *separate* files, we combine the > operator from the previous section with a new operator for redirecting the standard error stream, 2>:

```
$ ls -l tb1.fasta leafy1.fasta > listing.txt 2> listing.stderr
$ cat listing.txt
-rw-r--r--  1 vinceb  staff  152 Jan 20 21:24 tb1.fasta
$ cat listing.stderr
ls: leafy1.fasta: No such file or directory
```

Additionally, 2> has 2>>, which is analogous to >> (it will append to a file rather than overwrite it).

File Descriptors

The 2> notation may seem rather cryptic (and difficult to memorize), but there's a reason why standard error's redirect operator has a 2 in it. All open files (including streams) on Unix systems are assigned a unique integer known as a *file descriptor*. Unix's three standard streams—standard input (which we'll see in a bit), *standard output*, and *standard error*—are given the file descriptors 0, 1, and 2, respectively. It's even possible to use 1> as the redirect operator for standard output, though this is not common in practice and may be confusing to collaborators.

Occasionally a program will produce messages we don't need or care about. Redirection can be a useful way to silence diagnostic information some programs write to standard out: we just redirect to a logfile like *stderr.txt*. However, in some cases, we don't need to save this output to a file and writing output to a physical disk can slow programs down. Luckily, Unix-like operating systems have a special "fake" disk (known as a *pseudodevice*) to redirect unwanted output to: */dev/null*. Output written to */dev/null* disappears, which is why it's sometimes jokingly referred to as a "blackhole" by nerds.

Using tail -f to Monitor Redirected Standard Error

We often need to redirect both the standard output and standard error for large bioinformatics programs that could run for days (or maybe weeks, or months!). With both these streams redirected, nothing will be printed to your terminal, including useful diagnostic messages you may want to keep an eye on during long-running tasks. If you wish to follow these messages, the program `tail` can be used to look at the last lines of an output file by calling `tail filename.txt`. For example, running `tail stderr.txt` will print the last 10 lines of the file *stderr.txt*. You can set the exact number of lines `tail` will print with the `-n` option.

Tail can also be used to constantly monitor a file with `-f` (`-f` for follow). As the monitored file is updated, `tail` will display the new lines to your terminal screen, rather than just display 10 lines and exiting as it would without this option. If you wish to stop the monitoring of a file, you can use Control-C to interrupt the `tail` process. The process writing to the file will not be interrupted when you close `tail`.

Using Standard Input Redirection

The Unix shell also provides a redirection operator for standard input. Normally standard input comes from your keyboard, but with the < redirection operator you can read standard input directly from a file. Though standard input redirection is less common than >, >>, and 2>, it is still occasionally useful:

```
$ program < inputfile > outputfile
```

In this example, the artificial file `inputfile` is provided to `program` through standard input, and all of `program`'s standard output is redirected to the file `outputfile`.

It's a bit more common to use Unix pipes (e.g., `cat inputfile | program > output file`) than <. Many programs we'll see later (like `grep`, `awk`, `sort`) also can take a file argument in addition to input through standard input. Other programs (common especially in bioinformatics) use a single dash argument (`-`) to indicate that they should use standard input, but this is a convention rather than a feature of Unix.

The Almighty Unix Pipe: Speed and Beauty in One

> We should have some ways of connecting programs like [a] garden hose—screw in another segment when it becomes necessary to massage data in another way.
>
> —Doug McIlory (1964)

In "Why Do We Use Unix in Bioinformatics? Modularity and the Unix Philosophy" on page 37 McIlroy's quote included the recommendation, "write programs to work together." This is easy thanks to the Unix pipe, a concept that McIlroy himself invented. Unix pipes are similar to the redirect operators we saw earlier, except rather than redirecting a program's standard output stream to a file, pipes redirect it to another program's standard input. Only standard output is piped to the next command; standard error still is printed to your terminal screen, as seen in Figure 3-2.

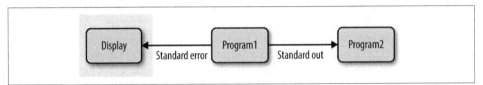

Figure 3-2. Piping standard output from program1 to program2; standard error is still printed to the user's terminal

You may be wondering why we would pipe a program's standard output *directly* into another program's standard input, rather than writing output to a file and then reading this file into the next program. In many cases, creating a file would be helpful in checking intermediate results and possibly debugging steps of your workflow—so why not do this every time?

The answer is that it often comes down to computational efficiency—reading and writing to the disk is very slow. We use pipes in bioinformatics (quite compulsively) not only because they are useful way of building pipelines, but because they're faster (in some cases, *much* faster). Modern disks are orders of magnitude slower than memory. For example, it only takes about 15 microseconds to read 1 megabyte of data from memory, but 2 milliseconds to read 1 megabyte from a disk. These 2 milliseconds are 2,000 microseconds, making reading from the disk more than 100 times slower (this is an estimate; actual numbers will vary according to your disk type and speed).

In practice, writing or reading from a disk (e.g., during redirection of standard output to a file) is often a bottleneck in data processing. For large next-generation sequencing data, this can slow things down quite considerably. If you implement a clever algorithm that's twice as fast as an older version, you may not even notice the difference if the true bottleneck is reading or writing to the disk. Additionally, unnecessarily redirecting output to a file uses up disk space. With large next-generation data and potentially many experimental samples, this can be quite a problem.

Passing the output of one program directly into the input of another program with pipes is a computationally efficient and simple way to interface Unix programs. This is another reason why bioinformaticians (and software engineers in general) like Unix. Pipes allow us to build larger, more complex tools from smaller modular parts. It doesn't matter what language a program is written in, either; pipes will work between anything as long as both programs understand the data passed between them. As the lowest common denominator between most programs, plain-text streams are often used—a point that McIlroy makes in his quote about the Unix philosophy.

Pipes in Action: Creating Simple Programs with Grep and Pipes

The Golden Rule of Bioinformatics is to not trust your tools or data. This skepticism requires constant sanity checking of intermediate results, which ensures your methods aren't biasing your data, or problems in your data aren't being exacerbated by your methods. However, writing custom scripts to check every bit of your intermediate data can be costly, even if you're a fast programmer who can write bug-free code the first go. Unix pipes allow us to quickly and iteratively build tiny command-line programs to inspect and manipulate data—an approach we'll explore in much more depth in Chapter 7. Pipes also are used extensively in larger bioinformatics workflows (Chapter 12), as they avoid latency issues from writing unnecessary files to disk. We'll learn the basics of pipes in this section, preparing you to use them throughout the rest of the book.

Let's look at how we can chain processes together with pipes. Suppose we're working with a FASTA file and a program warns that it contains non-nucleotide characters in sequences. You're surprised by this, as the sequences are only DNA. We can check for non-nucleotide characters easily with a Unix one-liner using pipes and the grep. The grep Unix tool searches files or standard input for strings that match patterns. These patterns can be simple strings, or regular expressions (there are actually two flavors of regular expressions, basic and extended; see man grep for more details). If you're unfamiliar with regular expressions, see the book's GitHub repository's *README* for resources.

Our pipeline would first remove all header lines (those that begin with >) from the FASTA files, as we only care if sequences have non-nucleotide characters. The remaining sequences of the FASTA file could then be *piped* to another instance of grep, which would only print lines containing non-nucleotide characters. To make these easier to spot in our terminal, we could also color these matching characters. The entire command would look like:

```
$ grep -v "^>" tb1.fasta | \    ❶
  grep --color -i "[^ATCG]"    ❷
CCCCAAAGACGGACCAATCCAGCAGCTTCTACTGCTAYCCATGCTCCCCTCCCTTCGCCGCCGCCGACGC
```

❶ First, we remove the FASTA header lines, which begin with the > character. Our regular expression pattern is ^>, which matches all lines that start with a > character. The caret symbol has two meanings in regular expressions, but in this context, it's used to anchor a pattern to the start of a line. Because we want to *exclude* lines starting with >, we *invert* the matching lines with the grep option -v. Finally, we pipe the standard output to the next command with the pipe character (|). The backslash (\) character simply means that we're continuing the command on the next line, and is used above to improve readability.

❷ Second, we want to find any characters that are *not* A, T, C, or G. It's easiest to construct a regular expression pattern that doesn't match A, T, C, or G. To do this, we use the caret symbol's second meaning in regular expressions. When used in brackets, a caret symbol matches anything that's *not* one of the characters in these brackets. So the pattern [^ATCG] matches any character that's not A, T, C, or G. Also, we ignore case with -i, because a, t, c, and g are valid nucleotides (lowercase characters are often used to indicate masked repeat or low-complexity sequences). Finally, we add grep's --color option to color the matching non-nucleotide characters.

When run in a terminal window, this would highlight "Y". Interestingly, Y is actually a valid extended ambiguous nucleotide code according to a standard set by IUPAC. Y represents the pYrimidine bases: C or T. Other single-letter IUPAC codes can represent uncertainty in sequence data. For example, puRine bases are represented by R, and a base that's either A, G, or T has the code D.

Let's discuss a few additional points about this simple Unix pipe. First, note that both regular expressions are in quotes, which is a good habit to get into. Also, if instead we had used grep -v > tb1.fasta, your shell would have interpreted the > as a redirect operator rather than a pattern supplied to grep. Unfortunately, this would mistakenly *overwrite* your *tb1.fasta* file! Most bioinformaticians have made this mistake at some point and learned the hard way (by losing the FASTA file they were hoping to grep), so beware.

This simple Unix one-liner takes only a few seconds to write and run, and works great for this particular task. We could have written a more complex program that explicitly parses the FASTA format, counts the number of occurrences, and outputs a list of sequence names with non-nucleotide characters. However, for our task—seeing why a program isn't working—building simple command-line tools on the fly is fast and sufficient. We'll see many more examples of how to build command line tools with Unix data programs and pipes in Chapter 7.

Combining Pipes and Redirection

Large bioinformatics programs like aligners, assemblers, and SNP callers will often use multiple streams simultaneously. Results (e.g., aligned reads, assembled contigs, or SNP calls) are output via the standard output stream while diagnostic messages, warnings, or errors are output to the standard error stream. In such cases, we need to combine pipes and redirection to manage all streams from a running program.

For example, suppose we have two imaginary programs: program1 and program2. Our first program, program1, does some processing on an input file called *input.txt* and outputs results to the standard output stream and diagnostic messages to the standard error stream. Our second program, program2, takes standard output from program1 as input and processes it. program2 also outputs its own diagnostic messages to standard error, and results to standard output. The tricky part is that we now have *two* processes outputting to both standard error and standard output. If we didn't capture both program1's and program2's standard error streams, we'd have a jumbled mess of diagnostic messages on our screen that scrolls by too quickly for us to read. Luckily, we can can combine pipes and redirects easily:

```
$ program1 input.txt 2> program1.stderr  | \     ❶
     program2 2> program2.stderr > results.txt ❷
```

❶ program1 processes the *input.txt* input file and then outputs its results to standard output. program1's standard error stream is redirected to the *program1.stderr* logfile. As before, the backslash is used to split these commands across multiple lines to improve readability (and is optional in your own work).

❷ Meanwhile, program2 uses the standard output from program1 as its standard input. The shell redirects program2's standard error stream to the *program2.stderr* logfile, and program2's standard output to *results.txt*.

Occasionally, we need to redirect a standard error stream to standard output. For example, suppose we wanted to use grep to search for "error" in both the standard output and standard error streams of program1. Using pipes wouldn't work, because pipes only link the standard output of one program to the standard input of the next. Pipes ignore standard error. We can get around this by first redirecting standard error to standard output, and then piping this merged stream to grep:

```
$ program1 2>&1 | grep "error"
```

The 2>&1 operator is what redirects standard error to the standard output stream.

Even More Redirection: A tee in Your Pipe

As mentioned earlier, pipes prevent unnecessary disk writing and reading operations by connecting the standard output of one process to the standard input of another. However, we do occasionally need to write intermediate files to disk in Unix pipelines. These intermediate files can be useful when debugging a pipeline or when you wish to store intermediate files for steps that take a long time to complete. Like a plumber's T joint, the Unix program tee diverts a copy of your pipeline's standard output stream to an intermediate file while still passing it through its standard output:

```
$ program1 input.txt | tee intermediate-file.txt | program2 > results.txt
```

Here, `program1`'s standard output is both written to *intermediate-file.txt* and piped directly into `program2`'s standard input.

Managing and Interacting with Processes

When we run programs through the Unix shell, they become *processes* until they successfully finish or terminate with an error. There are multiple processes running on your machine simultaneously—for example, system processes, as well as your web browser, email application, bioinformatics programs, and so on. In bioinformatics, we often work with processes that run for a large amount of time, so it's important we know how to work with and manage processes from the Unix shell. In this section, we'll learn the basics of manipulating processes: running and managing processes in the background, killing errant processes, and checking process exit status.

Background Processes

When we type a command in the shell and press Enter, we lose access to that shell prompt for however long the command takes to run. This is fine for short tasks, but waiting for a long-running bioinformatics program to complete before continuing work in the shell would kill our productivity. Rather than running your programs in the foreground (as you do normally when you run a command), the shell also gives you the option to run programs in the *background*. Running a process in the background frees the prompt so you can continue working.

We can tell the Unix shell to run a program in the background by appending an ampersand (&) to the end of our command. For example:

```
$ program1 input.txt > results.txt &
[1] 26577
```

The number returned by the shell is the *process ID* or PID of `program1`. This is a unique ID that allows you to identify and check the status of `program1` later on. We can check what processes we have running in the background with `jobs`:

```
$ jobs
[1]+  Running        program1 input.txt > results.txt
```

To bring a background process into the foreground again, we can use `fg` (for foreground). `fg` will bring the most recent process to the foreground. If you have many processes running in the background, they will all appear in the list output by the program `jobs`. The numbers like [1] are *job IDs* (which are different than the process IDs your system assigns your running programs). To return a specific background job to the foreground, use `fg %<num>` where `<num>` is its number in the job list. If we wanted to return `program1` to the foreground, both `fg` and `fg %1` would do the same thing, as there's only one background process:

```
$ fg
program1 input.txt > results.txt
```

Background Processes and Hangup Signals

There's a slight gotcha with background processes: although they run in the background and seem disconnected from our terminal, closing our terminal window would cause these processes to be killed. Unfortunately, a lot of long-running important jobs have been accidentally killed this way.

Whenever our terminal window closes, it sends a *hangup* signal. Hangup signals (also know as SIGHUP) are from the era in which network connections were much less reliable. A dropped connection could prevent a user from stopping an aberrant, resource-hungry process. To address this, the hangup signal is sent to all processes started from closed terminal. Nearly all Unix command-line programs stop running as soon as they receive this signal.

So beware—running a process in the background does not guarantee that it won't die when your terminal closes. To prevent this, we need to use the tool nohup or run it from within Tmux, two topics we'll cover in much more detail in Chapter 4.

It's also possible to place a process already running in the foreground into the background. To do this, we first need to *suspend* the process, and then use the bg command to run it in the background. Suspending a process temporarily pauses it, allowing you to put it in the background. We can suspend processes by sending a stop signal through the key combination Control-z. With our imaginary program1, we would accomplish this as follows:

```
$ program1 input.txt > results.txt # forgot to append ampersand
$ # enter control-z
[1]+  Stopped                 program1 input.txt > results.txt
$ bg
[1]+ program1 input.txt > results.txt
```

As with fg earlier, we could also use jobs to see the suspended process's job ID. If we have multiple running processes, we can specify which one to move to the background with bg %<num> (where <num> is the job ID).

Killing Processes

Occasionally we need to kill a process. It's not uncommon for a process to demand too many of our computer's resources or become nonresponsive, requiring that we send a special signal to kill the process. Killing a process ends it for good, and unlike suspending it with a stop signal, it's unrecoverable. If the process is currently running in your shell, you can kill it by entering Control-C. This only works if this process is running in the foreground, so if it's in the background you'll have to use the `fg` discussed earlier.

More advanced process management (including monitoring and finding processes with `top` and `ps`, and killing them with the `kill` command) is out of the scope of this chapter. However, there's *a lot* of information about process and resource management in this chapter's *README* on GitHub.

Exit Status: How to Programmatically Tell Whether Your Command Worked

One concern with long-running processes is that you're probably not going to wait around to monitor them. How do you know when they complete? How do you know if they successfully finished without an error? Unix programs exit with an *exit status*, which indicates whether a program terminated without a problem or with an error. By Unix standards, an exit status of 0 indicates the process ran successfully, and any nonzero status indicates some sort of error has occurred (and hopefully the program prints an understandable error message, too).

Warning Exit Statuses

Unfortunately, whether a program returns a nonzero status when it encounters an error is up to program developers. Occasionally, programmers forget to handle errors well (and this does indeed happen in bioinformatics programs), and programs can error out and still return a zero-exit status. This is yet another reason why it's crucial to follow The Golden Rule (i.e., don't trust your tools) and to always check your intermediate data.

The exit status isn't printed to the terminal, but your shell will set its value to a variable in your shell (aptly named a shell variable) named $?. We can use the `echo` command to look at this variable's value after running a command:

```
$ program1 input.txt > results.txt
$ echo $?
0
```

Exit statuses are incredibly useful because they allow us to programmatically chain commands together in the shell. A subsequent command in a chain is run conditionally on the last command's exit status. The shell provides two operators that implement this: one operator that runs the subsequent command only if the first command completed *successfully* (&&), and one operator that runs the next command only if the first completed *unsuccessfully* (||). If you're familiar with concept of short-circuit evaluation, you'll understand that these operators are short-circuiting *and* and *or*, respectively.

It's best to see these operators in an example. Suppose we wanted to run program1, have it write its output to file, and then have program2 read this output. To avoid the problem of program2 reading a file that's not complete because program1 terminated with an error, we want to start program2 only after program1 returns a zero (successful) exit code. The shell operator && executes subsequent commands only if previous commands have completed with a nonzero exit status:

```
$ program1 input.txt > intermediate-results.txt && \
    program2 intermediate-results.txt > results.txt
```

Using the || operator, we can have the shell execute a command only if the previous command has failed (exited with a nonzero status). This is useful for warning messages:

```
$ program1 input.txt > intermediate-results.txt || \
    echo "warning: an error occurred"
```

If you want to test && and ||, there are two Unix commands that do nothing but return either exit success (true) or exit failure (false). For example, think about why the following lines are printed:

```
$ true
$ echo $?
0
$ false
$ echo $?
1
$ true && echo "first command was a success"
first command was a success
$ true || echo "first command was not a success"
$ false || echo "first command was not a success"
first command was not a success
$ false && echo "first command was a success"
```

Additionally, if you don't care about the exit status and you just wish to execute two commands *sequentially*, you can use a single semicolon (;):

```
$ false; true; false; echo "none of the previous mattered"
none of the previous mattered
```

If you've only known the shell as something that you interacted with through a terminal, you may start to notice that it has many elements of a full programming language. This is because it is! In fact, you can write and execute shell scripts just as you do Python scripts. Keeping your bioinformatics shell work in a commented shell script kept under version control is the best way to ensure that your work is reproducible. We will discuss shell scripts in Chapter 12.

Command Substitution

Unix users like to have the Unix shell do work for them—this is why shell expansions like wildcards and brace expansion exist (if you need a refresher, refer back to Chapter 2). Another type of useful shell expansion is *command substitution*. Command substitution runs a Unix command inline and returns the output as a string that can be used in another command. This opens up a lot of useful possibilities.

A good example of when this is useful is in those early days of the New Year when we haven't yet adjusted to using the new date. For example, five days into 2013, I shared with my collaborators a directory of new results named *snp-sim-01-05-2012* (in *mm-dd-yyyy* format). After the embarrassment subsided, the Unix solution presented itself: the date command is there to programmatically return the current date as a string. We can use this string to automatically give our directories names containing the current date. We use command substitution to run the date program and replace this command with its output (the string). This is easier to understand through a simpler example:

```
$ grep -c '^>' input.fasta ❶
416
$ echo "There are $(grep -c '^>' input.fasta) entries in my FASTA file." ❷
There are 416 entries in my FASTA file.
```

❶ This command uses grep to count (the -c option stands for count) the number of lines matching the pattern. In this case, our pattern ^> matches FASTA header lines. Because each FASTA file entry has a header like ">sequence-a" that begins with ">", this command matches each of these headers and counts the number of FASTA entries.

❷ Now suppose we wanted to take the output of the grep command and insert it into *another* command—this is what command substitution is all about. In this case, we want echo to print a message containing how many FASTA entries there are to standard output. Using command substitution, we can calculate and return the number of FASTA entries directly into this string!

Using this command substitution approach, we can easily create dated directories using the command date +%F, where the argument +%F simply tells the date program to output the date in a particular format. date has multiple formatting options, so your European colleagues can specify a date as "19 May 2011" whereas your American colleagues can specify "May 19, 2011:"

```
$ mkdir results-$(date +%F)
$ ls results-2015-04-13
```

In general, the format returned by `date +%F` is a really good one for dated directories, because when results are sorted by name, directories in this format also sort chronologically:

```
$ ls -l
drwxr-xr-x  2 vinceb  staff  68 Feb  3 23:23 1999-07-01
drwxr-xr-x  2 vinceb  staff  68 Feb  3 23:22 2000-12-19
drwxr-xr-x  2 vinceb  staff  68 Feb  3 23:22 2011-02-03
drwxr-xr-x  2 vinceb  staff  68 Feb  3 23:22 2012-02-13
drwxr-xr-x  2 vinceb  staff  68 Feb  3 23:23 2012-05-26
drwxr-xr-x  2 vinceb  staff  68 Feb  3 23:22 2012-05-27
drwxr-xr-x  2 vinceb  staff  68 Feb  3 23:23 2012-07-04
drwxr-xr-x  2 vinceb  staff  68 Feb  3 23:23 2012-07-05
```

The cleverness behind this is what makes this date format, known as ISO 8601, useful.

Storing Your Unix Tricks

In Chapter 2, we made a project directory with `mkdir -p` and brace expansions. If you find yourself making the same project structures a lot, it's worthwhile to store it rather than typing it out each time. Why repeat yourself?

Early Unix users were a clever (or lazy) bunch and devised a tool for storing repeated command combinations: `alias`. If you're running a clever one-liner over and over again, use add `alias` to add it to your ~/.bashrc (or ~/.profile if on OS X). `alias` simply aliases your command to a shorter name alias. For example, if you always create project directories with the same directory structure, add a line like the following:

```
alias mkpr="mkdir -p {data/seqs,scripts,analysis}"
```

For small stuff like this, there's no point writing more complex scripts; adopt the Unix way and keep it simple. Another example is that we could alias our `date +%F` command to `today`:

```
alias today="date +%F"
```

Now, entering `mkdir results-$(today)` will create a dated results directory.

A word of warning, though: do not use your aliased command in project-level shell scripts! These reside in your shell's startup file (e.g., ~/.profile or ~/.bashrc), which is *outside* of your project directory. If you distribute your project directory, any shell programs that require alias definitions will not work. In computing, we say that such a practice is not portable—if it moves off your system, it breaks. Writing code to be portable, even if it's not going to run elsewhere, will help in keeping projects reproducible.

Like all good things, Unix tricks like command substitution are best used in moderation. Once you're a bit more familiar with these tricks, they're very quick and easy solutions to routine annoyances. In general, however, it's best to keep things simple and know when to reach for the quick Unix solution and when to use another tool like Python or R. We'll discuss this in more depth in Chapters 7 and 8.

CHAPTER 4
Working with Remote Machines

Most data-processing tasks in bioinformatics require more computing power than we have on our workstations, which means we must work with large servers or computing clusters. For some bioinformatics projects, it's likely you'll work predominantly over a network connection with remote machines. Unsurprisingly, working with remote machines can be quite frustrating for beginners and can hobble the productivity of experienced bioinformaticians. In this chapter, we'll learn how to make working with remote machines as effortless as possible so you can focus your time and efforts on the project itself.

Connecting to Remote Machines with SSH

There are many ways to connect to another machine over a network, but by far the most common is through the *secure shell* (*SSH*). We use SSH because it's encrypted (which makes it secure to send passwords, edit private files, etc.), and because it's on every Unix system. How your server, SSH, and your user account are configured is something you or your system administrator determines; this chapter won't cover these system administration topics. The material covered in this section should help you answer common SSH questions a sysadmin may ask (e.g., "Do you have an SSH public key?"). You'll also learn all of the basics you'll need as a bioinformatician to SSH into remote machines.

To initialize an SSH connection to a host (in this case, *biocluster.myuniversity.edu*), we use the ssh command:

```
$ ssh biocluster.myuniversity.edu
Password: ❶
Last login: Sun Aug 11 11:57:59 2013 from fisher.myisp.com
wsobchak@biocluster.myuniversity.edu$ ❷
```

❶ When connecting to a remote host with SSH, you'll be prompted for your remote user account's password.

❷ After logging in with your password, you're granted a shell prompt on the remote host. This allows you to execute commands on the remote host just as you'd execute them locally.

SSH also works with IP addresses—for example, you could connect to a machine with `ssh 192.169.237.42`. If your server uses a different port than the default (port 22), or your username on the remote machine is different from your local username, you'll need to specify these details when connecting:

```
$ ssh -p 50453 cdarwin@biocluster.myuniversity.edu
```

Here, we've specified the port with the flag `-p` and the username by using the syntax `user@domain`. If you're unable to connect to a host, using `ssh -v` (`-v` for verbose) can help you spot the issue. You can increase the verbosity by using `-vv` or `-vvv`; see `man ssh` for more details.

Storing Your Frequent SSH Hosts

Bioinformaticians are constantly having to SSH to servers, and typing out IP addresses or long domain names can become quite tedious. It's also burdensome to remember and type out additional details like the remote server's port or your remote username. The developers behind SSH created a clever alternative: the SSH config file. SSH config files store details about hosts you frequently connect to. This file is easy to create, and hosts stored in this file work not only with `ssh`, but also with two programs we'll learn about in Chapter 6: `scp` and `rsync`.

To create a file, just create and edit the file at *~/.ssh/config*. Each entry takes the following form:

```
Host bio_serv
    HostName 192.168.237.42
    User cdarwin
    Port 50453
```

You won't need to specify `Port` and `User` unless these differ from the remote host's defaults. With this file saved, you can SSH into *192.168.236.42* using the alias `ssh bio_serv` rather than typing out `ssh -p 50453 cdarwin@192.169.237.42`.

If you're working with many remote machine connections in many terminal tabs, it's sometimes useful to be make sure you're working on the host you think you are. You can always access the hostname with the command `hostname`:

```
$ hostname
biocluster.myuniversity.edu
```

Similarly, if you maintain multiple accounts on a server (e.g., a user account for anal-ysis and a more powerful administration account for sysadmin tasks), it can be useful to check which account you're using. The command whoami returns your username:

```
$ whoami
cdarwin
```

This is especially useful if you do occasionally log in with an administrator account with more privileges—the potential risks associated with making a mistake on an account with administrator privileges are much higher, so you should always be away when you're on this account (and minimize this time as much as possible).

Quick Authentication with SSH Keys

SSH requires that you type your password for the account on the remote machine. However, entering a password each time you login can get tedious, and not always safe (e.g., keyboard input could be monitored). A safer, easier alternative is to use an *SSH public key*. Public key cryptography is a fascinating technology, but the details are outside the scope of this book. To use SSH keys to log in into remote machines without passwords, we first need to generate a public/private key pair. We do this with the command ssh-keygen. It's very important that you note the difference between your public and private keys: you can distribute your public key to other servers, but your private key must be kept safe and secure and never shared.

Let's generate an SSH key pair using ssh-keygen:

```
$ ssh-keygen -b 2048
Generating public/private rsa key pair.
Enter file in which to save the key (/Users/username/.ssh/id_rsa):
Enter passphrase (empty for no passphrase):
Enter same passphrase again:
Your identification has been saved in /Users/username/.ssh/id_rsa.
Your public key has been saved in /Users/username/.ssh/id_rsa.pub.
The key fingerprint is:
e1:1e:3d:01:e1:a3:ed:2b:6b:fe:c1:8e:73:7f:1f:f0
The key's randomart image is:
+--[ RSA 2048]----+
|.o... ...        |
| .  .  o         |
| .       *       |
| .     o +       |
| .     S .       |
| o    . E        |
|  +    .         |
|oo+..  . .       |
|+=oo... o.       |
+-----------------+
```

This creates a private key at *~/.ssh/id_rsa* and a public key at *~/.ssh/id_rsa.pub*. ssh-keygen gives you the option to use an empty password, but it's generally recommended that you use a real password. If you're wondering, the random art ssh-keygen creates is a way of validating your keys (there are more details about this in man ssh if you're curious).

To use password-less authentication using SSH keys, first SSH to your remote host and log in with your password. Change directories to *~/.ssh*, and append the contents of your public key file (*id_rsa.pub, not* your private key!) to *~/.ssh/authorized_keys* (note that the ~ may be expanded to */home/username* or */Users/username* depending on the remote operating system). You can append this file by copying your public key from your local system, and pasting it to the *~/.ssh/authorized_keys* file on the remote system. Some systems have an ssh-copy-id command that automatically does this for you.

Again, be sure you're using your public key, and *not* the private key. If your private key ever is accidentally distributed, this compromises the security of the machines you've set up key-based authentication on. The *~/.ssh/id_rsa* private key has read/write permissions only for the creator, and these restrictive permissions should be kept this way.

After you've added your public key to the remote host, try logging in a few times. You'll notice that you keep getting prompted for your SSH key's password. If you're scratching your head wondering how this saves time, there's one more trick to know: ssh-agent. The ssh-agent program runs in the background on your local machine, and manages your SSH key(s). ssh-agent allows you to use your keys without entering their passwords each time—exactly what we want when we frequently connect to servers. SSH agent is usually already running on Unix-based systems, but if not, you can use eval ssh-agent to start it. Then, to tell ssh-agent about our key, we use ssh-add:

```
$ ssh-add
Enter passphrase for /Users/username/.ssh/id_rsa:
Identity added: /Users/username/.ssh/id_rsa
```

Now, the background ssh-agent process manages our key for us, and we won't have to enter our password each time we connect to a remote machine. I once calculated that I connect to different machines about 16 times a day, and it takes me about two seconds to enter my password on average (accounting for typing mistakes). If we were to assume I didn't work on weekends, this works out to about 8,320 seconds, or 2.3 hours a year of just SSH connections. After 10 years, this translates to nearly an entire day wasted on just connecting to machines. Learning these tricks may take an hour or so, but over the course of a career, this really saves time.

Maintaining Long-Running Jobs with nohup and tmux

In Chapter 3, we briefly discussed how processes (whether running in the foreground or background) will be terminated when we close our terminal window. Processes are also terminated if we disconnect from our servers or if our network connection temporarily drops out. This behavior is intentional—your program will receive the *hangup* signal (referred to more technically as SIGHUP), which will in almost all cases cause your application to exit immediately. Because we're perpetually working with remote machines in our daily bioinformatics work, we need a way to prevent hangups from stopping long-running applications. Leaving your local terminal's connection to a remote machine open while a program runs is a fragile solution—even the most reliable networks can have short outage blips. We'll look at two preferable solutions: nohup and Tmux. If you use a cluster, there are better ways to deal with hangups (e.g., submitting batch jobs to your cluster's software), but these depend on your specific cluster configuration. In this case, consult your system administrator.

nohup

nohup is simple command that executes a command and catches hangup signals sent from the terminal. Because the nohup command is catching and ignoring these hangup signals, the program you're running won't be interrupted. Running a command with nohup is as easy as adding nohup before your command:

```
$ nohup program1 > output.txt & ❶
[1] 10900 ❷
```

❶ We run the command with all options and arguments as we would normally, but by adding nohup this program will not be interrupted if your terminal were to close or the remote connection were to drop. Additionally, it's a good idea to redirect standard output and standard error just as we did in Chapter 3 so you can check output later.

❷ nohup returns the process ID number (or PID), which is how you can monitor or terminate this process if you need to (covered in "Killing Processes" on page 51). Because we lose access to this process when we run it through nohup, our only way of terminating it is by referring to it by its process ID.

Working with Remote Machines Through Tmux

An alternative to nohup is to use a *terminal multiplexer*. In addition to solving the hangup problem, using a terminal multiplexer will greatly increase your productivity when working over a remote connection. We'll use a terminal multiplexer called

Tmux, but a popular alternative is GNU Screen. Tmux and Screen have similar functionality, but Tmux is more actively developed and has some additional nice features.

Tmux (and terminal multiplexers in general) allow you to create a session containing multiple windows, each capable of running their own processes. Tmux's sessions are persistent, meaning that all windows and their processes can easily be restored by reattaching the session.

When run on a remote machine, Tmux allows you to maintain a persistent session that won't be lost if your connection drops or you close your terminal window to go home (or even quit your terminal program). Rather, all of Tmux's sessions can be reattached to whatever terminal you're currently on—just SSH back into the remote host and reattach the Tmux session. All windows will be undisturbed and all processes still running.

Installing and Configuring Tmux

Tmux is available through most package/port managers. On OS X, Tmux can be installed through Homebrew and on Ubuntu it's available through `apt-get`. After installing Tmux, I strongly suggest you go to this chapter's directory on GitHub and download the *.tmux.conf* file to your home directory. Just as your shell loads configurations from *~/.profile* or *~/.bashrc*, Tmux will load its configurations from *~/.tmux.conf*. The minimal settings in *.tmux.conf* make it easier to learn Tmux by giving you a useful display bar at the bottom and changing some of Tmux's key bindings to those that are more common among Tmux users.

Creating, Detaching, and Attaching Tmux Sessions

Tmux allows you to have multiple *sessions*, and within each session have multiple windows. Each Tmux session is a separate environment. Normally, I use a session for each different project I'm working on; for example, I might have a session for maize SNP calling, one for developing a new tool, and another for writing R code to analyze some *Drosophila* gene expression data. Within each of these sessions, I'd have multiple windows. For example, in my maize SNP calling project, I might have three windows open: one for interacting with the shell, one with a project notebook open in my text editor, and another with a Unix manual page open. Note that all of these windows are *within* Tmux; your terminal program's concept of tabs and windows is entirely different from Tmux's. Unlike Tmux, your terminal cannot maintain persistent sessions.

Let's create a new Tmux session. To make our examples a bit easier, we're going to do this on our local machine. However, to manage sessions on a remote host, we'd need to start Tmux on that remote host (this is often confusing for beginners). Running Tmux on a remote host is no different; we just SSH in to our host and start Tmux there. Suppose we wanted to create a Tmux session corresponding to our earlier Maize SNP calling example:

```
$ tmux new-session -s maize-snps
```

Tmux uses subcommands; the new-session subcommand just shown creates new sessions. The -s option simply gives this session a name so it's easier to identify later. If you're following along and you've correctly placed the .tmux.conf file in your home directory, your Tmux session should look like Figure 4-1.

```
bash-3.2$ echo "hello, tmux"
hello, tmux
bash-3.2$ 
```

```
maize-snps 0:bash*                                              "stebbins" 21:54 10-Feb-14
```

Figure 4-1. Tmux using the provided .tmux.conf file

Tmux looks just like a normal shell prompt except for the status bar it has added at the bottom of the screen (we'll talk more about this in a bit). When Tmux is open, we interact with Tmux through keyboard shortcuts. These shortcuts are all based on first pressing Control and *a*, and then adding a specific key after (releasing Control-*a* first). By default, Tmux uses Control-*b* rather than Control-*a*, but this is a change we've made in our .tmux.conf to follow the configuration preferred by most Tmux users.

The most useful feature of Tmux (and terminal multiplexers in general) is the ability to detach and reattach sessions without losing our work. Let's see how this works in Tmux. Let's first enter something in our blank shell so we can recognize this session later: echo "hello, tmux". To detach a session, we use Control-*a*, followed by *d* (for detach). After entering this, you should see Tmux close and be returned to your regular shell prompt.

After detaching, we can see that Tmux has kept our session alive by calling `tmux` with the `list-sessions` subcommand:

```
$ tmux list-sessions
maize-snps: 1 windows (created Mon Feb 10 00:06:00 2014) [180x41]
```

Now, let's reattach our session. We reattach sessions with the `attach-session` subcommand, but the shorter `attach` also works:

```
$ tmux attach
```

Note that because we only have one session running (our `maize-snps` session) we don't have to specify which session to attach. Had there been more than one session running, all session names would have been listed when we executed `list-sessions` and we could reattach a particular session using `-t <session-name>`. With only one Tmux session running, `tmux attach` is equivalent to `tmux attach-session -t maize-snps`.

Managing remote sessions with Tmux is no different than managing sessions locally as we did earlier. The only difference is that we create our sessions on the remote host after we connect with SSH. Closing our SSH connection (either intentionally or unintentionally due to a network drop) will cause Tmux to detach any active sessions.

Working with Tmux Windows

Each Tmux session can also contain multiple windows. This is especially handy when working on remote machines. With Tmux's windows, a single SSH connection to a remote host can support multiple activities in different windows. Tmux also allows you to create multiple *panes* within a window that allow you to split your windows into parts, but to save space I'll let the reader learn this functionality on their own. Consult the Tmux manual page (e.g., with `man tmux`) or read one of the many excellent Tmux tutorials on the Web.

Like other Tmux key sequences, we create and switch windows using Control-*a* and then another key. To create a window, we use Control-*a c*, and we use Control-*a n* and Control-*a p* to go to the next and previous windows, respectively. Table 4-1 lists the most commonly used Tmux key sequences. See `man tmux` for a complete list, or press Control-*a ?* from within a Tmux session.

Table 4-1. Common Tmux key sequences

Key sequence	Action
Control-*a d*	Detach
Control-*a c*	Create new window

Key sequence	Action
Control-*a n*	Go to next window
Control-*a p*	Go to previous window
Control-*a &*	Kill current window (`exit` in shell also works)
Control-*a ,*	Rename current window
Control-*a ?*	List all key sequences

Table 4-2 lists the most commonly used Tmux subcommands.

Table 4-2. Common Tmux subcommands

Subcommand	Action
`tmux list-sessions`	List all sessions.
`tmux new-session -s session-name`	Create a new session named "session-name".
`tmux attach-session -t session-name`	Attach a session named "session-name".
`tmux attach-session -d -t session-name`	Attach a session named "session-name", detaching it first.

If you use Emacs as your text editor, you'll quickly notice that the key binding Control-*a* may get in the way. To enter a literal Control-*a* (as used to go to the beginning of the line in Emacs or the Bash shell), use Control-*a a*.

## CHAPTER 5

Git for Scientists

In Chapter 2, we discussed organizing a bioinformatics project directory and how this helps keep your work tidy during development. Good organization also facilitates automating tasks, which makes our lives easier and leads to more reproducible work. However, as our project changes over time and possibly incorporates the work of our collaborators, we face an additional challenge: managing different file versions.

It's likely that you already use some sort of versioning system in your work. For example, you may have files with names such as *thesis-vers1.docx*, *thesis-vers3_CD_edits.docx*, *analysis-vers6.R*, and *thesis-vers8_CD+GM+SW_edits.docx*. Storing these past versions is helpful because it allows us to go back and restore whole files or sections if we need to. File versions also help us differentiate our copies of a file from those edited by a collaborator. However, this ad hoc file versioning system doesn't scale well to complicated bioinformatics projects—our otherwise tidy project directories would be muddled with different versioned scripts, R analyses, *README* files, and papers.

Project organization only gets more complicated when we work collaboratively. We could share our entire directory with a colleague through a service like Dropbox or Google Drive, but we run the risk of something getting deleted or corrupted. It's also not possible to drop an entire bioinformatics project directory into a shared directory, as it likely contains gigabytes (or more) of data that may be too large to share across a network. These tools are useful for sharing small files, but aren't intended to manage large collaborative projects involving changing code and data.

Luckily, software engineers have faced these same issues in modern collaborative software development and developed *version control systems* (VCS) to manage different versions of collaboratively edited code. The VCS we'll use in this chapter was written by Linus Torvalds and is called *Git*. Linus wrote Git to manage the Linux kernel (which he also wrote), a large codebase with thousands of collaborators simultaneously changing and working on files. As you can imagine, Git is well suited for project version control and collaborative work.

Admittedly, Git can be tricky to learn at first. I highly recommend you take the time to learn Git in this chapter, but be aware that understanding Git (like most topics in this book, and arguably everything in life) will take time and practice. Throughout this chapter, I will indicate when certain sections are especially advanced; you can revisit these later without problems in continuity with the rest of the book. Also, I recommend you practice Git with the example projects and code from the book to get the basic commands in the back of your head. After struggling in the beginning with Git, you'll soon see how it's the best version control system out there.

Why Git Is Necessary in Bioinformatics Projects

As a longtime proponent of Git, I've suggested it to many colleagues and offered to teach them the basics. In most cases, I find the hardest part is actually in convincing scientists they should adopt version control in their work. Because you may be wondering whether working through this chapter is worth it, I want to discuss why learning Git is definitely worth the effort. If you're already 100% convinced, you can dive into learning Git in the next section.

Git Allows You to Keep Snapshots of Your Project

With version control systems, you create snapshots of your current project at specific points in its development. If anything goes awry, you can rewind to a past snapshot of your project's state (called a commit) and restore files. In the fast pace of bioinformatics work, having this safeguard is very useful.

Git also helps fix a frustrating type of bug known as software regression, where a piece of code that was once working mysteriously stops working or gives different results. For example, suppose that you're working on an analysis of SNP data. You find in your analysis that 14% of your SNPs fall in coding regions in one stretch of a chromosome. This is relevant to your project, so you cite this percent in your paper and make a commit.

Two months later, you've forgotten the details of this analysis, but need to revisit the 14% statistic. Much to your surprise, when you rerun the analysis code, this changes to 26%! If you've been tracking your project's development by making commits (e.g., taking snapshots), you'll have an entire history of all of your project's changes and can pinpoint when your results changed.

Git commits allow you to easily reproduce and rollback to past versions of analysis. It's also easy to look at every commit, when it was committed, what has changed across commits, and even compare the difference between any two commits. Instead of redoing months of work to find a bug, Git can give you line-by-line code differences across versions.

In addition to simplifying bug finding, Git is an essential part of proper documentation. When your code produces results, it's essential that this version of code is fully documented for reproducibility. A good analogy comes from my friend and colleague Mike Covington: imagine you keep a lab notebook in pencil, and each time you run a new PCR you erase your past results and jot down the newest ones. This may sound extreme, but is functionally no different than changing code and not keeping a record of past versions.

Git Helps You Keep Track of Important Changes to Code

Most software changes over time as new features are added or bugs are fixed. It's important in scientific computing to follow the development of software we use, as a fixed bug could mean the difference between correct and incorrect results in our own work. Git can be very helpful in helping you track changes in code—to see this, let's look at a situation I've run into (and I suspect happens in labs all over the world).

Suppose a lab has a clever bioinformatician who has written a script that trims poor quality regions from reads. This bioinformatician then distributes this to all members of his lab. Two members of his lab send it to friends in other labs. About a month later, the clever bioinformatician realizes there's a bug that leads to incorrect results in certain cases. The bioinformatician quickly emails everyone in his lab the new version and warns them of the potential for incorrect results. Unfortunately, members of the other lab may not get the message and could continue using the older buggy version of the script.

Git helps solve this problem by making it easy to stay up to date with software development. With Git, it's easy to both track software changes and download new software versions. Furthermore, services like GitHub and Bitbucket host Git repositories on the Web, which makes sharing and collaborating on code across labs easy.

Git Helps Keep Software Organized and Available After People Leave

Imagine another situation: a postdoc moves to start her own lab, and all of her different software tools and scripts are scattered in different directories, or worse, completely lost. Disorderedly code disrupts and inconveniences other lab members; lost code leads to irreproducible results and could delay future research.

Git helps maintain both continuity in work and a full record of a project's history. Centralizing an entire project into a repository keeps it organized. Git stores every committed change, so the entire history of a project is available even if the main developer leaves and isn't around for questions. With the ability to roll back to past versions, modifying projects is less risky, making it easier to build off existing work.

Installing Git

If you're on OS X, install Git through Homebrew (e.g., `brew install git`); on Linux, use `apt-get` (e.g., `apt-get install git`). If your system does not have a package manager, the Git website (*http://git-scm.com/*) has both source code and executable versions of Git.

Basic Git: Creating Repositories, Tracking Files, and Staging and Committing Changes

Now that we've seen some Git concepts and how Git fits into your bioinformatics workflow, let's explore the most basic Git concepts of creating repositories, telling Git which files to track, and staging and committing changes.

Git Setup: Telling Git Who You Are

Because Git is meant to help with collaborative editing of files, you need to tell Git who you are and what your email address is. To do this, use:

```
$ git config --global user.name "Sewall Wright"
$ git config --global user.email "swright@adaptivelandscape.org"
```

Make sure to use your own name and email, or course. We interact with Git through *subcommands*, which are in the format `git <subcommand>`. Git has loads of subcommands, but you'll only need a few in your daily work.

Another useful Git setting to enable now is terminal colors. Many of Git's subcommands use terminal colors to visually indicate changes (e.g., red for deletion and green for something new or modified). We can enable this with:

```
$ git config --global color.ui true
```

git init and git clone: Creating Repositories

To get started with Git, we first need to initialize a directory as a Git *repository*. A *repository* is a directory that's under version control. It contains both your current working files and snapshots of the project at certain points in time. In version control lingo, these snapshots are known as *commits*. Working with Git is fundamentally about creating and manipulating these commits: creating commits, looking at past commits, sharing commits, and comparing different commits.

With Git, there are two primary ways to create a repository: by initializing one from an existing directory, or cloning a repository that exists elsewhere. Either way, the result is a directory that Git treats as a repository. Git only manages the files and subdirectories inside the repository directory—it cannot manage files outside of your repository.

Let's start by initializing the *zmays-snps/* project directory we created in Chapter 2 as a Git repository. Change into the *zmays-snps/* directory and use the Git subcommand git init:

```
$ git init
Initialized empty Git repository in /Users/vinceb/Projects/zmays-snps/.git/
```

git init creates a hidden directory called *.git/* in your *zmays-snps/* project directory (you can see it with ls -a). This *.git/* directory is how Git manages your repository behind the scenes. However, don't modify or remove anything in this directory—it's meant to be manipulated by Git only. Instead, we interact with our repository through Git subcommands like git init.

The other way to create a repository is through cloning an existing repository. You can clone repositories from anywhere: somewhere else on your filesystem, from your local network, or across the Internet. Nowadays, with repository hosting services like GitHub and Bitbucket, it's most common to clone Git repositories from the Web.

Let's practice cloning a repository from GitHub. For this example, we'll clone the Seqtk code from Heng Li's GitHub page. Seqtk is short for SEQuence ToolKit, and contains a well-written and useful set of tools for processing FASTQ and FASTA files. First, visit the GitHub repository (*http://github.com/lh3/seqtk*) and poke around a bit. All of GitHub's repositories have this URL syntax: *user/repository*. Note on this repository's page that clone URL on the righthand side—this is where you can copy the link to clone this repository.

Now, let's switch to a directory outside of *zmays-snps/*. Whichever directory you choose is fine; I use a *~/src/* directory for cloning and compiling other developers' tools. From this directory, run:

```
$ git clone git://github.com/lh3/seqtk.git
Cloning into 'seqtk'...
remote: Counting objects: 92, done.
remote: Compressing objects: 100% (47/47), done.
remote: Total 92 (delta 56), reused 80 (delta 44)
Receiving objects: 100% (92/92), 32.58 KiB, done.
Resolving deltas: 100% (56/56), done.
```

git clone clones seqtk to your local directory, mirroring the original repository on GitHub. Note that you won't be able to directly modify Heng Li's original GitHub repository—cloning this repository only gives you access to retrieve new updates from the GitHub repository as they're released.

Now, if you cd into *seqtk/* and run ls, you'll see seqtk's source:

```
$ cd seqtk
$ ls
Makefile  README.md khash.h   kseq.h    seqtk.c
```

Despite originating through different methods, both *zmays-snps/* and *seqtk/* are Git repositories.

Tracking Files in Git: git add and git status Part I

Although you've initialized the *zmays-snps/* as a Git repository, Git doesn't automatically begin tracking every file in this directory. Rather, you need to tell Git which files to *track* using the subcommand `git add`. This is actually a useful feature of Git—bioinformatics projects contain many files we don't want to track, including large data files, intermediate results, or anything that could be easily regenerated by rerunning a command.

Before tracking a file, let's use the command `git status` to check Git's status of the files in our repository (switch to the *zmays-snps/* directory if you are elsewhere):

```
$ git status
# On branch master ❶
#
# Initial commit
#
# Untracked files: ❷
#   (use "git add <file>..." to include in what will be committed)
#
#       README
#       data/
nothing added to commit but untracked files present (use "git add" to track)
```

`git status` tell us:

❶ We're on branch *master*, which is the default Git *branch*. Branches allow you to work on and switch between different versions of your project simultaneously. Git's simple and powerful branches are a primary reason it's such a popular version control system. We're only going to work with Git's default *master* branch for now, but we'll learn more about branches later in this chapter.

❷ We have a list of "Untracked files," which include everything in the root project directory. Because we haven't told Git to track anything, Git has nothing to put in a commit if we were to try.

It's good to get `git status` under your fingers, as it's one of the most frequently used Git commands. `git status` describes the current state of your project repository: what's changed, what's ready to be included in the next commit, and what's not being tracked. We'll use it extensively throughout the rest of this chapter.

Let's use `git add` to tell Git to track the *README* and *data/README* files in our *zmays-snps/* directory:

```
$ git add README data/README
```

Now, Git is tracking both *data/README* and *README*. We can verify this by running `git status` again:

```
$ ls
README    analysis data      scripts
$ git status
# On branch master
#
# Initial commit
#
# Changes to be committed:
#   (use "git rm --cached <file>..." to unstage)
#
#       new file:   README ❶
#       new file:   data/README
#
# Untracked files:
#   (use "git add <file>..." to include in what will be committed)
#
#       data/seqs/ ❷
```

❶ Note now how Git lists *README* and *data/README* as new files, under the section "changes to be committed." If we made a commit now, our commit would take a snapshot of the exact version of these files as they were when we added them with `git add`.

❷ There are also untracked directories like *data/seqs/*, as we have not told Git to track these yet. Conveniently, `git status` reminds us we could use `git add` to add these to a commit.

The *scripts/* and *analysis/* directories are not included in `git status` because they are empty. The *data/seqs/* directory is included because it contains the empty sequence files we created with touch in Chapter 2.

Staging Files in Git: git add and git status Part II

With Git, there's a difference between tracked files and files *staged* to be included in the next commit. This is a subtle difference, and one that often causes a lot of confusion for beginners learning Git. A file that's tracked means Git knows about it. A staged file is not only tracked, but its latest changes are staged to be included in the next commit (see Figure 5-1).

A good way to illustrate the difference is to consider what happens when we change one of the files we started tracking with `git add`. Changes made to a tracked file will not automatically be included in the next commit. To include these new changes, we would need to explicitly *stage* them—using `git add` again. Part of the confusion lies in the fact that `git add` both tracks new files and stages the changes made to tracked files. Let's work through an example to make this clearer.

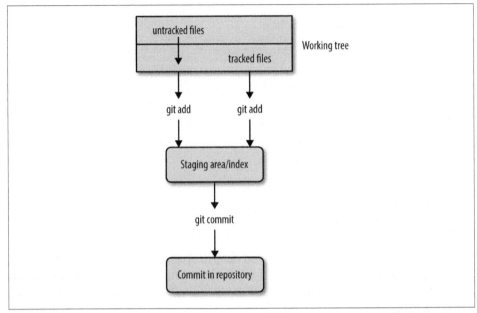

Figure 5-1. Git's separation of the working tree (all files in your repository), the staging area (files to be included in the next commit), and committed changes (a snapshot of a version of your project at some point in time); git add on an untracked file begins tracking it and stages it, while git add on a tracked file just stages it for the next commit

From the `git status` output from the last section, we see that both the *data/README* and *README* files are ready to be committed. However, look what happens when we make a change to one of these tracked files and then call `git status`:

```
$ echo "Zea Mays SNP Calling Project" >> README    # change file README
$ git status
# On branch master
#
# Initial commit
#
# Changes to be committed:
#   (use "git rm --cached <file>..." to unstage)
#
#       new file:   README
#       new file:   data/README
#
# Changes not staged for commit:
#   (use "git add <file>..." to update what will be committed)
#   (use "git checkout -- <file>..." to discard changes in working directory)
#
```

```
#         modified:    README
#
# Untracked files:
#    (use "git add <file>..." to include in what will be committed)
#
#         data/seqs/
```

After modifying *README*, `git status` lists *README* under "changes not staged for commit." This is because we've made changes to this file since initially tracking and staging *README* with `git add` (when first tracking a file, its current version is also staged). If we were to make a commit now, our commit would include the previous version of *README*, *not* this newly modified version.

To add these recent modifications to *README* in our next commit, we stage them using `git add`. Let's do this now and see what `git status` returns:

```
$ git add README
$ git status
# On branch master
#
# Initial commit
#
# Changes to be committed:
#    (use "git rm --cached <file>..." to unstage)
#
#         new file:    README
#         new file:    data/README
#
# Untracked files:
#    (use "git add <file>..." to include in what will be committed)
#
#         data/seqs/
#         notebook.md
```

Now, *README* is listed under "Changes to be committed" again, because we've staged these changes with `git add`. Our next commit will include the most recent version.

Again, don't fret if you find this confusing. The difference is subtle, and it doesn't help that we use `git add` for both operations. Remember the two roles of `git add`:

- Alerting Git to start tracking untracked files (this also stages the current version of the file to be included in the next commit)
- Staging changes made to an already tracked file (staged changes will be included in the next commit)

It's important to be aware that any modifications made to a file since the last time it was staged will *not* be included in the next commit unless they are staged with `git add`. This extra step may seem like an inconvenience but actually has many benefits.

Suppose you've made changes to many files in a project. Two of these files' changes are complete, but everything else isn't quite ready. Using Git's staging, you can stage and commit only these two complete files and keep other incomplete files out of your commit. Through planned staging, your commits can reflect meaningful points in development rather than random snapshots of your entire project directory (which would likely include many files in a state of disarray). When we learn about committing in the next section, we'll see a shortcut to stage and commit all modified files.

git commit: Taking a Snapshot of Your Project

We've spoken a lot about commits, but haven't actually made one yet. When first learning Git, the trickiest part of making a commit is understanding staging. Actually committing your staged commits is quite easy:

```
$ git commit -m "initial import"
 2 files changed, 1 insertion(+)
 create mode 100644 README
 create mode 100644 data/README
```

This command commits your staged changes to your repository with the *commit message* "initial import." Commit messages are notes to your collaborators (and yourself in the future) about what a particular commit includes. Optionally, you can omit the -m option, and Git will open up your default text editor. If you prefer to write commit messages in a text editor (useful if they are multiline messages), you can change the default editor Git uses with:

```
$ git config --global core.editor emacs
```

where emacs can be replaced by vim (the default) or another text editor of your choice.

Some Advice on Commit Messages

Commit messages may seem like an inconvenience, but it pays off in the future to have a description of how a commit changes code and what functionality is affected. In three months when you need to figure out why your SNP calling analyses are returning unexpected results, it's much easier to find relevant commits if they have messages like "modifying SNP frequency function to fix singleton bug, refactored coverage calculation" rather than "cont" (that's an actual commit I've seen in a public project). For an entertaining take on this, see xkcd's "Git Commit" comic (*http://xkcd.com/ 1296/*).

Earlier, we staged our changes using git add. Because programmers like shortcuts, there's an easy way to stage all tracked files' changes and commit them in one command: git commit -a -m "your commit message". The option -a tells git commit to automatically stage all modified tracked files in this commit. Note that while this saves time, it also will throw *all* changes to tracked files in this commit. Ideally commits should reflect helpful snapshots of your project's development, so including every slightly changed file may later be confusing when you look at your repository's history. Instead, make frequent commits that correspond to discrete changes to your project like "new counting feature added" or "fixed bug that led to incorrect translation."

We've included all changes in our commit, so our working directory is now "clean": no tracked files differ from the version in the last commit. Until we make modifications, `git status` indicates there's nothing to commit:

```
$ git status
# On branch master
# Untracked files:
#   (use "git add <file>..." to include in what will be committed)
#
#       data/seqs/
```

Untracked files and directories will still remain untracked (e.g., *data/seqs/*), and any unstaged changes to tracked files will not be included in the next commit unless added. Sometimes a working directory with unstaged changes is referred to as "messy," but this isn't a problem.

Seeing File Differences: git diff

So far we've seen the Git tools needed to help you stage and commit changes in your repository. We've used the `git status` subcommand to see which files are tracked, which have changes, and which are staged for the next commit. Another subcommand is quite helpful in this process: `git diff`.

Without any arguments, `git diff` shows you the difference between the files in your working directory and what's been staged. If none of your changes have been staged, `git diff` shows us the difference between your last commit and the current versions of your files. For example, if I add a line to *README.md* and run `git diff`:

```
$ echo "Project started 2013-01-03" >> README
$ git diff
diff --git a/README b/README
index 5483cfd..ba8d7fc 100644
--- a/README ❶
+++ b/README
@@ -1 +1,2 @@ ❷
 Zea Mays SNP Calling Project
+Project started 2013-01-03 ❸
```

This format (called a *unified diff*) may seem a bit cryptic at first. When Git's terminal colors are enabled, `git diff`'s output is easier to read, as added lines will be green and deleted lines will be red.

❶ This line (and the one following it) indicate there are two versions of the *README* file we are comparing, a and b. The `---` indicates the original file—in our case, the one from our last commit. `+++` indicates the changed version.

❷ This denotes the start of a changed hunk (hunk is diff's term for a large changed block), and indicates which line the changes start on, and how long they are. Diffs try to break your changes down into hunks so that you can easily identify the parts that have been changed. If you're curious about the specifics, see Wikipedia's page on the diff utility (*http://bit.ly/diff-utility*).

❸ Here's the meat of the change. Spaces before the line (e.g., the line that begins Zea Mays… indicates nothing was changed (and just provide context). Plus signs indicate a line addition (e.g., the line that begins Project…), and negative signs indicate a line deletion (not shown in this diff because we've only added a line). Changes to a line are represented as a deletion of the original line and an addition of the new line.

After we stage a file, `git diff` won't show any changes, because `git diff` compares the version of files in your working directory to the last staged version. For example:

```
$ git add README
$ git diff # shows nothing
```

If we wanted to compare what's been staged to our last commit (which will show us exactly what's going into the next commit), we can use `git diff --staged` (in old versions of Git this won't work, so upgrade if it doesn't). Indeed, we can see the change we just staged:

```
$ git diff --staged
diff --git a/README b/README
index 5483cfd..ba8d7fc 100644
--- a/README
+++ b/README
@@ -1 +1,2 @@
 Zea Mays SNP Calling Project
+Project started 2013-01-03
```

`git diff` can also be used to compare arbitrary objects in our Git commit history, a topic we'll see in "More git diff: Comparing Commits and Files" on page 100.

Seeing Your Commit History: git log

Commits are like chains (more technically, directed acyclic graphs), with each commit pointing to its parent (as in Figure 5-2).

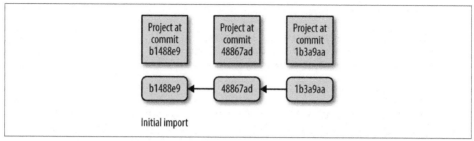

Initial import

Figure 5-2. Commits in Git take discrete snapshots of your project at some point in time, and each commit (except the first) points to its parent commit; this chain of commits is your set of connected snapshots that show how your project repository evolves

We can use `git log` to visualize our chain of commits:

```
$ git log
commit 3d7ffa6f0276e607dcd94e18d26d21de2d96a460 ❶
Author: Vince Buffalo <vsbuffaloAAAAAA@gmail.com>
Date:   Mon Sep 23 23:55:08 2013 -0700

    initial import
```

❶ This strange looking mix of numbers and characters is a *SHA-1* checksum. Each commit will have one of these, and they will depend on your repository's past commit history and the current files. SHA-1 hashes act as a unique ID for each commit in your repository. You can always refer to a commit by its SHA-1 hash.

git log and Your Terminal Pager

`git log` opens up your repository's history in your default *pager* (usually either the program `more` or `less`). If you're unfamiliar with pagers, `less`, and `more`, don't fret. To exit and get back to your prompt, hit the letter *q*. You can move forward by pressing the space bar, and move backward by pressing *b*. We'll look at `less` in more detail in Chapter 7.

Let's commit the change we made in the last section:

```
$ git commit -a -m "added information about project to README"
[master 94e2365] added information about project to README
 1 file changed, 1 insertion(+)
```

Now, if we look at our commit history with `git log`, we see:

```
$ git log
commit 94e2365dd66701a35629d29173d640fdae32fa5c
Author: Vince Buffalo <vsbuffaloAAAAAA@gmail.com>
Date:   Tue Sep 24 00:02:11 2013 -0700

    added information about project to README

commit 3d7ffa6f0276e607dcd94e18d26d21de2d96a460
Author: Vince Buffalo <vsbuffaloAAAAAA@gmail.com>
Date:   Mon Sep 23 23:55:08 2013 -0700

    initial import
```

As we continue to make and commit changes to our repository, this chain of commits will grow. If you want to see a nice example of a longer Git history, change directories to the *seqtk* repository we cloned earlier and call git log.

Moving and Removing Files: git mv and git rm

When Git tracks your files, it wants to be in charge. Using the command mv to move a tracked file will confuse Git. The same applies when you remove a file with rm. To move or remove tracked files in Git, we need to use Git's version of mv and rm: git mv and git rm.

For example, our *README* file doesn't have an extension. This isn't a problem, but because the *README* file might later contain Markdown, it's not a bad idea to change its extension to *.md*. You can do this using git mv:

```
$ git mv README README.md
$ git mv data/README data/README.md
```

Like all changes, this isn't stored in your repository until you commit it. If you ls your files, you can see your working copy has been renamed:

```
$ ls
README.md   analysis   data        notebook.md scripts
```

Using git status, we see this change is staged and ready to be committed:

```
$ git status
# On branch master
# Changes to be committed:
#   (use "git reset HEAD <file>..." to unstage)
#
#       renamed:    README -> README.md
#       renamed:    data/README -> data/README.md
#
# Untracked files:
#   (use "git add <file>..." to include in what will be committed)
#
#       data/seqs/
```

git `mv` already staged these commits for us; git `add` is only necessary for staging modifications to the contents of files, not moving or removing files. Let's commit these changes:

```
$ git commit -m "added markdown extensions to README files"
[master e4feb22] added markdown extensions to README files
 2 files changed, 0 insertions(+), 0 deletions(-)
 rename README => README.md (100%)
 rename data/{README => README.md} (100%)
```

Note that even if you change or remove a file and commit it, it still exists in past snapshots. Git does its best to make everything recoverable. We'll see how to recover files later on in this chapter.

Telling Git What to Ignore: .gitignore

You may have noticed that git `status` keeps listing which files are not tracked. As the number of files in your bioinformatics project starts to increase (this happens quickly!) this long list will become a burden.

Many of the items in this untracked list may be files we never want to commit. Sequencing data files are a great example: they're usually much too large to include in a repository. If we were to commit these large files, collaborators cloning your repository would have to download these enormous data files. We'll talk about other ways of managing these later, but for now, let's just ignore them.

Suppose we wanted to ignore all FASTQ files (with the extension *.fastq*) in the *data/ seqs/* directory. To do this, create and edit the file *.gitignore* in your *zmays-snps/* repository directory, and add:

```
data/seqs/*.fastq
```

Now, git `status` gives us:

```
$ git status
# On branch master
# Untracked files:
#   (use "git add <file>..." to include in what will be committed)
#
#       .gitignore
```

It seems we've gotten rid of one annoyance (the *data/seqs/* directory in "Untracked files") but added another (the new *.gitignore*). Actually, the best way to resolve this is to add and commit your *.gitignore* file. It may seem counterintuitive to contribute a file to a project that's merely there to tell Git what to ignore. However, this is a good practice; it saves collaborators from seeing a listing of untracked files Git should ignore. Let's go ahead and stage the *.gitignore* file, and commit this and the filename changes we made earlier:

```
$ git add .gitignore
$ git commit -m "added .gitignore"
[master c509f63] added .gitignore
 1 file changed, 1 insertion(+)
 create mode 100644 .gitignore
```

What should we tell *.gitignore* to ignore? In the context of a bioinformatics project, here are some guidelines:

Large files
> These should be ignored and managed by other means, as Git isn't designed to manage really large files. Large files slow creating, pushing, and pulling commits. This can lead to quite a burden when collaborators clone your repository.

Intermediate files
> Bioinformatics projects are often filled with intermediate files. For example, if you align reads to a genome, this will create SAM or BAM files. Even if these aren't large files, these should probably be ignored. If a data file can easily be reproduced by rerunning a command (or better yet, a script), it's usually preferable to just store how it was created. Ultimately, recording and storing how you created an intermediate file in Git is more important than the actual file. This also ensures reproducibility.

Text editor temporary files
> Text editors like Emacs and Vim will sometimes create temporary files in your directory. These can look like *textfile.txt~* or *#textfile.txt#*. There's no point in storing these in Git, and they can be an annoyance when viewing progress with `git status`. These files should always be added to *.gitignore*. Luckily, *.gitignore* takes wildcards, so these can be ignored with entries like *~ and \#*\#.

Temporary code files
> Some language interpreters (e.g., Python) produce temporary files (usually with some sort of optimized code). With Python, these look like *overlap.pyc*.

We can use a *global .gitignore* file to universally ignore a file across all of our projects. Good candidates of files to globally ignore are our text editor's temporary files or files your operating system creates (e.g., OS X will sometimes create hidden files named *.DS_Store* in directories to store details like icon position). GitHub maintains a useful repository of global *.gitignore* suggestions (*http://bit.ly/git_ignore*).

You can create a global *.gitignore* file in *~/.gitignore_global* and then configure Git to use this with the following:

```
git config --global core.excludesfile ~/.gitignore_global
```

A repository should store everything required to replicate a project except large datasets and external programs. This includes all scripts, documentation, analysis, and possibly even a final manuscript. Organizing your repository this way means that all of your project's *dependencies* are in one place and are managed by Git. In the long run, it's far easier to have Git keep track of your project's files, than try to keep track of them yourself.

Undoing a Stage: git reset

Recall that one nice feature of Git is that you don't have to include messy changes in a commit—just don't stage these files. If you accidentally stage a messy file for a commit with git add, you can unstage it with git reset. For example, suppose you add a change to a file, stage it, but decide it's not ready to be committed:

```
$ echo "TODO: ask sequencing center about adapters" >> README.md
$ git add README.md
$ git status
# On branch master
# Changes to be committed:
#   (use "git reset HEAD <file>..." to unstage)
#
#       new file:   README.md
#
```

With git status, we can see that our change to *README.md* is going to be included in the next commit. To unstage this change, follow the directions git status provides:

```
$ git reset HEAD README.md
$ git status
# On branch master
# Changes not staged for commit:
#   (use "git add <file>..." to update what will be committed)
#   (use "git checkout -- <file>..." to discard changes in working
directory)
#
#       modified:   README.md
#
```

The syntax seems a little strange, but all we're doing is resetting our staging area (which Git calls the *index*) to the version at HEAD for our *README.md* file. In Git's lingo, *HEAD* is an alias or pointer to the last commit on the *current* branch (which is, as mentioned earlier, the default Git branch called *master*). Git's reset command is a powerful tool, but its default action is to just reset your index. We'll see additional ways to use git reset when we learn about working with commit histories.

Collaborating with Git: Git Remotes, git push, and git pull

Thus far, we've covered the very basics of Git: tracking and working with files, staging changes, making commits, and looking at our commit history. Commits are the foundation of Git—they are the snapshots of our project as it evolves. Commits allow you to go back in time and look at, compare, and recover past versions, which are all topics we look at later in this chapter. In this section, we're going to learn how to collaborate with Git, which at its core is just about sharing commits between your repository and your collaborators' repositories.

The basis of sharing commits in Git is the idea of a *remote repository*, which is just a version of your repository hosted elsewhere. This could be a shared departmental server, your colleague's version of your repository, or on a repository hosting service like GitHub or Bitbucket. Collaborating with Git first requires we configure our local repository to work with our remote repositories. Then, we can retrieve commits from a remote repository (a *pull*) and send commits to a remote repository (a *push*).

Note that Git, as a *distributed* version control system, allows you to work with remote repositories any way you like. These *workflow* choices are up to you and your collaborators. In this chapter, we'll learn an easy common workflow to get started with: collaborating over a *shared central repository*.

Let's take a look at an example: suppose that you're working on a project you wish to share with a colleague. You start the project in your local repository. After you've made a few commits, you want to share your progress by sharing these commits with your collaborator. Let's step through the entire workflow before seeing how to execute it with Git:

1. You create a shared central repository on a server that both you and your collaborator have access to.

2. You push your project's initial commits to this repository (seen in (a) in Figure 5-3).

3. Your collaborator then retrieves your initial work by cloning this central repository (seen in (b) in Figure 5-3).

4. Then, your collaborator makes her changes to the project, commits them to her local repository, and then pushes these commits to the central repository (seen in (a) in Figure 5-4).

5. You then pull in the commits your collaborator pushed to the central repository (seen in (b) in Figure 5-4). The commit history of your project will be a mix of both you and your collaborator's commits.

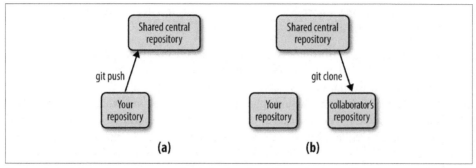

Figure 5-3. After creating a new shared central repository, you push your project's commits (a); your collaborator can retrieve your project and its commits by cloning this central repository (b)

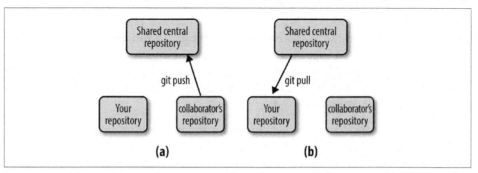

Figure 5-4. After making and committing changes, your collaborator pushes them to the central repository (a); to retrieve your collaborator's new commits, you pull them from the central repository (b)

This process then repeats: you and your collaborator work independently in your own local repositories, and when either of you have commits to share, you push them to the central repository. In many cases, if you and your collaborator work on different files or different sections of the same file, Git can automatically figure out how best to merge these changes. This is an amazing feature of collaborating with Git: you and your collaborator can work on the same project simultaneously. Before we dive into how to do this, there is one caveat to discuss.

It's important to note that Git cannot always automatically merge code and documents. If you and your collaborator are both editing the same section of the same file and both of you commit these changes, this will create a *merge conflict*. Unfortunately, one of you will have to resolve the conflicting files manually. Merge conflicts occur when you (or your collaborator) pull in commits from the central repository and your collaborator's (or your) commits conflict with those in the local repository. In these cases, Git just isn't smart enough to figure out how to reconcile you and your collaborator's conflicting versions.

Most of us are familiar with this process when we collaboratively edit manuscripts in a word processor. If you write a manuscript and send it to all your collaborators to edit, you will need to manually settle sections with conflicting edits. Usually, we get around this messy situation through planning and prioritizing which coauthors will edit first, and gradually incorporating changes. Likewise, good communication and planning with your collaborators can go far in preventing Git merge conflicts, too. Additionally, it's helpful to frequently push and pull commits to the central repository; this keeps all collaborators synced so everyone's working with the newest versions of files.

Creating a Shared Central Repository with GitHub

The first step of our workflow is to create a shared central repository, which is what you and your collaborator(s) share commits through. In our examples, we will use GitHub, a web-based Git repository hosting service. Bitbucket (*http://bitbucket.org/*) is another Git repository hosting service you and your collaborators could use. Both are excellent; we'll use GitHub because it's already home to many large bioinformatics projects like Biopython and Samtools.

Navigate to *http://github.com* and sign up for an account. After your account is set up, you'll be brought to the GitHub home page, which is a newsfeed for you to follow project development (this newsfeed is useful to follow how bioinformatics software you use changes over time). On the main page, there's a link to create a new repository. After you've navigated to the Create a New Repository page (*https://github.com/new*), you'll see you need to provide a repository name, and you'll have the choice to initialize with a *README.md* file (GitHub plays well with Markdown), a *.gitignore* file, and a license (to license your software project). For now, just create a repository named *zmays-snps*. After you've clicked the "Create repository" button, GitHub will forward you to an empty repository page—the public frontend of your project.

There are a few things to note about GitHub:

- Public repositories are free, but private repositories require you to pay. Luckily, GitHub has a special program for educational users (*http://github.com/edu*). If you need private repositories without cost, Bitbucket has a different pricing scheme and provides some for free. Or, you can set up your own internal Git repository on your network if you have shared server space. Setting up your own Git server is out of the scope of this book, but see "Git on the Server - Setting Up the Server" in Scott Chacon and Ben Straub's free online book *Pro Git* (*http://bit.ly/setup-server*) for more information. If your repository is public, anyone can see the source (and even clone and develop their own versions of your repository). However, other users don't have access to modify your GitHub repository unless you grant it to them.

- If you're going to use GitHub for collaboration, all participating collaborators need a GitHub account.

- By default, you are the only person who has write (push) access to the repository you created. To use your remote repository as a shared central repository, you'll have to add *collaborators* in your GitHub repository's settings. Collaborators are GitHub users who have access to push their changes to your repository on GitHub (which modifies it).

- There are other common GitHub workflows. For example, if you manage a lab or other group, you can set up an organization account. You can create repositories and share them with collaborators under the organization's name. We'll discuss other GitHub workflows later in this chapter.

Authenticating with Git Remotes

GitHub uses SSH keys to authenticate you (the same sort we generated in "Quick Authentication with SSH Keys" on page 59). SSH keys prevent you from having to enter a password each time you push or pull from your remote repository. Recall in "Quick Authentication with SSH Keys" on page 59 we generated two SSH keys: a public key and a private key. Navigate to your account settings on GitHub, and in account settings, find the SSH keys tab. Here, you can enter your public SSH key (remember, don't share your private key!) by using cat ~/.ssh/id_rsa.pub to view it, copying it to your clipboard, and pasting it into GitHub's form. You can then try out your SSH public key by using:

```
$ ssh -T git@github.com
Hi vsbuffalo! You've successfully authenticated, but
 GitHub does not provide shell access.
```

If you're having troubles with this, consult GitHub's "Generating SSH Keys" article (*http://bit.ly/genkeys*).

GitHub allows you to use to HTTP as a protocol, but this is typically only used if your network blocks SSH. By default, HTTP asks for passwords each time you try to pull and push (which gets tiresome quickly), but there are ways around this—see GitHub's "Caching Your GitHub Password in Git" (*http://bit.ly/cache-pw*) article.

Connecting with Git Remotes: git remote

Now, let's configure our local repository to use the GitHub repository we've just created as a remote repository. We can do this with git remote add:

```
$ git remote add origin git@github.com:username/zmays-snps.git
```

In this command, we specify not only the address of our Git repository (*git@github.com:username/zmays-snps.git*), but also a name for it: *origin*. By convention, *origin* is the name of your primary remote repository. In fact, earlier when we cloned Seqtk from GitHub, Git automatically added the URL we cloned from as a remote named *origin*.

Now if you enter `git remote -v` (the -v makes it more verbose), you see that our local Git repository knows about the remote repository:

```
$ git remote -v
origin  git@github.com:username/zmays-snps.git (fetch)
origin  git@github.com:username/zmays-snps.git (push)
```

Indeed, *origin* is now a repository we can push commits to and fetch commits from. We'll see how to do both of these operations in the next two sections.

It's worth noting too that you can have multiple remote repositories. Earlier, we mentioned that Git is a distributed version control system; as a result, we can have many remote repositories. We'll come back to how this is useful later on. For now, note that you can add other remote repositories with different names. If you ever need to delete an unused remote repository, you can with `git remote rm <repository-name>`.

Pushing Commits to a Remote Repository with git push

With our remotes added, we're ready to share our work by pushing our commits to a remote repository. Collaboration on Git is characterized by repeatedly pushing your work to allow your collaborators to see and work on it, and pulling their changes into your own local repository. As you start collaborating, remember you only share the commits you've made.

Let's push our initial commits from *zmays-snps* into our remote repository on GitHub. The subcommand we use here is `git push <remote-name> <branch>`. We'll talk more about using branches later, but recall from "Tracking Files in Git: git add and git status Part I" on page 72 that our default branch name is *master*. Thus, to push our *zmays-snps* repository's commits, we do this:

```
$ git push origin master
Counting objects: 14, done.
Delta compression using up to 2 threads.
Compressing objects: 100% (9/9), done.
Writing objects: 100% (14/14), 1.24 KiB | 0 bytes/s, done.
Total 14 (delta 0), reused 0 (delta 0)
To git@github.com:vsbuffalo/zmays-snps.git
 * [new branch]      master -> master
```

That's it—your collaborator now has access to all commits that were on your *master* branch through the central repository. Your collaborator retrieves these commits by pulling them from the central repository into her own local repository.

Pulling Commits from a Remote Repository with git pull

As you push new commits to the central repository, your collaborator's repository will go out of date, as there are commits on the shared repository she doesn't have in her own local repository. She'll need to pull these commits in before continuing with her work. Collaboration on Git is a back-and-forth exchange, where one person pushes their latest commits to the remote repository, other collaborators pull changes into their local repositories, make their own changes and commits, and then push these commits to the central repository for others to see and work with.

To work through an example of this exchange, we will clone our own repository to a different directory, mimicking a collaborator's version of the project. Let's first clone our remote repository to a local directory named *zmay-snps-barbara/*. This directory name reflects that this local repository is meant to represent our colleague Barbara's repository. We can clone *zmays-snps* from GitHub to a local directory named *zmays-snps-barbara/* as follows:

```
$ git clone git@github.com:vsbuffalo/zmays-snps.git zmays-snps-barbara
Cloning into 'zmays-snps-barbara'...
remote: Counting objects: 14, done.
remote: Compressing objects: 100% (9/9), done.
remote: Total 14 (delta 0), reused 14 (delta 0)
Receiving objects: 100% (14/14), done.
Checking connectivity... done
```

Now, both repositories have the same commits. You can verify this by using `git log` and seeing that both have the same commits. Now, in our original *zmay-snps/* local repository, let's modify a file, make a commit, and push to the central repository:

```
$ echo "Samples expected from sequencing core 2013-01-10" >> README.md
$ git commit -a -m "added information about samples"
[master 46f0781] added information about samples
 1 file changed, 1 insertion(+)
$ git push origin master
Counting objects: 5, done.
Delta compression using up to 2 threads.
Compressing objects: 100% (3/3), done.
Writing objects: 100% (3/3), 415 bytes | 0 bytes/s, done.
Total 3 (delta 0), reused 0 (delta 0)
To git@github.com:vsbuffalo/zmays-snps.git
   c509f63..46f0781  master -> master
```

Now, Barbara's repository (*zmays-snps-barbara*) is a commit behind both our local *zmays-snps* repository and the central shared repository. Barbara can pull in this change as follows:

```
$ # in zmays-snps-barbara/
$ git pull origin master
remote: Counting objects: 5, done.
remote: Compressing objects: 100% (3/3), done.
```

```
remote: Total 3 (delta 0), reused 3 (delta 0)
Unpacking objects: 100% (3/3), done.
From github.com:vsbuffalo/zmays-snps
 * branch            master      -> FETCH_HEAD
   c509f63..46f0781  master      -> origin/master
Updating c509f63..46f0781
Fast-forward
 README.md | 1 +
 1 file changed, 1 insertion(+)
```

We can verify that Barbara's repository contains the most recent commit using `git log`. Because we just want a quick image of the last few commits, I will use `git log` with some helpful formatting options:

```
$ # in zmays-snps-barbara/
$ git log --pretty=oneline --abbrev-commit
46f0781 added information about samples
c509f63 added .gitignore
e4feb22 added markdown extensions to README files
94e2365 added information about project to README
3d7ffa6 initial import
```

Now, our commits are in both the central repository and Barbara's repository.

Working with Your Collaborators: Pushing and Pulling

Once you grow a bit more acquainted with pushing and pulling commits, it will become second nature. I recommend practicing this with fake repositories with a lab-mate or friend to get the hang of it. Other than merge conflicts (which we cover in the next section), there's nothing tricky about pushing and pulling. Let's go through a few more pushes and pulls so it's extremely clear.

In the last section, Barbara pulled our new commit into her repository. But she will also create and push her own commits to the central repository. To continue our example, let's make a commit from Barbara's local repository and push it to the central repository. Because there is no Barbara (Git is using the account we made at the beginning of this chapter to make commits), I will modify `git log`'s output to show Barbara as the collaborator. Suppose she adds the following line to the *README.md*:

```
$ # in zmays-snps-barbara/ -- Barbara's version
$ echo "\n\nMaize reference genome version: refgen3" >> README.md
$ git commit -a -m "added reference genome info"
[master 269aa09] added reference genome info
 1 file changed, 3 insertions(+)
$ git push origin master
Counting objects: 5, done.
Delta compression using up to 2 threads.
Compressing objects: 100% (3/3), done.
Writing objects: 100% (3/3), 390 bytes | 0 bytes/s, done.
Total 3 (delta 1), reused 0 (delta 0)
```

```
To git@github.com:vsbuffalo/zmays-snps.git
   46f0781..269aa09  master -> master
```

Now Barbara's local repository and the central repository are two commits ahead of
our local repository. Let's switch to our *zmays-snps* repository, and pull these new
commits in. We can see how Barbara's commits changed *README.md* with `cat`:

```
$ # in zmays-snps/ -- our version
$ git pull origin master
From github.com:vsbuffalo/zmays-snps
 * branch            master      -> FETCH_HEAD
Updating 46f0781..269aa09
Fast-forward
 README.md | 3 +++
 1 file changed, 3 insertions(+)

$ cat README.md
Zea Mays SNP Calling Project
Project started 2013-01-03
Samples expected from sequencing core 2013-01-10


Maize reference genome version: refgen3
```

If we were to look at the last two log entries, they would look as follows:

```
$ git log -n 2
commit 269aa09418b0d47645c5d077369686ff04b16393
Author: Barbara <barbara@barbarasmaize.com>
Date:   Sat Sep 28 22:58:55 2013 -0700

    added reference genome info

commit 46f0781e9e081c6c9ee08b2d83a8464e9a26ae1f
Author: Vince Buffalo <vsbuffaloAAAAAA@gmail.com>
Date:   Tue Sep 24 00:31:31 2013 -0700

    added information about samples
```

This is what collaboration looks like in Git's history: a set of sequential commits made
by different people. Each is a snapshot of their repository and the changes they made
since the last commit. All commits, whether they originate from your collaborator's
or your repository, are part of the same history and point to their parent commit.

Because new commits build on top of the commit history, it's helpful to do the following to avoid problems:

- When pulling in changes, it helps to have your project's changes committed. Git
  will error out if a pull would change a file that you have uncommitted changes to,
  but it's still helpful to commit your important changes before pulling.

- Pull often. This complements the earlier advice: planning and communicating what you'll work on with your collaborators. By pulling in your collaborator's changes often, you're in a better position to build on your collaborators' changes. Avoid working on older, out-of-date commits.

Merge Conflicts

Occasionally, you'll pull in commits and Git will warn you there's a merge conflict. Resolving merge conflicts can be a bit tricky—if you're struggling with this chapter so far, you can bookmark this section and return to it when you encounter a merge conflict in your own work.

Merge conflicts occur when Git can't automatically merge your repository with the commits from the latest pull—Git needs your input on how best to resolve a conflict in the version of the file. Merge conflicts seem scary, but the strategy to solve them is always the same:

1. Use `git status` to find the conflicting file(s).

2. Open and edit those files manually to a version that fixes the conflict.

3. Use `git add` to tell Git that you've resolved the conflict in a particular file.

4. Once all conflicts are resolved, use `git status` to check that all changes are staged. Then, commit the resolved versions of the conflicting file(s). It's also wise to immediately push this merge commit, so your collaborators see that you've resolved a conflict and can continue their work on this new version accordingly.

As an example, let's create a merge conflict between our *zmays-snps* repository and Barbara's *zmays-snps-barbara* repository. One common situation where merge conflicts arise is to pull in a collaborator's changes that affect a file you've made and committed changes to. For example, suppose that Barbara changed *README.md* to something like the following (you'll have to do this in your text editor if you're following along):

```
Zea Mays SNP Calling Project
Project started 2013-01-03
Samples expected from sequencing core 2013-01-10


Maize reference genome version: refgen3, downloaded 2013-01-04 from
http://maizegdb.org into `/share/data/refgen3/`.
```

After making these edits to *README.md*, Barbara commits and pushes these changes. Meanwhile, in your repository, you also changed the last line:

```
Zea Mays SNP Calling Project
Project started 2013-01-03
```

```
Samples expected from sequencing core 2013-01-10
```

```
We downloaded refgen3 on 2013-01-04.
```

You commit this change, and then try to push to the shared central repository. To your surprise, you get the following error message:

```
$ git push origin master
To git@github.com:vsbuffalo/zmays-snps.git
 ! [rejected]        master -> master (fetch first)
error: failed to push some refs to 'git@github.com:vsbuffalo/zmays-snps.git'
hint: Updates were rejected because the remote contains work that you do
hint: not have locally. This is usually caused by another repository pushing
hint: to the same ref. You may want to first integrate the remote changes
hint: (e.g., 'git pull ...') before pushing again.
hint: See the 'Note about fast-forwards' in 'git push --help' for details.
```

Git rejects your push attempt because Barbara has already updated the central repository's *master* branch. As Git's message describes, we need to resolve this by integrating the commits Barbara has pushed into our own local repository. Let's pull in Barbara's commit, and then try pushing as the message suggests (note that this error is not a merge conflict—rather, it just tells us we can't push to a remote that's one or more commits ahead of our local repository):

```
$ git pull origin master
remote: Counting objects: 5, done.
remote: Compressing objects: 100% (2/2), done.
remote: Total 3 (delta 1), reused 3 (delta 1)
Unpacking objects: 100% (3/3), done.
From github.com:vsbuffalo/zmays-snps
 * branch            master     -> FETCH_HEAD
   269aa09..dafce75  master     -> origin/master
Auto-merging README.md
CONFLICT (content): Merge conflict in README.md
Automatic merge failed; fix conflicts and then commit the result.
```

This is the merge conflict. This message isn't very helpful, so we follow the first step of the merge strategy by checking everything with `git status`:

```
$ git status
# On branch master
# You have unmerged paths.
#   (fix conflicts and run "git commit")
#
# Unmerged paths:
#   (use "git add <file>..." to mark resolution)
#
#       both modified:      README.md
#
no changes added to commit (use "git add" and/or "git commit -a")
```

`git status` tells us that there is only one file with a merge conflict, *README.md* (because we both edited it). The second step of our strategy is to look at our conflicting file(s) in our text editor:

```
Zea Mays SNP Calling Project
Project started 2013-01-03
Samples expected from sequencing core 2013-01-10


<<<<<<< HEAD ❶
We downloaded refgen3 on 2013-01-04.
======= ❷
Maize reference genome version: refgen3, downloaded 2013-01-04 from
http://maizegdb.org into `/share/data/refgen3/`.
>>>>>>> dafce75dc531d123922741613d8f29b894e605ac ❸
```

Notice Git has changed the content of this file in indicating the conflicting lines.

❶ This is the start of our version, the one that's HEAD in our repository. HEAD is Git's lingo for the latest commit (technically, HEAD points to the latest commit on the current branch).

❷ Indicates the end of HEAD and beginning of our collaborator's changes.

❸ This final delimiter indicates the end of our collaborator's version, and the different conflicting chunk. Git does its best to try to isolate the conflicting lines, so there can be many of these chunks.

Now we use step two of our merge conflict strategy: edit the conflicting file to resolve all conflicts. Remember, Git raises merge conflicts when it can't figure out what to do, so you're the one who has to manually resolve the issue. Resolving merge conflicts in files takes some practice. After resolving the conflict in *README.md*, the edited file would appear as follows:

```
Zea Mays SNP Calling Project
Project started 2013-01-03
Samples expected from sequencing core 2013-01-10


We downloaded the B73 reference genome (refgen3) on 2013-01-04 from
http://maizegdb.org into `/share/data/refgen3/`.
```

I've edited this file so it's a combination of both versions. We're happy now with our changes, so we continue to the third step of our strategy—using `git add` to declare this conflict has been resolved:

```
$ git add README.md
```

Now, the final step in our strategy—check `git status` to ensure all conflicts are resolved and ready to be merged, and commit them:

```
$ git status
git status
# On branch master
# All conflicts fixed but you are still merging.
#   (use "git commit" to conclude merge)
#
# Changes to be committed:
#
#       modified:   README.md
#

$ git commit -a -m "resolved merge conflict in README.md"
[master 20041ab] resolved merge conflict in README.md
```

That's it: our merge conflict is resolved! With our local repository up to date, our last step is to share our merge commit with our collaborator. This way, our collaborators know of the merge and can continue their work from the new merged version of the file.

After pushing our merge commit to the central repository with `git push`, let's switch to Barbara's local repository and pull in the merge commit:

```
$ git pull origin master
remote: Counting objects: 10, done.
remote: Compressing objects: 100% (4/4), done.
remote: Total 6 (delta 2), reused 5 (delta 2)
Unpacking objects: 100% (6/6), done.
From github.com:vsbuffalo/zmays-snps
 * branch            master      -> FETCH_HEAD
   dafce75..20041ab  master      -> origin/master
Updating dafce75..20041ab
Fast-forward
 README.md | 2 +-
 1 file changed, 1 insertion(+), 1 deletion(-)
```

Using `git log` we see that this is a special commit—a merge commit:

```
commit cd72acf0a81cdd688cb713465cb774320caeb2fd
Merge: f9114a1 d99121e
Author: Vince Buffalo <vsbuffaloAAAAAA@gmail.com>
Date:   Sat Sep 28 20:38:01 2013 -0700

    resolved merge conflict in README.md
```

Merge commits are special, in that they have two parents. This happened because both Barbara and I committed changes to the same file with the same parent commit. Graphically, this situation looks like Figure 5-5.

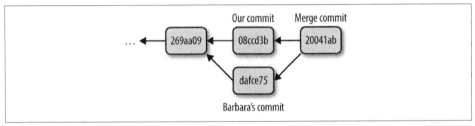

Figure 5-5. A merge commit has two parents—in this case, Barbara's version and our version; merge commits resolve conflicts between versions

We can also see the same story through `git log` with the option `--graph`, which draws a plain-text graph of your commits:

```
*   commit 20041abaab156c39152a632ea7e306540f89f706
|\  Merge: 08ccd3b dafce75
| | Author: Vince Buffalo <vsbuffaloAAAAAA@gmail.com>
| | Date:   Sat Sep 28 23:13:07 2013 -0700
| |
| |     resolved merge conflict in README.md
| |
| * commit dafce75dc531d123922741613d8f29b894e605ac
| | Author: Vince Buffalo <vsbuffaloAAAAAA@gmail.com>
| | Date:   Sat Sep 28 23:09:01 2013 -0700
| |
| |     added ref genome download date and link
| |
* | commit 08ccd3b056785513442fc405f568e61306d62d4b
|/  Author: Vince Buffalo <vsbuffaloAAAAAA@gmail.com>
|   Date:   Sat Sep 28 23:10:39 2013 -0700
|
|       added reference download date
```

Merge conflicts are intimidating at first, but following the four-step strategy introduced at the beginning of this section will get you through it. Remember to repeatedly check `git status` to see what needs to be resolved, and use `git add` to stage edited files as you resolve the conflicts in them. At any point, if you're overwhelmed, you can abort a merge with `git merge --abort` and start over (but beware: you'll lose any changes you've made).

There's one important caveat to merging: if your project's code is spread out across a few files, resolving a merge conflict does not guarantee that your code *works*. Even if Git can fast-forward your local repository after a pull, it's still possible your collaborator's changes may break something (such is the danger when working with collaborators!). It's always wise to do some sanity checks after pulling in code.

For complex merge conflicts, you may want to use a merge tool. Merge tools help visualize merge conflicts by placing both versions side by side, and pointing out what's different (rather than using Git's inline notation that uses inequality and equal signs). Some commonly used merge tools include Meld (*http://meldmerge.org/*) and Kdiff (*http://kdiff3.sourceforge.net/*).

More GitHub Workflows: Forking and Pull Requests

While the shared central repository workflow is the easiest way to get started collaborating with Git, GitHub suggests a slightly different workflow based on *forking* repositories. When you visit a repository owned by another user on GitHub, you'll see a "fork" link. Forking is an entirely GitHub concept—it is not part of Git itself. By forking another person's GitHub repository, you're copying their repository to your own GitHub account. You can then clone your forked version and continue development in your own repository. Changes you push from your local version to your remote repository do not interfere with the main project. If you decide that you've made changes you want to share with the main repository, you can request that your commits are pulled using a `pull request` (another feature of GitHub).

This is the workflow GitHub is designed around, and it works very well with projects with many collaborators. Development primarily takes place in contributors' own repositories. A developer's contributions are only incorporated into the main project when pulled in. This is in contrast to a shared central repository workflow, where collaborators can push their commits to the main project at their will. As a result, lead developers can carefully control what commits are added to the project, which prevents the hassle of new changes breaking functionality or creating bugs.

Using Git to Make Life Easier: Working with Past Commits

So far in this chapter we've created commits in our local repository and shared these commits with our collaborators. But our commit history allows us to do much more than collaboration—we can compare different versions of files, retrieve past versions, and tag certain commits with helpful messages.

After this point, the material in this chapter becomes a bit more advanced. Readers can skip ahead to Chapter 6 without a loss of continuity. If you do skip ahead, bookmark this section, as it contains many tricks used to get out of trouble (e.g., restoring files, stashing your working changes, finding bugs by comparing versions, and editing and undoing commits). In the final section, we'll also cover branching, which is a more advanced Git workflow—but one that can make your life easier.

Getting Files from the Past: git checkout

Anything in Git that's been committed is easy to recover. Suppose you accidentally overwrite your current version of *README.md* by using > instead of >>. You see this change with `git status`:

```
$ echo "Added an accidental line" > README.md
$ cat README.md
Added an accidental line
$ git status
# On branch master
# Changes not staged for commit:
#   (use "git add <file>..." to update what will be committed)
#   (use "git checkout -- <file>..." to discard changes in working
directory)
#
#       modified:   README.md
#
no changes added to commit (use "git add" and/or "git commit -a")
```

This mishap accidentally wiped out the previous contents of *README.md*! However, we can restore this file by checking out the version in our last commit (the commit HEAD points to) with the command `git checkout -- <file>`. Note that you don't need to remember this command, as it's included in `git status` messages. Let's restore *README.md*:

```
$ git checkout -- README.md
$ cat README.md
Zea Mays SNP Calling Project
Project started 2013-01-03
Samples expected from sequencing core 2013-01-10


We downloaded the B72 reference genome (refgen3) on 2013-01-04 from
http://maizegdb.org into `/share/data/refgen3/`.
```

But beware: restoring a file this way erases all changes made to that file since the last commit! If you're curious, the cryptic -- indicates to Git that you're checking out a file, not a branch (`git checkout` is also used to check out branches; commands with multiple uses are common in Git).

By default, `git checkout` restores the file version from HEAD. However, `git checkout` can restore any arbitrary version from commit history. For example, suppose we want to restore the version of *README.md* one commit before HEAD. The past three commits from our history looks like this (using some options to make `git log` more concise):

```
$ git log --pretty=oneline --abbrev-commit -n 3
20041ab resolved merge conflict in README.md
08ccd3b added reference download date
dafce75 added ref genome download date and link
```

Thus, we want to restore *README.md* to the version from commit 08ccd3b. These SHA-1 IDs (even the abbreviated one shown here) function as *absolute* references to your commits (similar to absolute paths in Unix like */some/dir/path/file.txt*). We can always refer to a specific commit by its SHA-1 ID. So, to restore *README.md* to the version from commit `08ccd3b`, we use:

```
$ git checkout 08ccd3b -- README.md
$ cat README.md
Zea Mays SNP Calling Project
Project started 2013-01-03
Samples expected from sequencing core 2013-01-10


We downloaded refgen3 on 2013-01-04.
```

If we restore to get the most recent commit's version, we could use:

```
$ git checkout 20041ab -- README.md
$ git status
# On branch master
nothing to commit, working directory clean
```

Note that after checking out the latest version of the *README.md* file from commit `20041ab`, nothing has effectively changed in our working directory; you can verify this using `git status`.

Stashing Your Changes: git stash

One very useful Git subcommand is `git stash`, which saves any working changes you've made since the last commit and restores your repository to the version at HEAD. You can then reapply these saved changes later. `git stash` is handy when we want to save our messy, partial progress before operations that are best performed with a clean working directory—for example, `git pull` or branching (more on branching later).

Let's practice using `git stash` by first adding a line to *README.md*:

```
$ echo "\\nAdapter file: adapters.fa" >> README.md
$ git status
# On branch master
# Changes not staged for commit:
#   (use "git add <file>..." to update what will be committed)
#   (use "git checkout -- <file>..." to discard changes in working
directory)
#
#       modified:   README.md
#
no changes added to commit (use "git add" and/or "git commit -a")
```

Then, let's stash this change using `git stash`:

```
$ git stash
Saved working directory and index state WIP on master: 20041ab
resolved merge conflict in README.md
HEAD is now at 20041ab resolved merge conflict in README.md
```

Stashing our working changes sets our directory to the same state it was in at the last commit; now our project directory is clean.

To reapply the changes we stashed, use `git stash pop`:

```
$ git stash pop
# On branch master
# Changes not staged for commit:
#   (use "git add <file>..." to update what will be committed)
#   (use "git checkout -- <file>..." to discard changes in working
# directory)
#
#       modified:   README.md
#
no changes added to commit (use "git add" and/or "git commit -a")
Dropped refs/stash@{0} (785dad46104116610d5840b317f05465a5f07c8b)
```

Note that the changes stored with `git stash` are not committed; `git stash` is a separate way to store changes outside of your commit history. If you start using `git stash` a lot in your work, check out other useful stash subcommands like `git stash apply` and `git stash list`.

More git diff: Comparing Commits and Files

Earlier, we used `git diff` to look at the difference between our working directory and our staging area. But `git diff` has many other uses and features; in this section, we'll look at how we can use `git diff` to compare our current working tree to other commits.

One use for `git diff` is to compare the difference between two arbitrary commits. For example, if we wanted to compare what we have now (at HEAD) to commit dafce75:

```
$ git diff dafce75
diff --git a/README.md b/README.md
index 016ed0c..9491359 100644
--- a/README.md
+++ b/README.md
@@ -3,5 +3,7 @@ Project started 2013-01-03
 Samples expected from sequencing core 2013-01-10


-Maize reference genome version: refgen3, downloaded 2013-01-04 from
+We downloaded the B72 reference genome (refgen3) on 2013-01-04 from
 http://maizegdb.org into `/share/data/refgen3/`.
+
+Adapter file: adapters.fa
```

Specifying Revisions Relative to HEAD

Like writing out absolute paths in Unix, referring to commits by their full SHA-1 ID is tedious. While we can reduce typing by using the abbreviated commits (the first seven characters of the full SHA-1 ID), there's an easier way: relative ancestry references. Similar to using relative paths in Unix like ./ and ../, Git allows you to specify commits relative to HEAD (or any other commit, with SHA-1 IDs).

The caret notation (^) represents the *parent* commit of a commit. For example, to refer to the parent of the most recent commit on the current branch (HEAD), we'd use HEAD^ (commit 08ccd3b in our examples).

Similarly, if we'd wanted to refer to our parent's parent commit (dafce75 in our example), we use HEAD^^. Our example repository doesn't have enough commits to refer to the parent of *this* commit, but if it did, we could use HEAD^^^. At a certain point, using this notation is no easier than copying and pasting a SHA-1, so a succinct alternative syntax exists: git HEAD~<n>, where <n> is the number of commits back in the ancestry of HEAD (including the last one). Using this notation, HEAD^^ is the same as HEAD~2.

Specifying revisions becomes more complicated with merge commits, as these have *two* parents. Git has an elaborate language to specify these commits. For a full specification, enter git rev-parse --help and see the "Specifying Revisions" section of this manual page.

Using git diff, we can also view all changes made to a file between two commits. To do this, specify both commits and the file's path as arguments (e.g., git diff <commit> <commit> <path>). For example, to compare our version of *README.md* across commits 269aa09 and 46f0781, we could use either:

```
$ git diff 46f0781 269aa09 README.md
# or
$ git diff HEAD~3 HEAD~2 README.md
```

This second command utilizes the relative ancestry references explained in "Specifying Revisions Relative to HEAD" on page 101.

How does this help? Git's ability to compare the changes between two commits allows you to find where and how a bug entered your code. For example, if a modified script produces different results from an earlier version, you can use git diff to see exactly which lines differ across versions. Git also has a tool called git bisect to help developers find where exactly bugs entered their commit history. git bisect is out of the scope of this chapter, but there are some good examples in git bisect --help.

Undoing and Editing Commits: git commit --amend

At some point, you're bound to accidentally commit changes you didn't mean to or make an embarrassing typo in a commit message. For example, suppose we were to make a mistake in a commit message:

```
$ git commit -a -m "added adpters file to readme"
[master f4993e3] added adpters file to readme
 1 file changed, 2 insertions(+)
```

We could easily amend our commit with:

```
$ git commit --amend
```

`git commit --amend` opens up your last commit message in your default text editor, allowing you to edit it. Amending commits isn't limited to just changing the commit message though. You can make changes to your file, stage them, and then amend these staged changes with `git commit --amend`. In general, unless you've made a mistake, it's best to just use separate commits.

It's also possible to undo commits using either `git revert` or the more advanced `git reset` (which if used improperly can lead to data loss). These are more advanced topics that we don't have space to cover in this chapter, but I've listed some resources on this issue in this chapter's *README* file on GitHub.

Working with Branches

Our final topic is probably Git's greatest feature: branching. If you're feeling overwhelmed so far by Git (I certainly did when I first learned it), you can move forward to Chapter 6 and work through this section later.

Branching is much easier with Git than in other version control systems—in fact, I switched to Git after growing frustrated with another version control system's branching model. Git's branches are virtual, meaning that branching doesn't require actually copying files in your repository. You can create, merge, and share branches effortlessly with Git. Here are some examples of how branches can help you in your bioinformatics work:

- Branches allow you to experiment in your project without the risk of adversely affecting the main branch, *master*. For example, if in the middle of a variant calling project you want to experiment with a new pipeline, you can create a new branch and implement the changes there. If the new variant caller doesn't work out, you can easily switch back to the *master* branch—it will be unaffected by your experiment.

- If you're developing software, branches allow you to develop new features or bug fixes without affecting the working production version, the *master* branch. Once the feature or bug fix is complete, you can merge it into the *master* branch, incorporating the change into your production version.

- Similarly, branches simplify working collaboratively on repositories. Each collaborator can work on their own separate branches, which prevents disrupting the *master* branch for other collaborators. When a collaborator's changes are ready to be shared, they can be merged into the *master* branch.

Creating and Working with Branches: git branch and git checkout

As a simple example, let's create a new branch called *readme-changes*. Suppose we want to make some edits to *README.md*, but we're not sure these changes are ready to be incorporated into the main branch, *master*.

To create a Git branch, we use `git branch <branchname>`. When called without any arguments, `git branch` lists all branches. Let's create the *readme-changes* branch and check that it exists:

```
$ git branch readme-changes
$ git branch
* master
  readme-changes
```

The asterisk next to `master` is there to indicate that this is the branch we're currently on. To switch to the *readme-changes* branch, use `git checkout readme-changes`:

```
$ git checkout readme-changes
Switched to branch 'readme-changes'
$ git branch
  master
* readme-changes
```

Notice now that the asterisk is next to `readme-changes`, indicating this is our current branch. Now, suppose we edit our *README.md* section extensively, like so:

```
# Zea Mays SNP Calling Project
Project started 2013-01-03.

## Samples
Samples downloaded 2013-01-11 into `data/seqs`:

        data/seqs/zmaysA_R1.fastq
        data/seqs/zmaysA_R2.fastq
        data/seqs/zmaysB_R1.fastq
        data/seqs/zmaysB_R2.fastq
        data/seqs/zmaysC_R1.fastq
        data/seqs/zmaysC_R2.fastq


## Reference
```

```
We downloaded the B72 reference genome (refgen3) on 2013-01-04 from
http://maizegdb.org into `/share/data/refgen3/`.
```

Now if we commit these changes, our commit is added to the *readme-changes* branch. We can verify this by switching back to the *master* branch and seeing that this commit doesn't exist:

```
$ git commit -a -m "reformatted readme, added sample info" ❶
[readme-changes 6e680b6] reformatted readme, added sample info
 1 file changed, 12 insertions(+), 3 deletions(-)
$ git log --abbrev-commit --pretty=oneline -n 3 ❷
6e680b6 reformatted readme, added sample info
20041ab resolved merge conflict in README.md
08ccd3b added reference download date
$ git checkout master ❸
Switched to branch 'master'
$ git log --abbrev-commit --pretty=oneline -n 3 ❹
20041ab resolved merge conflict in README.md
08ccd3b added reference download date
dafce75 added ref genome download date and link
```

❶ Our commit, made on the branch *readme-changes*.

❷ The commit we just made (6e680b6).

❸ Switching back to our *master* branch.

❹ Our last commit on *master* is 20041ab. Our changes to *README.md* are only on the *readme-changes* branch, and when we switch back to master, Git swaps our files out to those versions on *that* branch.

Back on the *master* branch, suppose we add the *adapters.fa* file, and commit this change:

```
$ git branch
* master
  readme-changes
$ echo ">adapter-1\\nGATGATCATTCAGCGACTACGATCG" >> adapters.fa
$ git add adapters.fa
$ git commit -a -m "added adapters file"
[master dd57e33] added adapters file
 1 file changed, 2 insertions(+)
 create mode 100644 adapters.fa
```

Now, both branches have new commits. This situation looks like Figure 5-6.

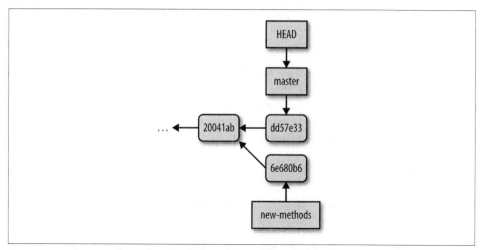

Figure 5-6. Our two branches (within Git, branches are represented as pointers at commits, as depicted here), the master and readme-changes branches have diverged, as they point to different commits (our HEAD points to master, indicating this is the current branch we're on)

Another way to visualize this is with `git log`. We'll use the `--branches` option to specify we want to see all branches, and `-n 2` to only see these last commits:

```
$ git log --abbrev-commit --pretty=oneline --graph --branches -n2
* dd57e33 added adapters file
| * 6e680b6 reformatted readme, added sample info
|/
```

Merging Branches: git merge

With our two branches diverged, we now want to merge them together. The strategy to merge two branches is simple. First, use `git checkout` to switch to the branch we want to merge the other branch into. Then, use `git merge <otherbranch>` to merge the other branch into the current branch. In our example, we want to merge the *readme-changes* branch into *master*, so we switch to *master* first. Then we use:

```
$ git merge readme-changes
Merge made by the 'recursive' strategy.
 README.md | 15 +++++++++++---
 1 file changed, 12 insertions(+), 3 deletions(-)
```

There wasn't a merge conflict, so `git merge` opens our text editor and has us write a merge commit message. Once again, let's use `git log` to see this:

```
$ git log --abbrev-commit --pretty=oneline --graph --branches -n 3
*   e9a81b9 Merge branch 'readme-changes'
|\
| * 6e680b6 reformatted readme, added sample info
```

```
* | dd57e33 added adapters file
| /
```

Bear in mind that merge conflicts can occur when merging branches. In fact, the merge conflict we encountered in "Merge Conflicts" on page 92 when pulling in remote changes was a conflict between two branches; git pull does a merge between a remote branch and a local branch (more on this in "Branches and Remotes" on page 106). Had we encountered a merge conflict when running git merge, we'd follow the same strategy as in "Merge Conflicts" on page 92 to resolve it.

When we've used git log to look at our history, we've only been looking a few commits back—let's look at the entire Git history now:

```
$ git log --abbrev-commit --pretty=oneline --graph --branches
*   e9a81b9 Merge branch 'readme-changes'
|\
| * 6e680b6 reformatted readme, added sample info
* | dd57e33 added adapters file
|/
*   20041ab resolved merge conflict in README.md
|\
| * dafce75 added ref genome download date and link
* | 08ccd3b added reference download date
|/
* 269aa09 added reference genome info
* 46f0781 added information about samples
* c509f63 added .gitignore
* e4feb22 added markdown extensions to README files
* 94e2365 added information about project to README
* 3d7ffa6 initial import
```

Note that we have two bubbles in our history: one from the merge conflict we resolved after git pull, and the other from our recent merge of the *readme-changes* branch.

Branches and Remotes

The branch we created in the previous section was entirely local—so far, our collaborators are unable to see this branch or its commits. This is a nice feature of Git: you can create and work with branches to fit your workflow needs without having to share these branches with collaborators. In some cases, we do want to share our local branches with collaborators. In this section, we'll see how Git's branches and remote repositories are related, and how we can share work on local branches with collaborators.

Remote branches are a special type of local branch. In fact, you've already interacted with these remote branches when you've pushed to and pulled from remote repositories. Using git branch with the option --all, we can see these hidden remote branches:

```
$ git branch --all
* master
  readme-changes
  remotes/origin/master
```

remotes/origin/master is a remote branch—we can't do work on it, but it can be synchronized with the latest commits from the remote repository using `git fetch`. Interestingly, a `git pull` is nothing more than a `git fetch` followed by a `git merge`. Though a bit technical, understanding this idea will greatly help you in working with remote repositories and remote branches. Let's step through an example.

Suppose that Barbara is starting a new document that will detail all the bioinformatics methods of your project. She creates a *new-methods* branch, makes some commits, and then pushes these commits on this branch to our central repository:

```
$ git push origin new-methods
Counting objects: 4, done.
Delta compression using up to 2 threads.
Compressing objects: 100% (2/2), done.
Writing objects: 100% (3/3), 307 bytes | 0 bytes/s, done.
Total 3 (delta 1), reused 0 (delta 0)
To git@github.com:vsbuffalo/zmays-snps.git
 * [new branch]      new-methods -> new-methods
```

Back in our repository, we can fetch Barbara's latest branches using `git fetch <remotename>`. This creates a new remote branch, which we can see with `git branch --all`:

```
$ git fetch origin
remote: Counting objects: 4, done.
remote: Compressing objects: 100% (1/1), done.
remote: Total 3 (delta 1), reused 3 (delta 1)
Unpacking objects: 100% (3/3), done.
From github.com:vsbuffalo/zmays-snps
 = [up to date]      master      -> origin/master
 * [new branch]      new-methods -> origin/new-methods
$ git branch --all
* master
  new-methods
  remotes/origin/master
  remotes/origin/new-methods
```

`git fetch` doesn't change any of your local branches; rather, it just synchronizes your remote branches with the newest commits from the remote repositories. If after a `git fetch` we wanted to incorporate the new commits on our remote branch into our local branch, we would use a `git merge`. For example, we could merge Barbara's *new-methods* branch into our *master* branch with `git merge origin/new-methods`, which emulates a `git pull`.

However, Barbara's branch is just getting started—suppose we want to develop on the *new-methods* branch before merging it into our *master* branch. We cannot develop on remote branches (e.g., our *remotes/origin/new-methods*), so we need to make a new branch that *starts* from this branch:

```
$ git checkout -b new-methods origin/new-methods
Branch new-methods set up to track remote branch new-methods from origin.
Switched to a new branch 'new-methods'
```

Here, we've used `git checkout` to simultaneously create and switch a new branch using the -b option. Note that Git notified us that it's *tracking* this branch. This means that this local branch knows which remote branch to push to and pull from if we were to just use `git push` or `git pull` without arguments. If we were to commit a change on this branch, we could then push it to the remote with `git push`:

```
$ echo "\\n(1) trim adapters\\n(2) quality-based trimming" >> methods.md
$ git commit -am "added quality trimming steps"
[new-methods 5f78374] added quality trimming steps
 1 file changed, 3 insertions(+)
$ git push
Counting objects: 5, done.
Delta compression using up to 2 threads.
Compressing objects: 100% (3/3), done.
Writing objects: 100% (3/3), 339 bytes | 0 bytes/s, done.
Total 3 (delta 1), reused 0 (delta 0)
To git@github.com:vsbuffalo/zmays-snps.git
    6364ebb..9468e38  new-methods -> new-methods
```

Development can continue on *new-methods* until you and your collaborator decide to merge these changes into *master*. At this point, this branch's work has been incorporated into the main part of the project. If you like, you can delete the remote branch with `git push origin :new-methods` and your local branch with `git branch -d new-methods`.

Continuing Your Git Education

Git is a massively powerful version control system. This chapter has introduced the basics of version control and collaborating through pushing and pulling, which is enough to apply to your daily bioinformatics work. We've also covered some basic tools and techniques that can get you out of trouble or make working with Git easier, such as `git checkout` to restore files, `git stash` to stash your working changes, and `git branch` to work with branches. After you've mastered all of these concepts, you may want to move on to more advanced Git topics such as rebasing (`git rebase`), searching revisions (`git grep`), and submodules. However, none of these topics are required in daily Git use; you can search out and learn these topics as you need them. A great resource for these advanced topics is Scott Chacon and Ben Straub's *Pro Git* book (*http://git-scm.com/book/en/v2*).

## CHAPTER 6

Bioinformatics Data

Thus far, we've covered many of the preliminaries to get started in bioinformatics: organizing a project directory, intermediate Unix, working with remote machines, and using version control. However, we've ignored an important component of a new bioinformatics project: data.

Data is a requisite of any bioinformatics project. We further our understanding of complex biological systems by refining a large amount of data to a point where we can extract meaning from it. Unfortunately, many tasks that are simple with small or medium-sized datasets are a challenge with the large and complex datasets common in genomics. These challenges include:

Retrieving data

> Whether downloading large sequencing datasets or accessing a web application hundreds of times to download specific files, retrieving data in bioinformatics can require special tools and skills.

Ensuring data integrity

> Transferring large datasets across networks creates more opportunities for data corruption, which can later lead to incorrect analyses. Consequently, we need to ensure data integrity with tools before continuing with analysis. The same tools can also be used to verify we're using the correct version of data in an analysis.

Compression

> The data we work with in bioinformatics is large enough that it often needs to be compressed. Consequently, working with compressed data is an essential skill in bioinformatics.

Retrieving Bioinformatics Data

Suppose you've just been told the sequencing for your project has been completed: you have six lanes of Illumina data to download from your sequencing center. Downloading this amount of data through your web browser is not feasible: web browsers are not designed to download such large datasets. Additionally, you'd need to download this sequencing data to your server, not the local workstation where you browse the Internet. To do this, you'd need to SSH into your data-crunching server and download your data directly to this machine using command-line tools. We'll take a look at some of these tools in this section.

Downloading Data with wget and curl

Two common command-line programs for downloading data from the Web are wget and curl. Depending on your system, these may not be already installed; you'll have to install them with a package manager (e.g., Homebrew or apt-get). While curl and wget are similar in basic functionality, their relative strengths are different enough that I use both frequently.

wget

wget is useful for quickly downloading a file from the command line—for example, human chromosome 22 from the GRCh37 (also known as hg19) assembly version:

```
$ wget http://hgdownload.soe.ucsc.edu/goldenPath/hg19/chromosomes/chr22.fa.gz
--2013-06-30 00:15:45--  http://[...]/goldenPath/hg19/chromosomes/chr22.fa.gz
Resolving hgdownload.soe.ucsc.edu... 128.114.119.163
Connecting to hgdownload.soe.ucsc.edu|128.114.119.163|:80... connected.
HTTP request sent, awaiting response... 200 OK
Length: 11327826 (11M) [application/x-gzip]
Saving to: 'chr22.fa.gz'

17% [======>                            ] 1,989,172    234KB/s  eta 66s
```

wget downloads this file to your current directory and provides a useful progress indicator. Notice that the link to chromosome 22 begins with "http" (short for Hyper-Text Transfer Protocol). wget can also handle FTP links (which start with "ftp," short for File Transfer Protocol). In general, FTP is preferable to HTTP for large files (and is often recommended by websites like the UCSC Genome Browser).

Because UCSC generously provides the human reference genome publicly, we don't need any authentication to gain access to this file. However, if you're downloading data from a Lab Information Management System (LIMS), you may need to first authenticate with a username and password. For simple HTTP or FTP authentication, you can authenticate using wget's --user= and --ask-password options. Some sites use more elaborate authentication procedures; in these cases, you'll need to contact the site administrator.

One of wget's strengths is that it can download data *recursively*. When run with the recursive option (`--recursive` or `-r`), wget will also follow and download the pages linked to, and even follow and download links on these pages, and so forth. By default (to avoid downloading the entire Internet), wget limits itself to only follow five links deep (but this can be customized using the option `--level` or `-l`).

Recursive downloading can be useful for downloading all files of a certain type from a page. For example, if another lab has a web page containing many GTF files we wish to download, we could use:

```
$ wget --accept "*.gtf" --no-directories --recursive --no-parent \
    http://genomics.someuniversity.edu/labsite/annotation.html
```

But beware! wget's recursive downloading can be quite aggressive. If not constrained, wget will download everything it can reach within the maximum depth set by `--level`. In the preceding example, we limited wget in two ways: with `--no-parent` to prevent wget from downloading pages higher in the directory structure, and with `--accept "*.gtf"`, which only allows wget to download filenames matching this pattern.

Exercise caution when using wget's recursive option; it can utilize a lot of network resources and overload the remote server. In some cases, the remote host may block your IP address if you're downloading too much too quickly. If you plan on downloading a lot of data from a remote host, it's best to inquire with the website's sysadmin about recommended download limits so your IP isn't blocked. wget's `--limit-rate` option can be used to limit how quickly wget downloads files.

wget is an incredibly powerful tool; the preceding examples have only scraped the surface of its capabilities. See Table 6-1 for some commonly used options, or man wget for an exhaustive list.

Table 6-1. Useful wget options

Option	Values	Use
-A, --accept	Either a suffix like ".fastq" or a pattern with *, ?, or [and], optionally comma-separated list	Only download files matching this criteria.
-R, --reject	Same as with --accept	Don't download files matching this; for example, to download all files on a page except Excel files, use --reject ".xls".

Option	Values	Use
-nd, --no-directory	No value	Don't place locally downloaded files in same directory hierarchy as remote files.
-r, --recursive	No value	Follow and download links on a page, to a maximum depth of five by default.
-np, --no-parent	No value	Don't move above the parent directory.
--limit-rate	number of bytes to allow per second	Throttle download bandwidth.
--user=user	FTP or HTTP username	Username for HTTP or FTP authentication.
--ask-password	No value	Prompt for password for HTTP of FTP authentication; --password= could also be used, but then your password is in your shell's history.
-O	Output filename	Download file to filename specified; useful if link doesn't have an informative name (e.g., http://lims.sequencingcenter.com/seqs.html?id=sample_A_03).

Curl

curl serves a slightly different purpose than wget. wget is great for downloading files via HTTP or FTP and scraping data from a web page using its recursive option. curl behaves similarly, although by default writes the file to standard output. To download chromosome 22 as we did with wget, we'd use:

```
$ curl http://[...]/goldenPath/hg19/chromosomes/chr22.fa.gz > chr22.fa.gz
  % Total    % Received % Xferd  Average Speed   Time    Time     Time  Current
                                 Dload  Upload   Total   Spent    Left  Speed
 14 10.8M   14 1593k    0     0   531k      0  0:00:20  0:00:02  0:00:18  646k
```

Note that I've had to truncate the URL so this example fits within a page; the URL is the same as we used with wget earlier.

If you prefer not to redirect curl's output, use -O <filename> to write the results to *<filename>*. If you omit the filename argument, curl will use same filename as the remote host.

curl has the advantage that it can transfer files using more protocols than wget, including SFTP (secure FTP) and SCP (secure copy). One especially nice feature of curl is that it can follow page redirects if the -L/--location option is enabled. With this enabled, curl will download the ultimate page the link redirects to, not the redirect page itself. Finally, Curl itself is also a library, meaning in addition to the command-line `curl` program, Curl's functionality is wrapped by software libraries like RCurl (*http://bit.ly/cran-rcurl*) and pycurl (*http://pycurl.sourceforge.net/doc/*).

Rsync and Secure Copy (scp)

`wget` and `curl` are appropriate for quickly downloading files from the command line, but are not optimal for some heavier-duty tasks. For example, suppose a colleague needs all large sequencing datasets in your project directory that are ignored by Git (e.g., in your *.gitignore*). A better tool for synchronizing these entire directories across a network is Rsync.

Rsync is a superior option for these types of tasks for a few reasons. First, Rsync is often faster because it only sends the *difference* between file versions (when a copy already exists or partially exists) and it can compress files during transfers. Second, Rsync has an archive option that preserves links, modification timestamps, permissions, ownership, and other file attributes. This makes Rsync an excellent choice for network backups of entire directories. Rsync also has numerous features and options to handle different backup scenarios, such as what to do if a file exists on the remote host.

`rsync`'s basic syntax is `rsync source destination`, where `source` is the source of the files or directories you'd like to copy, and `destination` is the destination you'd like to copy these files to. Either `source` or `destination` can be a remote host specified in the format `user@host:/path/to/directory/`.

Let's look at an example of how we can use `rsync` to copy over an entire directory to another machine. Suppose you'd like to copy all of your project's data in *zea_mays/data* to your colleague's directory */home/deborah/zea_mays/data* on the host *192.168.237.42*. The most common combination of `rsync` options used to copy an entire directory are `-avz`. The option `-a` enables `wrsync`'s archive mode, `-z` enables file transfer compression, and `-v` makes `rsync`'s progress more verbose so you can see what's being transferred. Because we'll be connecting to the remote host through SSH, we also need to use `-e ssh`. Our directory copying command would look as follows:

```
$ rsync -avz -e ssh zea_mays/data/ vinceb@[...]:/home/deborah/zea_mays/data
building file list ... done
zmaysA_R1.fastq
zmaysA_R2.fastq
zmaysB_R1.fastq
zmaysB_R2.fastq
zmaysC_R1.fastq
zmaysC_R2.fastq
sent 2861400 bytes  received 42 bytes  107978.94 bytes/sec
total size is 8806085  speedup is 3.08
```

One subtle yet important behavior of `rsync` is that trailing slashes (e.g., *data/* versus *data*) are meaningful when specifying paths in `rsync`. A trailing slash in the source path means copy the *contents* of the source directory, whereas no trailing slash means copy the *entire directory itself*. Because we'd like to copy all contents of *zea_mays/data/* to */home/deborah/zea_mays/data* in our example, we use a trailing slash. If the *data/* directory didn't already exist on the remote destination host, we'd want to copy it and its contents by using *zea_mays/data* (e.g., omitting the trailing slash).

Because rsync only transmits files if they don't exist or they've changed, you can (and should) run rsync again after your files have transferred. This operates as a simple check to ensure everything is synchronized between the two directories. It's also a good idea to check the exit status of rsync when calling it in scripts; rsync will exit with a nonzero status if it runs into problems transferring files. Lastly, rsync can use host aliases specified in an SSH config file (see the first Tip in "Connecting to Remote Machines with SSH" on page 57). You can omit -e ssh if you connect to a host through an SSH host alias.

Occasionally, we just need to quickly copy a single file over SSH—for tasks where Unix's cp would be sufficient, but needs to work over an SSH connection. rsync would work, but it's a bit overkill. Secure copy (scp) is perfect for this purpose. Secure copy works just like cp, except we need to specify both host and path (using the same user@host:/path/to/file notation as wget). For example, we could transfer a single GTF file to *192.168.237.42:/home/deborah/zea_mays/data/* using:

```
$ scp Zea_mays.AGPv3.20.gtf 192.168.237.42:/home/deborah/zea_mays/data/

Zea_mays.AGPv3.20.gtf                          100%   55      0.1KB/s   00:00
```

Data Integrity

Data we download into our project directory is the starting point of all future analyses and conclusions. Although it may seem improbable, the risk of data corruption during transfers is a concern when transferring large datasets. These large files take a long time to transfer, which translates to more opportunities for network connections to drop and bits to be lost. In addition to verifying your transfer finished without error, it's also important to explicitly check the transferred data's integrity with *checksums*. Checksums are very compressed summaries of data, computed in a way that even if just one bit of the data is changed, the checksum will be different.

Data integrity checks are also helpful in keeping track of data versions. In collaborative projects, our analyses may depend on our colleagues' intermediate results. When these intermediate results change, all downstream analyses that depend on these results need to be rerun. With many intermediate files, it's not always clear which data has changed and which steps need to be rerun. The checksums would differ if the data changed even the tiniest bit, so we can use them to calculate the version of the data. Checksums also facilitate reproducibility, as we can link a particular analysis and set of results to an exact version of data summarized by the data's checksum value.

SHA and MD5 Checksums

The two most common checksum algorithms for ensuring data integrity are MD5 and SHA-1. We've already encountered SHA-1 in Chapter 4, as this is what Git uses for its commit IDs. MD5 is an older checksum algorithm, but one that is still commonly used. Both MD5 and SHA-1 behave similarly, but SHA-1 is newer and generally preferred. However, MD5 is more common; it's likely to be what you encounter if a server has precomputed checksums on a set of files.

Let's get acquainted with checksums using SHA-1. We can pass arbitrary strings to the program shasum (on some systems, it's sha1sum) through standard in:

```
$ echo "bioinformatics is fun" | shasum
f9b70d0d1b0a55263f1b012adab6abf572e3030b  -
$ echo "bioinformatic is fun" | shasum
e7f33eedcfdc9aef8a9b4fec07e58f0cf292aa67  -
```

The long string of numbers and letters is the SHA-1 checksum. Checksums are reported in hexadecimal format, where each digit can be one of 16 characters: digits 0 through 9, and the letters a, b, c, d, e, and f. The trailing dash indicates this is the SHA-1 checksum of input from standard in. Note that when we omitted the "s" in "bioinformatics" and calculate the SHA-1 checksum, the checksum value has entirely changed. This is the strength of using checksums: they change when the tiniest part of the input changes. Checksums are completely deterministic, meaning that regardless of the time the checksum is calculated or the system used, checksum values will only differ if the input differs.

We can also use checksums with file input (note that the content of *Csyrichta_TAGGACT_L008_R1_001.fastq* is fake example data):

```
$ shasum Csyrichta_TAGGACT_L008_R1_001.fastq
fea7d7a582cdfb64915d486ca39da9ebf7ef1d83  Csyrichta_TAGGACT_L008_R1_001.fastq
```

If our sequencing center says the checksum of the *Csyrichta_TAG-GACT_L008_R1_001.fastq.gz* sequencing file is "069bf5894783db241e26f4e44201bd12f2d5aa42" and our local SHA checksum is "fea7d7a582cdfb64915d486ca39da9ebf7ef1d83," we know our file differs somehow from the original version.

When downloading many files, it can get rather tedious to check each checksum individually. The program shasum has a convenient solution—it can create and validate against a file containing the checksums of files. We can create a SHA-1 checksum file for all FASTQ files in the *data/* directory as follows:

```
$ shasum data/*fastq > fastq_checksums.sha
$ cat fastq_checksums.sha
524d9a057c51b1[...]d8b1cbe2eaf92c96a9    data/Csyrichta_TAGGACT_L008_R1_001.fastq
d2940f444f00c7[...]4f9c9314ab7e1a1b16    data/Csyrichta_TAGGACT_L008_R1_002.fastq
623a4ca571d572[...]1ec51b9ecd53d3aef6    data/Csyrichta_TAGGACT_L008_R1_003.fastq
f0b3a4302daf7a[...]7bf1628dfcb07535bb    data/Csyrichta_TAGGACT_L008_R1_004.fastq
53e2410863c36a[...]4c4c219966dd9a2fe5    data/Csyrichta_TAGGACT_L008_R1_005.fastq
e4d0ccf541e90c[...]5db75a3bef8c88ede7    data/Csyrichta_TAGGACT_L008_R1_006.fastq
```

Then, we can use shasum's check option (-c) to validate that these files match the
original versions:

```
$ shasum -c fastq_checksums.sha
data/Csyrichta_TAGGACT_L008_R1_001.fastq: OK
data/Csyrichta_TAGGACT_L008_R1_002.fastq: OK
data/Csyrichta_TAGGACT_L008_R1_003.fastq: OK
data/Csyrichta_TAGGACT_L008_R1_004.fastq: OK
data/Csyrichta_TAGGACT_L008_R1_005.fastq: OK
data/Csyrichta_TAGGACT_L008_R1_006.fastq: FAILED
shasum: WARNING: 1 computed checksum did NOT match
```

In the event that the checksums of a file disagree, shasum will show you which file
failed validation and exit with a nonzero error status.

The program md5sum (or md5 on OS X) calculates MD5 hashes and is similar in operation to shasum. However, note that on OS X, the md5 command doesn't have the -c
option, so you'll need to install the GNU version for this option. Also, some servers
use an antiquated checksum implementation such as sum or chsum. How we use these
older command-line checksum programs is similar to using shasum and md5sum.

Finally, you may be curious how all files can be summarized by a 40-character-long
SHA-1 checksum. They can't—there are only 16^{40} possible different checksums. However, 16^{40} is a huge number so the probability of a checksum collision is very, very
small. For the purposes of checking data integrity, the risk of a collision is negligible.

Looking at Differences Between Data

While checksums are a great method to check if files are different, they don't tell us
how files differ. One approach to this is to compute the *diff* between two files using
the Unix tool diff. Unix's diff works line by line, and outputs blocks (called *hunks*)
that differ between files (resembling Git's git diff command we saw in Chapter 4).

Suppose you notice a collaborator was working with a different version of a file than
the one you're using. Her version is *gene-2.bed*, and your version is *gene-1.bed* (these
files are on GitHub if you'd like to follow along). Because downstream results depend
on this dataset, you want to check if the files are indeed different. After comparing
the SHA-1 checksums, you find the files aren't identical. Before rerunning your analy-

sis using your collaborator's version, you'd like to check whether the files differ signif-icantly. We can do this by computing the diff between *gene-1.bed* and *gene-2.bed*:

```
$ diff -u gene-1.bed gene-2.bed

--- gene-1.bed  2014-02-22 12:53:14.000000000 -0800 ❶
+++ gene-2.bed  2015-03-10 01:55:01.000000000 -0700
@@ -1,22 +1,19 @@ ❷
 1      6206197 6206270 GENE00000025907
 1      6223599 6223745 GENE00000025907 ❸
 1      6227940 6228049 GENE00000025907
+1      6222341 6228319 GENE00000025907 ❹
 1      6229959 6230073 GENE00000025907
-1      6230003 6230005 GENE00000025907 ❺
 1      6233961 6234087 GENE00000025907
 1      6234229 6234311 GENE00000025907
 1      6206227 6206270 GENE00000025907
 1      6227940 6228049 GENE00000025907
 1      6229959 6230073 GENE00000025907
-1      6230003 6230073 GENE00000025907 ❻
+1      6230133 6230191 GENE00000025907
 1      6233961 6234087 GENE00000025907
 1      6234229 6234399 GENE00000025907
 1      6238262 6238384 GENE00000025907
-1      6214645 6214957 GENE00000025907
 1      6227940 6228049 GENE00000025907
 1      6229959 6230073 GENE00000025907
-1      6230003 6230073 GENE00000025907
 1      6233961 6234087 GENE00000025907
 1      6234229 6234399 GENE00000025907
-1      6238262 6238464 GENE00000025907
 1      6239952 6240378 GENE00000025907
```

The option -u tells diff to output in *unified diff format*, which is a format nearly identical to the one used by git diff. I've chosen to use unified diffs rather than diff's default diff format because unified diffs provide more context.

Unified diffs are broken down into hunks that differ between the two files. Let's step through the key parts of this format:

❶ These two lines are the header of the unified diff. The original file *gene-1.bed* is prefixed by ---, and the modified file *gene-2.bed* is prefixed by +++. The date and time in these two lines are the modification times of these files.

❷ This line indicates the start of a changed hunk. The pairs of integers between @@ and @@ indicate where the hunk begins, and how long it is, in the original file (-1,22) and modified file (+1,19), respectively.

❸ Lines in the diff that begin with a space indicate the modified file's line hasn't changed.

❹ Lines in the diff that begin with a + indicate a line has been added to the modified file.

❺ Similarly, - indicates lines removed in the modified file.

❻ An adjacent line deletion and line addition indicates that this line was changed in the modified file.

Diff files appear very cryptic at first, but you'll grow familiar with them over time. diff's output can also be redirected to a file, which creates a *patch file*. Patch files act as instructions on how to update a plain-text file, making the changes contained in the diff file. The Unix tool patch can apply changes to a file needed to be patched. Patches are used more commonly in software development than bioinformatics, so we won't cover them in detail. Lastly, it's important to note that diffs can be computationally expensive to compute on large files, so be cautious when running diff on large datasets.

Compressing Data and Working with Compressed Data

Data compression, the process of condensing data so that it takes up less space (on disk drives, in memory, or across network transfers), is an indispensable technology in modern bioinformatics. For example, sequences from a recent Illumina HiSeq run when compressed with Gzip take up 21,408,674,240 bytes, which is a bit under 20 gigabytes. Uncompressed, this file is a whopping 63,203,414,514 bytes (around 58 gigabytes). This FASTQ file has 150 million 200bp reads, which is 10x coverage of the human genome, 190x coverage of the Arabidopsis genome, or a measly 2x coverage of the hexaploid wheat genome. The compression ratio (uncompressed size/ compressed size) of this data is approximately 2.95, which translates to a significant space saving of about 66%. Your own bioinformatics projects will likely contain much more data, especially as sequencing costs continue to drop and it's possible to sequence genomes to higher depth, include more biological replicates or time points in expression studies, or sequence more individuals in genotyping studies.

For the most part, data can remain compressed on the disk throughout processing and analyses. Most well-written bioinformatics tools can work natively with compressed data as input, without requiring us to decompress it to disk first. Using pipes and redirection (covered in Chapter 3), we can stream compressed data and write compressed files directly to the disk. Additionally, common Unix tools like cat, grep, and less all have variants that work with compressed data, and Python's gzip module allows us to read and write compressed data from within Python. So while working with large datasets in bioinformatics can be challenging, using the compression tools in Unix and software libraries make our lives much easier.

gzip

The two most common compression systems used on Unix are gzip and bzip2. Both have their advantages: gzip compresses and decompresses data faster than bzip2, but bzip2 has a higher compression ratio (the previously mentioned FASTQ file is only about 16 GB when compressed with bzip2). Generally, gzip is used in bioinformatics to compress most sizable files, while bzip2 is more common for long-term data archiving. We'll focus primarily on gzip, but bzip2's tools behave very similarly to gzip.

The command-line tool `gzip` allows you to compress files in a few different ways. First, `gzip` can compress results from standard input. This is quite useful, as we can compress results directly from another bioinformatics program's standard output. For example, suppose we have a program that removes low-quality bases from FASTQ files called `trimmer` (this is an imaginary program). Our `trimmer` program can handle gzipped input files natively, but writes uncompressed trimmed FASTQ results to standard output. Using `gzip`, we can compress `trimmer`'s output in place, before writing to the disk:

```
$ trimmer in.fastq.gz | gzip > out.fastq.gz
```

gzip takes input from standard in, compresses it, and writes this compressed output to standard out.

gzip also can compress files on disk in place. If our *in.fastq.gz* file weren't compressed, we could compress it as follows:

```
$ ls
in.fastq
$ gzip in.fastq
$ ls
in.fastq.gz
```

gzip will compress this file in place, replacing the original uncompressed version with the compressed file (appending the extension *.gz* to the original filename). Similarly, we can decompress files in place with the command `gunzip`:

```
$ gunzip in.fastq.gz
$ ls
in.fastq
```

Note that this replaces our *in.fastq.gz* file with the decompressed version (removing the *.gz* suffix, too). Both gzip and gunzip can also output their results to standard out (rather than changing files in place). This can be enabled using the -c option:

```
$ gzip -c in.fastq > in.fastq.gz
$ gunzip -c in.fastq.gz > duplicate_in.fastq
```

A nice feature of the gzip compression algorithm is that you can concatenate gzip compressed output directly to an existing gzip file. For example, if we wanted to compress the *in2.fastq* file and append it to our compressed *in.fastq.gz* file, we wouldn't have to decompress *in.fastq.gz* first, concatenate the two files, and then compress the concatenated file. Instead, we can do the following:

```
$ ls
in.fastq.gz in2.fastq
$ gzip -c in2.fastq >> in.fastq.gz
```

Importantly, note that the redirection operator we use is >>; had we used >, we would overwrite our compressed version of *in2.fastq* to *in.fastq.gz* (rather than append to it). Always exercise caution when using redirection, and make sure you're using the appropriate operator (and keep file backups!). You may get a slightly better compression ratio by compressing everything together (e.g., with `cat in.fastq in2.fastq | gzip > in.fastq.gz`), but the convenience of appending to an existing gzipped file is useful. Also, note that gzip does not separate these compressed files: files compressed together are *concatenated*. If you need to compress multiple separate files into a single archive, use the `tar` utility (see the examples section of `man tar` for details).

Working with Gzipped Compressed Files

Perhaps the greatest advantage of gzip (and bzip2) is that many common Unix and bioinformatics tools can work directly with compressed files. For example, we can search compressed files using `grep`'s analog for gzipped files, `zgrep`. Likewise, `cat` has `zcat` (on some systems like OS X, this is `gzcat`), `diff` has `zdiff`, and `less` has `zless`. If programs cannot handle compressed input, you can use `zcat` and pipe output directly to the standard input of another program.

These programs that handle compressed input behave exactly like their standard counterpart. For example, all options available in `grep` are available in `zgrep`:

```
$ zgrep --color -i -n "AGATAGAT" Csyrichta_TAGGACT_L008_R1_001.fastq.gz
2706: ACTTCGGAGAGCCCATATATACACACTAAGATAGATAGCGTTAGCTAATGTAGATAGATT
```

There can be a slight performance cost in working with gzipped files, as your CPU must decompress input first. Usually, the convenience of z-tools like `zgrep`, `zless`, and `zcat` and the saved disk space outweigh any potential performance hits.

Case Study: Reproducibly Downloading Data

Downloading data reproducibly can be deceptively complex. We usually download genomic resources like sequence and annotation files from remote servers over the Internet, which may change in the future. Furthermore, new versions of sequence and annotation data may be released, so it's imperative that we document everything about how data was acquired for full reproducibility. As a demonstration of this, let's step through a case study. We'll download a few genomic and sequence resources for mouse (*Mus musculus*) and document how we acquired them.

For this example, we'll download the GRCm38 mouse reference genome and accompanying annotation. Note that this case study involves downloading large files, so you may not want to follow along with these examples. The mouse, human (*Homo sapiens*), and zebrafish (*Danio rerio*) genomes releases are coordinated through the Genome Reference Consortium (*http://bit.ly/gene-ref-con*). The "GRC" prefix in GRCm38 refers to the Genome Reference Consortium. We can download GRCm38 from Ensembl (a member of the consortium) using wget. For this and other examples in this section, I've had to truncate the URLs so they fit within a book's page width; see this chapter's *README.md* on GitHub for the full links for copying and pasting if you're following along.

```
$ wget ftp://ftp.ensembl.org/[...]/Mus_musculus.GRCm38.74.dna.toplevel.fa.gz
```

Ensembl's website provides links to reference genomes, annotation, variation data, and other useful files for many organisms. This FTP link comes from navigating to *http://www.ensembl.org*, clicking the mouse project page, and then clicking the "Download DNA sequence" link. If we were to document how we downloaded this file, our Markdown *README.md* might include something like:

```
Mouse (*Mus musculus*) reference genome version GRCm38 (Ensembl
release 74) was downloaded on Sat Feb 22 21:24:42 PST 2014, using:

    wget ftp://ftp.ensembl.org/[...]/Mus_musculus.GRCm38.74.dna.toplevel.fa.gz
```

We might want to look at the chromosomes, scaffolds, and contigs this files contains as a sanity check. This file is a gzipped FASTA file, so we can take a quick peek at all sequence headers by grepping for the regular expression "^>", which matches all lines beginning with > (a FASTA header). We can use the zgrep program to extract the FASTA headers on this gzipped file:

```
$ zgrep "^>" Mus_musculus.GRCm38.74.dna.toplevel.fa.gz | less
```

Ensembl also provides a checksum file in the parent directory called *CHECKSUMS*. This checksum file contains checksums calculated using the older Unix tool sum. We can compare our checksum values with those in *CHECKSUMS* using the sum program:

```
$ wget ftp://ftp.ensembl.org/pub/release-74/fasta/mus_musculus/dna/CHECKSUMS
$ sum Mus_musculus.GRCm38.74.dna.toplevel.fa.gz
53504 793314
```

The checksum 53504 agrees with the entry in the *CHECKSUMS* file for the entry *Mus_musculus.GRCm38.74.dna.toplevel.fa.gz*. I also like to include the SHA-1 sums of all important data in my data *README.md* file, so future collaborators can verify their data files are exactly the same as those I used. Let's calculate the SHA-1 sum using shasum:

```
$ shasum Mus_musculus.GRCm38.74.dna.toplevel.fa.gz
01c868e22a981[...]c2154c20ae7899c5f  Mus_musculus.GRCm38.74.dna.toplevel.fa.gz
```

Then, we can copy and paste this SHA-1 sum into our *README.md*. Next, we can download an accompanying GTF from Ensembl and the *CHECKSUMS* file for this directory:

```
$ wget ftp://ftp.ensembl.org/[...]/Mus_musculus.GRCm38.74.gtf.gz
$ wget ftp://ftp.ensembl.org/[...]/CHECKSUMS
```

Again, let's ensure that our checksums match those in the *CHECKSUMS* file and run shasum on this file for our own documentation:

```
$ sum Mus_musculus.GRCm38.74.gtf.gz
00985 15074
$ shasum cf5bb5f8bda2803410bb04b708bff59cb575e379  Mus_musculus.GRCm38.74.gtf.gz
```

And again, we copy the SHA-1 into our *README.md*. So far, our *README.md* might look as follows:

```
## Genome and Annotation Data

Mouse (*Mus musculus*) reference genome version GRCm38 (Ensembl
release 74) was downloaded on Sat Feb 22 21:24:42 PST 2014, using:

    wget ftp://ftp.ensembl.org/[...]/Mus_musculus.GRCm38.74.dna.toplevel.fa.gz

Gene annotation data (also Ensembl release 74) was downloaded from Ensembl on
Sat Feb 22 23:30:27 PST 2014, using:

    wget ftp://ftp.ensembl.org/[...]/Mus_musculus.GRCm38.74.gtf.gz

## SHA-1 Sums

 - `Mus_musculus.GRCm38.74.dna.toplevel.fa.gz`: 01c868e22a9815c[...]c2154c20ae7899c5f
 - `Mus_musculus.GRCm38.74.gtf.gz`: cf5bb5f8bda2803[...]708bff59cb575e379
```

Although this isn't a lot of documentation, this is infinitely better than not documenting how data was acquired. As this example demonstrates, it takes very little effort to properly track the data that enters your project, and thereby ensure reproducibility. The most important step in documenting your work is that you're consistent and make it a habit.

## PART III

Practice: Bioinformatics Data Skills

## CHAPTER 7

Unix Data Tools

We often forget how science and engineering function. Ideas come from previous exploration more often than from lightning strokes.

—John W. Tukey

In Chapter 3, we learned the basics of the Unix shell: using streams, redirecting output, pipes, and working with processes. These core concepts not only allow us to use the shell to run command-line bioinformatics tools, but to leverage Unix as a modular work environment for working with bioinformatics data. In this chapter, we'll see how we can combine the Unix shell with command-line data tools to explore and manipulate data quickly.

Unix Data Tools and the Unix One-Liner Approach: Lessons from Programming Pearls

Understanding how to use Unix data tools in bioinformatics isn't only about learning what each tool does, it's about mastering the practice of connecting tools together—creating programs from *Unix pipelines*. By connecting data tools together with pipes, we can construct programs that parse, manipulate, and summarize data. Unix pipelines can be developed in shell scripts or as "one-liners"—tiny programs built by connecting Unix tools with pipes directly on the shell. Whether in a script or as a one-liner, building more complex programs from small, modular tools capitalizes on the design and philosophy of Unix (discussed in "Why Do We Use Unix in Bioinformatics? Modularity and the Unix Philosophy" on page 37). The pipeline approach to building programs is a well-established tradition in Unix (and bioinformatics) because it's a fast way to solve problems, incredibly powerful, and adaptable to a variety of a problems. An illustrative example of the power of simple Unix pipelines comes from a famous exchange between two brilliant computer scientists: Donald Knuth and Doug McIlroy (recall from Chapter 3 that McIlroy invented Unix pipes).

In a 1986 "Programming Pearls" column in the *Communications of the ACM* magazine, columnist Jon Bentley had computer scientist Donald Knuth write a simple program to count and print the k most common words in a file alongside their counts, in descending order. Knuth was chosen to write this program to demonstrate *literate programming*, a method of programming that Knuth pioneered. Literate programs are written as a text document explaining how to solve a programming problem (in plain English) with code interspersed throughout the document. Code inside this document can then be "tangled" out of the document using literate programming tools (this approach might be recognizable to readers familiar with R's knitr or Sweave—both are modern descendants of this concept). Knuth's literate program was seven pages long, and also highly customized to this particular programming problem; for example, Knuth implemented a custom data structure for the task of counting English words. Bentley then asked that McIlroy critique Knuth's seven-page-long solution. McIlroy commended Knuth's literate programming and novel data structure, but overall disagreed with his engineering approach. McIlroy replied with a six-line Unix script that solved the same programming problem:

```
tr -cs A-Za-z '\n' | ❶
tr A-Z a-z | ❷
sort | ❸
uniq -c | ❹
sort -rn | ❺
sed ${1}q ❻
```

While you shouldn't worry about fully understanding this now (we'll learn these tools in this chapter), McIlroy's basic approach was:

❶ Translate all nonalphabetical characters (-c takes the complement of the first argument) to newlines and squeeze all adjacent characters together (-s) after translating. This creates one-word lines for the entire input stream.

❷ Translate all uppercase letters to lowercase.

❸ Sort input, bringing identical words on consecutive lines.

❹ Remove all duplicate consecutive lines, keeping only one with a count of the occurrences (-c).

❺ Sort in reverse (-r) numeric order (-n).

❻ Print the first k number of lines supplied by the first argument of the script (${1}) and quit.

McIlroy's solution is a beautiful example of the Unix approach. McIlroy originally wrote this as a script, but it can easily be turned into a one-liner entered directly on

the shell (assuming *k* here is 10). However, I've had to add a line break here so that
the code does not extend outside of the page margins:

```
$ cat input.txt \
  | tr -cs A-Za-z '\n' | tr A-Z a-z | sort | uniq -c | sort -rn | sed 10q
```

McIlroy's script was doubtlessly much faster to implement than Knuth's program and
works just as well (and arguably better, as there were a few minor bugs in Knuth's solution). Also, his solution was built on reusable Unix data tools (or as he called them,
"Unix staples") rather than "programmed monolithically from scratch," to use McIlroy's phrasing. The speed and power of this approach is why it's a core part of bioinformatics work.

When to Use the Unix Pipeline Approach and How to Use It Safely

Although McIlroy's example is appealing, the Unix one-liner approach isn't appropriate for all problems. Many bioinformatics tasks are better accomplished through a
custom, well-documented script, more akin to Knuth's program in "Programming
Pearls." Knowing when to use a fast and simple engineering solution like a Unix pipeline and when to resort to writing a well-documented Python or R script takes experience. As with most tasks in bioinformatics, choosing the most suitable approach can
be half the battle.

Unix pipelines entered directly into the command line shine as a fast, low-level data
manipulation toolkit to explore data, transform data between formats, and inspect
data for potential problems. In this context, we're not looking for thorough, theory-
shattering answers—we usually just want a quick picture of our data. We're willing to
sacrifice a well-documented implementation that solves a specific problem in favor of
a quick rough picture built from modular Unix tools. As McIlroy explained in his
response:

> The simple pipeline ... will suffice to get answers right now, not next week or next
> month. It could well be enough to finish the job. But even for a production project ... it
> would make a handsome down payment, useful for *testing the value of the answers and
> for smoking out follow-on questions.*
>
> —Doug McIlroy (my emphasis)

Many tasks in bioinformatics are of this nature: we want to get a quick answer and
keep moving forward with our project. We could write a custom script, but for simple
tasks this might be overkill and would take more time than necessary. As we'll see
later in this chapter, building Unix pipelines is fast: we can iteratively assemble and
test Unix pipelines directly in the shell.

For larger, more complex tasks it's often preferable to write a custom script in a language like Python (or R if the work involves lots of data analysis). While shell

approaches (whether a one-liner or a shell script) are useful, these don't allow for the same level of flexibility in checking input data, structuring programs, use of data structures, code documentation, and adding assert statements and tests as languages like Python and R. These languages also have better tools for stepwise documentation of larger analyses, like R's knitr (introduced in the "Reproducibility with Knitr and Rmarkdown" on page 254) and iPython notebooks. In contrast, lengthy Unix pipelines can be fragile and less robust than a custom script.

So in cases where using Unix pipelines is appropriate, what steps can we take to ensure they're reproducible? As mentioned in Chapter 1, it's essential that everything that produces a result is documented. Because Unix one-liners are entered directly in the shell, it's particularly easy to lose track of which one-liner produced what version of output. Remembering to record one-liners requires extra diligence (and is often neglected, especially in bioinformatics work). Storing pipelines in scripts is a good approach—not only do scripts serve as documentation of what steps were performed on data, but they allow pipelines to be rerun and can be checked into a Git repository. We'll look at scripting in more detail in Chapter 12.

Inspecting and Manipulating Text Data with Unix Tools

In this chapter, our focus is on learning how to use core Unix tools to manipulate and explore plain-text data formats. Many formats in bioinformatics are simple tabular plain-text files delimited by a character. The most common tabular plain-text file format used in bioinformatics is tab-delimited. This is not an accident: most Unix tools such as cut and awk treat tabs as delimiters by default. Bioinformatics evolved to favor tab-delimited formats because of the convenience of working with these files using Unix tools. Tab-delimited file formats are also simple to parse with scripting languages like Python and Perl, and easy to load into R.

Tabular Plain-Text Data Formats

Tabular plain-text data formats are used extensively in computing. The basic format is incredibly simple: each row (also known as a record) is kept on its own line, and each column (also known as a field) is separated by some delimiter. There are three flavors you will encounter: tab-delimited, comma-separated, and variable space-delimited.

Of these three formats, tab-delimited is the most commonly used in bioinformatics. File formats such as BED, GTF/GFF, SAM, tabular BLAST output, and VCF are all examples of tab-delimited files. Columns of a tab-delimited file are separated by a single tab character (which has the escape code \t). A common convention (but not a standard) is to include metadata on the first few lines of a tab-delimited file. These metadata lines begin with # to differentiate them from the tabular data records. Because tab-delimited files use a tab to delimit columns, tabs in data are not allowed.

Comma-separated values (CSV) is another common format. CSV is similar to tab-delimited, except the delimiter is a comma character. While not a common occurrence in bioinformatics, it is possible that the data stored in CSV format contain commas (which would interfere with the ability to parse it). Some variants just don't allow this, while others use quotes around entries that could contain commas. Unfortunately, there's no standard CSV format that defines how to handle this and many other issues with CSV—though some guidelines are given in RFC 4180 (*http://bit.ly/rfc-4180*).

Lastly, there are space-delimited formats. A few stubborn bioinformatics programs use a variable number of spaces to separate columns. In general, tab-delimited formats and CSV are better choices than space-delimited formats because it's quite common to encounter data containing spaces.

Despite the simplicity of tabular data formats, there's one major common headache: how lines are separated. Linux and OS X use a single linefeed character (with the escape code \n) to separate lines, while Windows uses a DOS-style line separator of a carriage return and a linefeed character (\r\n). CSV files generally use this DOS-style too, as this is specified in the CSV specification RFC-4180 (which in practice is loosely followed). Occasionally, you might encounter files separated by only carriage returns, too.

In this chapter, we'll work with very simple genomic feature formats: BED (three-column) and GTF files. These file formats store the positions of features such as genes, exons, and variants in tab-delimited format. Don't worry too much about the specifics of these formats; we'll cover both in more detail in Chapter 9. Our goal in this chapter is primarily to develop the skills to freely manipulate plain-text files or streams using Unix data tools. We'll learn each tool separately, and cumulatively work up to more advanced pipelines and programs.

Inspecting Data with Head and Tail

Many files we encounter in bioinformatics are much too long to inspect with cat—running cat on a file a million lines long would quickly fill your shell with text scrolling far too fast to make sense of. A better option is to take a look at the top of a file with head. Here, let's took a look at the file *Mus_musculus.GRCm38.75_chr1.bed*:

```
$ head Mus_musculus.GRCm38.75_chr1.bed
1       3054233 3054733
1       3054233 3054733
1       3054233 3054733
1       3102016 3102125
1       3102016 3102125
1       3102016 3102125
1       3205901 3671498
1       3205901 3216344
```

```
1        3213609 3216344
1        3205901 3207317
```

We can also control how many lines we see with head through the -n argument:

```
$ head -n 3 Mus_musculus.GRCm38.75_chr1.bed
1        3054233 3054733
1        3054233 3054733
1        3054233 3054733
```

head is useful for a quick inspection of files. head -n3 allows you to quickly inspect a file to see if a column header exists, how many columns there are, what delimiter is being used, some sample rows, and so on.

head has a related command designed to look at the end, or *tail* of a file. tail works just like head:

```
$ tail -n 3 Mus_musculus.GRCm38.75_chr1.bed
1        195240910       195241007
1        195240910       195241007
1        195240910       195241007
```

We can also use tail to remove the header of a file. Normally the -n argument specifies how many of the last lines of a file to include, but if -n is given a number x preceded with a + sign (e.g., +x), tail will start from the xth line. So to chop off a header, we start from the second line with -n +2. Here, we'll use the command seq to generate a file of 3 numbers, and chop of the first line:

```
$ seq 3 > nums.txt
$ cat nums.txt
1
2
3
$ tail -n +2 nums.txt
2
3
```

Sometimes it's useful to see both the beginning and end of a file—for example, if we have a sorted BED file and we want to see the positions of the first feature and last feature. We can do this using a trick from data scientist (and former bioinformatician) Seth Brown:

```
$ (head -n 2; tail -n 2) < Mus_musculus.GRCm38.75_chr1.bed
1        3054233 3054733
1        3054233 3054733
1        195240910       195241007
1        195240910       195241007
```

This is a useful trick, but it's a bit long to type. To keep it handy, we can create a shortcut in your shell configuration file, which is either ~/.bashrc or ~/.profile:

```
# inspect the first and last 3 lines of a file
i() { (head -n 2; tail -n 2) < "$1" | column -t}
```

Then, either run `source` on your shell configuration file, or start a new terminal session and ensure this works. Then we can use `i` (for inspect) as a normal command:

```
$ i Mus_musculus.GRCm38.75_chr1.bed
1  3054233    3054733
1  3054233    3054733
1  195240910  195241007
1  195240910  195241007
```

`head` is also useful for taking a peek at data resulting from a Unix pipeline. For example, suppose we want to `grep` the *Mus_musculus.GRCm38.75_chr1.gtf* file for rows containing the string `gene_id "ENSMUSG00000025907"` (because our GTF is well structured, it's safe to assume that these are all features belonging to this gene—but this may not always be the case!). We'll use `grep`'s results as the standard input for the next program in our pipeline, but first we want to check `grep`'s standard out to see if everything looks correct. We can pipe the standard out of `grep` directly to `head` to take a look:

```
$ grep 'gene_id "ENSMUSG00000025907"' Mus_musculus.GRCm38.75_chr1.gtf | head -n 1
1 protein_coding  gene  6206197 6276648 [...] gene_id "ENSMUSG00000025907" [...]
```

Note that for the sake of clarity, I've omitted the full line of this GTF, as it's quite long.

After printing the first few rows of your data to ensure your pipeline is working properly, the `head` process exits. This is an important feature that helps ensure your pipes don't needlessly keep processing data. When `head` exits, your shell catches this and stops the *entire* pipe, *including* the `grep` process too. Under the hood, your shell sends a signal to other programs in the pipe called `SIGPIPE`—much like the signal that's sent when you press Control-c (that signal is `SIGINT`). When building complex pipelines that process large amounts of data, this is extremely important. It means that in a pipeline like:

```
$ grep "some_string" huge_file.txt | program1 | program2 | head -n 5
```

`grep` won't continue searching *huge_file.txt*, and `program1` and `program2` don't continue processing input after `head` outputs 5 lines and exits. While `head` is a good illustration of this feature of pipes, `SIGPIPE` works with all programs (unless the program explicitly catches and ignore this symbol—a possibility, but not one we encounter with bioinformatics programs).

less

`less` is also a useful program for a inspecting files and the output of pipes. `less` is a *terminal pager*, a program that allows us to view large amounts of text in our terminals. Normally, if we `cat` a long file to screen, the text flashes by in an instant—`less` allows us to view and scroll through long files and standard output a screen at a time. Other applications can call the default terminal pager to handle displaying large amounts of output; this is how `git log` displays an entire Git repository's commit history. You might run across another common, but older terminal pager called `more`, but `less` has more features and is generally preferred (the name of `less` is a play on "less is more").

`less` runs more like an application than a command: once we start `less`, it will stay open until we quit it. Let's review an example—in this chapter's directory in the book's GitHub repository, there's a file called *contaminated.fastq*. Let's look at this with `less`:

```
$ less contaminated.fastq
```

This will open up the program `less` in your terminal and display a FASTQ file full of sequences. First, if you need to quit `less`, press *q*. At any time, you can bring up a help page of all of `less`'s commands with *h*.

Moving around in `less` is simple: press space to go down a page, and *b* to go up a page. You can use *j* and *k* to go down and up a line at a time (these are the same keys that the editor Vim uses to move down and up). To go back up to the top of a file, enter *g*; to go to the bottom of a file, press *G*. Table 7-1 lists the most commonly used `less` commands. We'll talk a bit about how this works when `less` is taking input from another program through a pipe in a bit.

Table 7-1. Commonly used less commands

Shortcut	Action
space bar	Next page
b	Previous page
g	First line
G	Last line
j	Down (one line at at time)
k	Up (one line at at time)
/<pattern>	Search down (forward) for string <pattern>
?<pattern>	Search up (backward) for string <pattern>
n	Repeat last search downward (forward)
N	Repeat last search upward (backward)

One of the most useful features of `less` is that it allows you to search text and highlights matches. Visually highlighting matches can be an extremely useful way to find potential problems in data. For example, let's use `less` to get a quick sense of whether there are 3' adapter contaminants in the *contaminated.fastq* file. In this case, we'll look for AGATCGGAAGAGCACACGTCTGAACTCCAGTCAC (a known adapter from the Illumina Tru-Seq® kit[1]). Our goal isn't to do an exhaustive test or remove these adapters—we just want to take a 30-second peek to check if there's any indication there could be contamination.

Searching for this entire string won't be very helpful, for the following reasons:

- It's likely that only part of the adapter will be in sequences
- It's common for there to be a few mismatches in sequences, making exact matching ineffective (especially since the base calling accuracy typically drops at the 3' end of Illumina sequencing reads)

To get around this, let's search for the first 11 bases, AGATCGGAAGA. First, we open *contaminated.fastq* in `less`, and then press / and enter AGATCGG. The results are in Figure 7-1, which passes the interocular test—the results hit you right between the eyes. Note the skew in match position toward the end of sequencing reads (where we expect contamination) and the high similarity in bases after the match. Although only a quick visual inspection, this is quite informative.

`less` is also extremely useful in debugging our command-line pipelines. One of the great beauties of the Unix pipe is that it's easy to debug at any point—just pipe the output of the command you want to debug to `less` and delete everything after. When you run the pipe, `less` will capture the output of the last command and pause so you can inspect it.

`less` is also crucial when iteratively building up a pipeline—which is the best way to construct pipelines. Suppose we have an imaginary pipeline that involves three programs, `step1`, `step2`, and `step3`. Our finished pipeline will look like `step1 input.txt | step2 | step3 > output.txt`. However, we want to build this up in pieces, running `step1 input.txt` first and checking its output, then adding in `step3` and checking that output, and so forth. The natural way to do this is with `less`:

```
$ step1 input.txt | less                    # inspect output in less
$ step1 input.txt | step2 | less
$ step1 input.txt | step2 | step3 | less
```

---

1 Courtesy of Illumina, Inc.

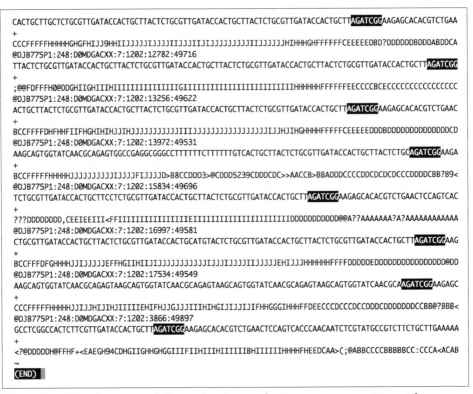

Figure 7-1. Using less to search for contaminant adapter sequences starting with "AGATCGG"; note how the nucleotides after the match are all very similar

A useful behavior of pipes is that the execution of a program with output piped to less will be paused when less has a full screen of data. This is due to how pipes *block* programs from writing to a pipe when the pipe is full. When you pipe a program's output to less and inspect it, less stops reading input from the pipe. Soon, the pipe becomes full and blocks the program putting data into the pipe from continuing. The result is that we can throw less after a complex pipe processing large data and not worry about wasting computing power—the pipe will block and we can spend as much time as needed to inspect the output.

Plain-Text Data Summary Information with wc, ls, and awk

In addition to peeking at a file with head, tail, or less, we may want other bits of summary information about a plain-text data file like the number of rows or columns. With plain-text data formats like tab-delimited and CSV files, the number of rows is usually the number of lines. We can retrieve this with the program wc (for word count):

```
$ wc Mus_musculus.GRCm38.75_chr1.bed
  81226  243678 1698545 Mus_musculus.GRCm38.75_chr1.bed
```

By default, wc outputs the number of words, lines, and characters of the supplied file. It can also work with many files:

```
$ wc Mus_musculus.GRCm38.75_chr1.bed Mus_musculus.GRCm38.75_chr1.gtf
  81226  243678 1698545 Mus_musculus.GRCm38.75_chr1.bed
  81231 2385570 26607149 Mus_musculus.GRCm38.75_chr1.gtf
 162457 2629248 28305694 total
```

Often, we only care about the number of lines. We can use option -l to just return the number of lines:

```
$ wc -l Mus_musculus.GRCm38.75_chr1.bed
  81226 Mus_musculus.GRCm38.75_chr1.bed
```

You might have noticed a discrepancy between the BED file and the GTF file for this chromosome 1 mouse annotation. What's going on? Using head, we can inspect the *Mus_musculus.GRCm38.75_chr1.gtf* file and see that the first few lines are comments:

```
$ head -n 5 Mus_musculus.GRCm38.75_chr1.gtf
#!genome-build GRCm38.p2
#!genome-version GRCm38
#!genome-date 2012-01
#!genome-build-accession NCBI:GCA_000001635.4
#!genebuild-last-updated 2013-09
```

The five-line discrepancy we see with wc -l is due to this header. Using a hash mark (#) as a comment field for metadata is a common convention; it is one we need to consider when using Unix data tools.

Another bit of information we usually want about a file is its size. The easiest way to do this is with our old Unix friend, ls, with the -l option:

```
$ ls -l Mus_musculus.GRCm38.75_chr1.bed
-rw-r--r-- 1 vinceb  staff  1698545 Jul 14 22:40 Mus_musculus.GRCm38.75_chr1.bed
```

In the fourth column (the one before the creation data) ls -l reports file sizes in bytes. If we wish to use human-readable sizes, we can use ls -lh:

```
$ ls -lh Mus_musculus.GRCm38.75_chr1.bed
-rw-r--r-- 1 vinceb  staff  1.6M Jul 14 22:40 Mus_musculus.GRCm38.75_chr1.bed
```

Here, "M" indicates megabytes; if a file is gigabytes in size, ls -lh will output results in gigabytes, "G."

Data Formats and Assumptions

Although `wc -l` is a quick way to determine how many rows there are in a plain-text data file (e.g., a TSV or CSV file), it makes the assumption that your data is well formatted. For example, imagine that a script writes data output like:

```
$ cat some_data.bed
1  3054233  3054733
1  3054233  3054733
1  3054233  3054733


$ wc -l data.txt
     5 data.txt
```

There's a subtle problem here: while there are only three rows of data, there are five lines. These two extra lines are empty newlines at the end of the file. So while `wc -l` is a quick and easy way to count the number of *lines* in a file, it isn't the most robust way to check how many *rows* of data are in a file. Still, `wc -l` will work well enough in most cases when we just need a rough idea how many rows there are. If we wish to exclude lines with just white-space (spaces, tabs, or newlines), we can use `grep`:

```
$ grep -c "[^ \\n\\t]" some_data.bed
3
```

We'll talk a lot more about `grep` later on.

There's one other bit of information we often want about a file: how many columns it contains. We could always manually count the number of columns of the first row with `head -n 1`, but a far easier way is to use `awk`. Awk is an easy, small programming language great at working with text data like TSV and CSV files. We'll introduce `awk` as a language in much more detail in "Text Processing with Awk" on page 157, but let's use an `awk` one-liner to return how many fields a file contains:

```
$ awk -F "\t" '{print NF; exit}' Mus_musculus.GRCm38.75_chr1.bed
3
```

`awk` was designed for tabular plain-text data processing, and consequently has a built-in variable NF set to the number of fields of the current dataset. This simple `awk` one-liner simply prints the number of fields of the first row of the *Mus_musculus.GRCm38.75_chr1.bed* file, and then exits. By default, `awk` treats white-space (tabs and spaces) as the field separator, but we could change this to just tabs by setting the `-F` argument of `awk` (because the examples in both BED and GTF formats we're working in are tab-delimited).

Finding how many columns there are in *Mus_musculus.GRCm38.75_chr1.gtf* is a bit trickier. Remember that our *Mus_musculus.GRCm38.75_chr1.gtf* file has a series of comments before it: five lines that begin with hash symbols (#) that contain helpful metadata like the genome build, version, date, and accession number. Because the first line of this file is a comment, our awk trick won't work—instead of reporting the number of data columns, it returns the number of columns of the first comment. To see how many columns of data there are, we need to first chop off the comments and then pass the results to our awk one-liner. One way to do this is with a tail trick we saw earlier:

```
$ tail -n +5 Mus_musculus.GRCm38.75_chr1.gtf | head -n 1 ❶
#!genebuild-last-updated 2013-09
$ tail -n +6 Mus_musculus.GRCm38.75_chr1.gtf | head ❷
1    pseudogene    gene    3054233    3054733    .    +    . [...]

$ tail -n +6 Mus_musculus.GRCm38.75_chr1.gtf | awk -F "\t" '{print NF; exit}' ❸
16
```

❶ Using tail with the -n +5 argument (note the preceding plus sign), we can chop off some rows. Before piping these results to awk, we pipe it to head to inspect what we have. Indeed, we see we've made a mistake: the first line returned is the last comment—we need to chop off one more line.

❷ Incrementing our -n argument to 6, and inspecting the results with head, we get the results we want: the first row of the standard output stream is the first row of the *Mus_musculus.GRCm38.75_chr1.gtf* GTF file.

❸ Now, we can pipe this data to our awk one-liner to get the number of columns in this file.

While removing a comment header block at the beginning of a file with tail does work, it's not very elegant and has weaknesses as an engineering solution. As you become more familiar with computing, you'll recognize a solution like this as brittle. While we've engineered a solution that does what we want, will it work on other files? Is it robust? Is this a *good* way to do this? The answer to all three questions is no. Recognizing when a solution is too fragile is an important part of developing Unix data skills.

The weakness with using tail -n +6 to drop commented header lines from a file is that this solution must be tailored to specific files. It's not a general solution, while removing comment lines from a file is a general problem. Using tail involves figuring out how many lines need to be chopped off and then hardcoding this value in our Unix pipeline. Here, a better solution would be to simply exclude all lines that match a comment line pattern. Using the program grep (which we'll talk more about in "The All-Powerful Grep" on page 140), we can easily exclude lines that begin with "#":

```
$ grep -v "^#" Mus_musculus.GRCm38.75_chr1.gtf | head -n 3
1  pseudogene                gene        3054233  3054733  .  +  .  [...]
1  unprocessed_pseudogene    transcript  3054233  3054733  .  +  .  [...]
1  unprocessed_pseudogene    exon        3054233  3054733  .  +  .  [...]
```

This solution is faster and easier (because we don't have to count how many commented header lines there are), in addition to being less fragile and more robust. Overall, it's a better engineered solution—an optimal balance of robustness, being generalizable, and capable of being implemented quickly. These are the types of solutions you should hunt for when working with Unix data tools: they get the job done and are neither over-engineered nor too fragile.

Working with Column Data with cut and Columns

When working with plain-text tabular data formats like tab-delimited and CSV files, we often need to extract specific columns from the original file or stream. For example, suppose we wanted to extract only the start positions (the second column) of the *Mus_musculus.GRCm38.75_chr1.bed* file. The simplest way to do this is with cut. This program cuts out specified columns (also known as fields) from a text file. By default, cut treats tabs as the delimiters, so to extract the second column we use:

```
$ cut -f 2 Mus_musculus.GRCm38.75_chr1.bed | head -n 3
3054233
3054233
3054233
```

The -f argument is how we specify which columns to keep. The argument -f also allows us to specify ranges of columns (e.g., -f 3-8) and sets of columns (e.g., -f 3,5,8). Note that it's *not* possible to reorder columns using using cut (e.g., -f 6,5,4,3 will not work, unfortunately). To reorder columns, you'll need to use awk, which is discussed later.

Using cut, we can convert our GTF for *Mus_musculus.GRCm38.75_chr1.gtf* to a three-column tab-delimited file of genomic ranges (e.g., chromosome, start, and end position). We'll chop off the metadata rows using the grep command covered earlier, and then use cut to extract the first, fourth, and fifth columns (chromosome, start, end):

```
$ grep -v "^#" Mus_musculus.GRCm38.75_chr1.gtf | cut -f1,4,5 | head -n 3
1  3054233  3054733
1  3054233  3054733
1  3054233  3054733
$ grep -v "^#" Mus_musculus.GRCm38.75_chr1.gtf | cut -f1,4,5 > test.txt
```

Note that although our three-column file of genomic positions *looks* like a BED-formatted file, it's not due to subtle differences in genomic range formats. We'll learn more about this in Chapter 9.

cut also allows us to specify the column delimiter character. So, if we were to come across a CSV file containing chromosome names, start positions, and end positions, we could select columns from it, too:

```
$ head -n 3 Mus_musculus.GRCm38.75_chr1_bed.csv
1,3054233,3054733
1,3054233,3054733
1,3054233,3054733
$ cut -d, -f2,3 Mus_musculus.GRCm38.75_chr1_bed.csv | head -n 3
3054233,3054733
3054233,3054733
3054233,3054733
```

Formatting Tabular Data with column

As you may have noticed when working with tab-delimited files, it's not always easy to see which elements belong to a particular column. For example:

```
$ grep -v "^#" Mus_musculus.GRCm38.75_chr1.gtf | cut -f1-8 | head -n3
1       pseudogene      gene    3054233 3054733 .       +       .
1       unprocessed_pseudogene  transcript      3054233 3054733 .       +       .
1       unprocessed_pseudogene  exon    3054233 3054733 .       +       .
```

While tabs are a terrific delimiter in plain-text data files, our variable width data leads our columns to not stack up well. There's a fix for this in Unix: program column -t (the -t option tells column to treat data as a table). column -t produces neat columns that are much easier to read:

```
$ grep -v "^#" Mus_musculus.GRCm38.75_chr1.gtf | cut -f 1-8 | column -t
| head -n 3
1  pseudogene              gene        3054233  3054733  . + .
1  unprocessed_pseudogene  transcript  3054233  3054733  . + .
1  unprocessed_pseudogene  exon        3054233  3054733  . + .
```

Note that you should only use columnt -t to visualize data in the terminal, not to reformat data to write to a file. Tab-delimited data is preferable to data delimited by a variable number of spaces, since it's easier for programs to parse.

Like cut, column's default delimiter is the tab character (\t). We can specify a different delimiter with the -s option. So, if we wanted to visualize the columns of the *Mus_musculus.GRCm38.75_chr1_bed.csv* file more easily, we could use:

```
$ column -s"," -t Mus_musculus.GRCm38.75_chr1_bed.csv | head -n 3
1  3054233  3054733
1  3054233  3054733
1  3054233  3054733
```

column illustrates an important point about how we should treat data: there's no reason to make data formats attractive at the expense of readable by programs. This relates to the recommendation, "write code for humans, write data for computers" ("Write Code for Humans, Write Data for Computers" on page 11). Although single-character delimited columns (like CSV or tab-delimited) can be difficult for humans to read, consider the following points:

- They work instantly with nearly all Unix tools.
- They are easy to convert to a readable format with `column -t`.

In general, it's easier to make computer-readable data attractive to humans than it is to make data in a human-friendly format readable to a computer. Unfortunately, data in formats that prioritize human readability over computer readability still linger in bioinformatics.

The All-Powerful Grep

Earlier, we've seen how `grep` is a useful tool for extracting lines of a file that match (or don't match) a pattern. `grep -v` allowed us to exclude the header rows of a GTF file in a more robust way than `tail`. But as we'll see in this section, this is just scratching the surface of `grep`'s capabilities; `grep` is one of the most powerful Unix data tools.

First, it's important to mention `grep` is fast. *Really fast.* If you need to find a pattern (fixed string or regular expression) in a file, `grep` will be faster than anything you could write in Python. Figure 7-2 shows the runtimes of four methods of finding exact matching lines in a file: `grep`, `sed`, `awk`, and a simple custom Python script. As you can see, `grep` dominates in these benchmarks: it's five times faster than the fastest alternative, Python. However, this is a bit of unfair comparison: `grep` is fast because it's tuned to do one task really well: find lines of a file that match a pattern. The other programs included in this benchmark are more versatile, but pay the price in terms of efficiency in this particular task. This demonstrates a point: if computational speed is our foremost priority (and there are many cases when it isn't as important as we think), Unix tools tuned to do certain tasks really often are the fastest implementation.

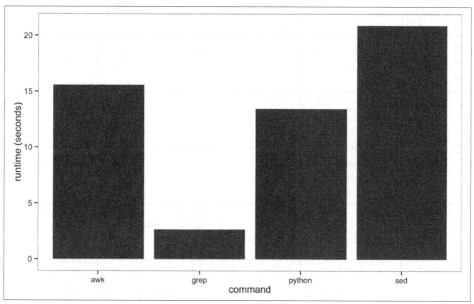

Figure 7-2. Benchmark of the time it takes to search the Maize genome for the exact string "AGATGCATG"

While we've seen `grep` used before in this book let's briefly review its basic usage. `grep` requires two arguments: the pattern (the string or basic regular expression you want to search for), and the file (or files) to search for it in. As a very simple example, let's use `grep` to find a gene, "Olfr418-ps1," in the file *Mus_musculus.GRCm38.75_chr1_genes.txt* (which contains all Ensembl gene identifiers and gene names for all protein-coding genes on chromosome 1):

```
$ grep "Olfr418-ps1" Mus_musculus.GRCm38.75_chr1_genes.txt
ENSMUSG00000049605      Olfr418-ps1
```

The quotes around the pattern aren't required, but it's safest to use quotes so our shells won't try to interpret any symbols. `grep` returns any lines that match the pattern, even ones that only partially match:

```
$ grep Olfr Mus_musculus.GRCm38.75_chr1_genes.txt | head -n 5
ENSMUSG00000067064      Olfr1416
ENSMUSG00000057464      Olfr1415
ENSMUSG00000042849      Olfr1414
ENSMUSG00000058904      Olfr1413
ENSMUSG00000046300      Olfr1412
```

One useful option when using `grep` is `--color=auto`. This option enables terminal colors, so the matching part of the pattern is colored in your terminal.

GNU, BSD, and the Flavors of Grep

Up until now, we've glossed over a very important detail: there are different *implementations* of Unix tools. Tools like `grep`, `cut`, and `sort` come from one of two flavors: BSD utils and GNU coreutils. Both of these implementations contain all standard Unix tools we use in this chapter, but their features may slightly differ from each other. BSD's tools are found on Max OS X and other Berkeley Software Distribution-derived operating systems like FreeBSD. GNU's coreutils are the standard set of tools found on Linux systems. It's important to know which implementation you're using (this is easy to tell by reading the man page). If you're using Mac OS X and would like to use GNU coreutils, you can install these through Homebrew with `brew install coreutils`. Each program will install with the prefix "g" (e.g., `cut` would be aliased to `gcut`), so as to not interfere with the system's default tools.

Unlike BSD's utils, GNU's coreutils are still actively developed. GNU's coreutils also have many more features and extensions than BSD's utils, some of which we use in this chapter. In general, I recommend you use GNU's coreutils over BSD utils, as the documentation is more thorough and the GNU extensions are helpful (and sometimes necessary). Throughout the chapter, I will indicate when a particular feature relies on the GNU version.

Earlier, we saw how `grep` could be used to only return lines that *do not* match the specified pattern—this is how we excluded the commented lines from our GTF file. The option we used was `-v`, for invert. For example, suppose you wanted a list of all genes that contain "Olfr," except "Olfr1413." Using `-v` and chaining together to calls to `grep` with pipes, we could use:

```
$ grep Olfr Mus_musculus.GRCm38.75_chr1_genes.txt | grep -v Olfr1413
```

But beware! What might go wrong with this? Partially matching may bite us here: while we wanted to exclude "Olfr1413," this command would *also* exclude genes like "Olfr1413a" and "Olfr14130." But we can get around this by using `-w`, which matches entire words (surrounded by whitespace). Let's look at how this works with a simpler toy example:

```
$ cat example.txt
bio
bioinfo
bioinformatics
computational biology
$ grep -v bioinfo example.txt
bio
computational biology
$ grep -v -w bioinfo example.txt
```

```
bio
bioinformatics
computational biology
```

By constraining our matches to be words, we're using a more restrictive pattern. In general, our patterns should always be as restrictive as possible to avoid unintentional matches caused by partial matching.

grep's default output often doesn't give us enough context of a match when we need to inspect results by eye; only the matching line is printed to standard output. There are three useful options to get around this context before (-B), context: after (-A), and context before and after (-C). Each of these arguments takes how many lines of context to provide:

```
$ grep -B1 "AGATCGG" contam.fastq | head -n 6 ❶
@DJB775P1:248:D0MDGACXX:7:1202:12362:49613
TGCTTACTCTGCGTTGATACCACTGCTTAGATCGGAAGAGCACACGTCTGAA
--
@DJB775P1:248:D0MDGACXX:7:1202:12782:49716
CTCTGCGTTGATACCACTGCTTACTCTGCGTTGATACCACTGCTTAGATCGG
--
$ grep -A2 "AGATCGG" contam.fastq | head -n 6 ❷
TGCTTACTCTGCGTTGATACCACTGCTTAGATCGGAAGAGCACACGTCTGAA
+
JJJJJIIJJJJJJHIHHHGHFFFFFFCEEEEEDBD?DDDDDDBDDDABDDCA
--
CTCTGCGTTGATACCACTGCTTACTCTGCGTTGATACCACTGCTTAGATCGG
+
```

❶ Print one line of context before (-B) the matching line.

❷ Print two lines of context after (-A) the matching line.

grep also supports a flavor of regular expression called *POSIX Basic Regular Expressions* (BRE). If you're familiar with the regular expressions in Perl or Python, you'll notice that grep's regular expressions aren't quite as powerful as the ones in these languages. Still, for many simple applications they work quite well. For example, if we wanted to find the Ensembl gene identifiers for both "Olfr1413" and "Olfr1411," we could use:

```
$ grep "Olfr141[13]" Mus_musculus.GRCm38.75_chr1_genes.txt
ENSMUSG00000058904      Olfr1413
ENSMUSG00000062497      Olfr1411
```

Here, we're using a shared prefix between these two gene names, and allowing the last single character to be either "1" or "3". However, this approach is less useful if we have more divergent patterns to search for. For example, constructing a BRE pattern to match both "Olfr218" and "Olfr1416" would be complex and error prone. For tasks like these, it's far easier to use grep's support for *POSIX Extended Regular Expressions*

(ERE). `grep` allows us to turn on ERE with the `-E` option (which on many systems is aliased to `egrep`). EREs allow us to use *alternation* (regular expression jargon for matching one of several possible patterns) to match either "Olfr218" or "Olfr1416." The syntax uses a pipe symbol (`|`):

```
$ grep -E "(Olfr1413|Olfr1411)" Mus_musculus.GRCm38.75_chr1_genes.txt
ENSMUSG00000058904      Olfr1413
ENSMUSG00000062497      Olfr1411
```

We're just scratching the surface of BRE and ERE now; we don't have the space to cover these two regular expression flavors in depth here (see "Assumptions This Book Makes" on page xvi for some resources on regular expressions). The important part is that you recognize there's a difference and know the terms necessary to find further help when you need it.

`grep` has an option to count how many lines match a pattern: `-c`. For example, suppose we wanted a quick look at how many genes start with "Olfr":

```
$ grep -c "\tOlfr" Mus_musculus.GRCm38.75_chr1_genes.txt
27
```

Alternatively, we could pipe the matching lines to `wc -l`:

```
$ grep "\tOlfr" Mus_musculus.GRCm38.75_chr1_genes.txt | wc -l
    27
```

Counting matching lines is extremely useful—especially with plain-text data where lines represent rows, and counting the number of lines that match a pattern can be used to count occurrences in the data. For example, suppose we wanted to know how many small nuclear RNAs are in our *Mus_musculus.GRCm38.75_chr1.gtf* file. snRNAs are annotated as `gene_biotype "snRNA"` in the last column of this GTF file. A simple way to count these features would be:

```
$ grep -c 'gene_biotype "snRNA"' Mus_musculus.GRCm38.75_chr1.gtf
315
```

Note here how we've used single quotes to specify our pattern, as our pattern includes the double-quote characters (`"`).

Currently, `grep` is outputting the entire matching line. In fact, this is one reason why `grep` is so fast: once it finds a match, it doesn't bother searching the rest of the line and just sends it to standard output. Sometimes, however, it's useful to use `grep` to extract only the matching part of the pattern. We can do this with `-o`:

```
$ grep -o "Olfr.*" Mus_musculus.GRCm38.75_chr1_genes.txt | head -n 3
Olfr1416
Olfr1415
Olfr1414
```

Or, suppose we wanted to extract all values of the "gene_id" field from the last column of our *Mus_musculus.GRCm38.75_chr1.gtf* file. This is easy with -o:

```
$ grep -E -o 'gene_id "\w+"' Mus_musculus.GRCm38.75_chr1.gtf | head -n 5
gene_id "ENSMUSG00000090025"
gene_id "ENSMUSG00000090025"
gene_id "ENSMUSG00000090025"
gene_id "ENSMUSG00000064842"
gene_id "ENSMUSG00000064842"
```

Here, we're using extended regular expressions to capture all gene names in the field. However, as you can see there's a great deal of redundancy: our GTF file has multiple features (transcripts, exons, start codons, etc.) that all have the same gene name. As a taste of what's to come in later sections, Example 7-1 shows how we could quickly convert this messy output from grep to a list of unique, sorted gene names.

Example 7-1. Cleaning a set of gene names with Unix data tools

```
$ grep -E -o 'gene_id "(\w+)"' Mus_musculus.GRCm38.75_chr1.gtf | \
    cut -f2 -d" " | \
    sed 's/"//g' | \
    sort | \
    uniq > mm_gene_id.txt
```

Even though it looks complex, this took less than one minute to write (and there are other possible solutions that omit cut, or only use awk). The length of this file (according to wc -l) is 2,027 line long—the same number we get when clicking around Ensembl's BioMart database interface for the same information. In the remaining sections of this chapter, we'll learn these tools so you can employ this type of quick pipeline in your work.

Decoding Plain-Text Data: hexdump

In bioinformatics, the plain-text data we work with is often encoded in *ASCII*. ASCII is a character encoding scheme that uses 7 bits to represent 128 different values, including letters (upper- and lowercase), numbers, and special nonvisible characters. While ASCII only uses 7 bits, nowadays computers use an 8-bit *byte* (a unit representing 8 bits) to store ASCII characters. More information about ASCII is available in your terminal through man ascii. Because plain-text data uses characters to encode information, our encoding scheme matters. When working with a plain-text file, 98% of the time you won't have to worry about the details of ASCII and how your file is encoded. However, the 2% of the time when encoding does matter—usually when an invisible non-ASCII character has entered data—it can lead to major headaches. In this section, we'll cover the basics of inspecting text data at a low level to solve these types of problems. If you'd like to skip this section for now, bookmark it in case you run into this issue at some point.

First, to look at a file's encoding use the program `file`, which infers what the encoding is from the file's content. For example, we see that many of the example files we've been working with in this chapter are ASCII-encoded:

```
$ file Mus_musculus.GRCm38.75_chr1.bed Mus_musculus.GRCm38.75_chr1.gtf
Mus_musculus.GRCm38.75_chr1.bed: ASCII text
Mus_musculus.GRCm38.75_chr1.gtf: ASCII text, with very long lines
```

Some files will have non-ASCII encoding schemes, and may contain special characters. The most common character encoding scheme is UTF-8, which is a superset of ASCII but allows for special characters. For example, the *utf8.txt* included in this chapter's GitHub directory is a UTF-8 file, as evident from `file`'s output:

```
$ file utf8.txt
utf8.txt: UTF-8 Unicode English text
```

Because UTF-8 is a superset of ASCII, if we were to delete the special characters in this file and save it, `file` would return that this file is ASCII-encoded.

Most files you'll download from data sources like Ensembl, NCBI, and UCSC's Genome Browser will not have special characters and will be ASCII-encoded (which again is simply UTF-8 without these special characters). Often, the problems I've run into are from data generated by humans, which through copying and pasting data from other sources may lead to unintentional special characters. For example, the *improper.fa* file in this chapter's directory in the GitHub repository looks like a regular FASTA file upon first inspection:

```
$ cat improper.fa
>good-sequence
AGCTAGCTACTAGCAGCTACTACGAGCATCTACGGCGCGATCTACG
>bad-sequence
GATCAGGCGACATCGAGCTATCACTACGAGCGAGAGATCAGCTATT
```

However, finding the reverse complement of these sequences using `bioawk` (don't worry about the details of this program yet—we'll cover it later) leads to strange results:

```
$ bioawk -cfastx '{print revcomp($seq)}' improper.fa
CGTAGATCGCGCCGTAGATGCTCGTAGTAGCTGCTAGTAGCTAGCT
AATAGCTGATC
```

What's going on? We have a non-ASCII character in our second sequence:

```
$ file improper.fa
improper.fa: UTF-8 Unicode text
```

Using the hexdump program, we can identify which letter is causing this problem. The hexdump program returns the hexadecimal values of each character. With the `-c` option, this also prints the character:

```
$ hexdump -c improper.fa
0000000   >   g   o   o   d   -   s   e   q   u   e   n   c   e  \n   A
```

```
0000010   G   C   T   A   G   C   T   A   C   T   A   G   C   A   G   C
0000020   T   A   C   T   A   C   G   A   G   C   A   T   C   T   A   C
0000030   G   G   C   G   C   G   A   T   C   T   A   C   G  \n   >   b
0000040   a   d   -   s   e   q   u   e   n   c   e  \n   G   A   T   C
0000050   A   G   G   C   G   A   C   A   T   C   G   A   G   C   T   A
0000060   T   C   A   C   T   A   C   G   A   G   C   G   A   G  221
0000070   G   A   T   C   A   G   C   T   A   T   T  \n
000007c
```

As we can see, the character after "CGAGCGAG" in the second sequence is clearly not an ASCII character. Another way to see non-ASCII characters is using grep. This command is a bit tricky (it searches for characters outside a hexadecimal range), but it's such a specific use case there's little reason to explain it in depth:

```
$ LC_CTYPE=C grep --color='auto' -P "[\x80-\xFF]" improper.fa
GATCAGGCGACATCGAGCTATCACTACGAGCGAG[m◆GATCAGCTATT
```

Note that this does *not* work with BSD grep, the version that comes with Mac OS X. Another useful grep option to add to this is -n, which adds line numbers to each matching line. On my systems, I have this handy line aliased to nonascii in my shell configuration file (often ~/.bashrc or ~/.profile):

```
$ alias nonascii="LC_CTYPE=C grep --color='auto' -n -P '[\x80-\xFF]'"
```

Overall, file, hexdump, and the grep command are useful for those situations where something isn't behaving correctly and you suspect a file's encoding may be to blame (which happened even during preparing this book's test data!). This is especially common with data curated by hand, by humans; always be wary of passing these files without inspection into an analysis pipeline.

Sorting Plain-Text Data with Sort

Very often we need to work with *sorted* plain-text data in bioinformatics. The two most common reasons to sort data are as follows:

- Certain operations are much more efficient when performed on sorted data.
- Sorting data is a prerequisite to finding all unique lines, using the Unix sort | uniq idiom.

We'll talk much more about sort | uniq in the next section; here we focus on how to sort data using sort.

First, like cut, sort is designed to work with plain-text data with columns. Running sort without any arguments simply sorts a file alphanumerically by line:

```
$ cat example.bed
chr1    26    39
chr1    32    47
chr3    11    28
```

```
chr1    40      49
chr3    16      27
chr1    9       28
chr2    35      54
chr1    10      19
$ sort example.bed
chr1    10      19
chr1    26      39
chr1    32      47
chr1    40      49
chr1    9       28
chr2    35      54
chr3    11      28
chr3    16      27
```

Because chromosome is the first column, sorting by line effectively groups chromosomes together, as these are "ties" in the sorted order. Grouped data is quite useful, as
we'll see.

Using Different Delimiters with sort

By default, sort treats blank characters (like tab or spaces) as field
delimiters. If your file uses another delimiter (such as a comma for
CSV files), you can specify the field separator with -t (e.g., -t",").

However, using sort's defaults of sorting alphanumerically by line doesn't handle tabular data properly. There are two new features we need:

- The ability to sort by particular columns

- The ability to tell sort that certain columns are numeric values (and not alphanumeric text; see the Tip "Leading Zeros and Sorting" in Chapter 2 for an example of the difference)

sort has a simple syntax to do this. Let's look at how we'd sort *example.bed* by chromosome (first column), and start position (second column):

```
$ sort -k1,1 -k2,2n example.bed
chr1    9       28
chr1    10      19
chr1    26      39
chr1    32      47
chr1    40      49
chr2    35      54
chr3    11      28
chr3    16      27
```

Here, we specify the columns (and their order) we want to sort by as -k arguments. In
technical terms, -k specifies the *sorting keys* and their order. Each -k argument takes a

range of columns as start,end, so to sort by a single column we use start,start. In the preceding example, we first sorted by the first column (chromosome), as the first -k argument was -k1,1. Sorting by the first column alone leads to many ties in rows with the same chromosomes (e.g., "chr1" and "chr3"). Adding a second -k argument with a different column tells sort how to break these ties. In our example, -k2,2n tells sort to sort by the second column (start position), treating this column as numerical data (because there's an n in -k2,2n).

The end result is that rows are grouped by chromosome and sorted by start position. We could then redirect the standard output stream of sort to a file:

```
$ sort -k1,1 -k2,2n example.bed > example_sorted.bed
```

If you need all columns to be sorted numerically, you can use the argument -n rather than specifying which particular columns are numeric with a syntax like -k2,2n.

Understanding the -k argument syntax is so important we're going to step through one more example. The *Mus_musculus.GRCm38.75_chr1_random.gtf* file is *Mus_musculus.GRCm38.75_chr1.gtf* with permuted rows (and without a metadata header). Let's suppose we wanted to again group rows by chromosome, and sort by position. Because this is a GTF file, the first column is chromosome and the fourth column is start position. So to sort this file, we'd use:

```
$ sort -k1,1 -k4,4n Mus_musculus.GRCm38.75_chr1_random.gtf > \
    Mus_musculus.GRCm38.75_chr1_sorted.gtf
```

Sorting Stability

There's one tricky technical detail about sorting worth being aware of: *sorting stability*. To understand stable sorting, we need to go back and think about how lines that have identical sorting keys are handled. If two lines are exactly identical according to all sorting keys we've specified, they are indistinguishable and equivalent when being sorted. When lines are equivalent, sort will sort them according to the entire line as a last-resort effort to put them in some order. What this means is that even if the two lines are identical according to the sorting keys, their *sorted order may be different from the order they appear in the original file*. This behavior makes sort an *unstable sort*.

If we don't want sort to change the order of lines that are equal according to our sort keys, we can specify the -s option. -s turns off this last-resort sorting, thus making sort a stable sorting algorithm.

Sorting can be computationally intensive. Unlike Unix tools, which operate on a single line a time, sort must compare multiple lines to sort a file. If you have a file that you suspect is already sorted, it's much cheaper to validate that it's indeed sorted rather than resort it. We can check if a file is sorted according to our -k arguments using -c:

```
$ sort -k1,1 -k2,2n -c example_sorted.bed ❶
$ echo $?
0
$ sort -k1,1 -k2,2n -c example.bed ❷
sort: example.bed:4: disorder: chr1      40      49
$ echo $?
1
```

❶ This file is already sorted by -k1,1 -k2,2n -c, so sort exits with exit status 0 (true).

❷ This file is not already sorted by -k1,1 -k2,2n -c, so sort returns the first out-of-order row it finds and exits with status 1 (false).

It's also possible to sort in reverse order with the -r argument:

```
$ sort -k1,1 -k2,2n -r example.bed
chr3    11    28
chr3    16    27
chr2    35    54
chr1    9     28
chr1    10    19
chr1    26    39
chr1    32    47
chr1    40    49
```

If you'd like to only reverse the sorting order of a single column, you can append r on that column's -k argument:

```
$ sort -k1,1 -k2,2nr example.bed
chr1    40    49
chr1    32    47
chr1    26    39
chr1    10    19
chr1    9     28
chr2    35    54
chr3    16    27
chr3    11    28
```

In this example, the effect is to keep the chromosomes sorted in alphanumeric ascending order, but sort the second column of start positions in descending numeric order.

There are a few other useful sorting options to discuss, but these are available for GNU sort only (not the BSD version as found on OS X). The first is -V, which is a clever alphanumeric sorting routine that understands numbers inside strings. To see why this is useful, consider the file *example2.bed*. Sorting with `sort -k1,1 -k2,2n` groups chromosomes but doesn't naturally order them as humans would:

```
cat example2.bed
chr2    15      19
chr22   32      46
chr10   31      47
chr1    34      49
chr11   6       16
chr2    17      22
chr2    27      46
chr10   30      42
$ sort -k1,1 -k2,2n example2.bed
chr1    34      49
chr10   30      42
chr10   31      47
chr11   6       16
chr2    15      19
chr2    17      22
chr2    27      46
chr22   32      46
```

Here, "chr2" is following "chr11" because the character "1" falls before "2"—sort isn't sorting by the number in the text. However, with V appended to -k1,1 we get the desired result:

```
$ sort -k1,1V -k2,2n example2.bed
chr1    34      49
chr2    15      19
chr2    17      22
chr2    27      46
chr10   30      42
chr10   31      47
chr11   6       16
chr22   32      46
```

In practice, Unix `sort` scales well to the moderately large text data we'll need to sort in bioinformatics. `sort` does this by using a sorting algorithm called *merge sort*. One nice feature of the merge sort algorithm is that it allows us to sort files larger than fit in our memory by storing sorted intermediate files on the disk. For large files, reading and writing these sorted intermediate files to the disk may be a bottleneck (remember: disk operations are very slow). Under the hood, `sort` uses a fixed-sized memory buffer to sort as much data in-memory as fits. Increasing the size of this buffer allows more data to be sorted in memory, which reduces the amount of temporary sorted files that need to be written and read off the disk. For example:

```
$ sort -k1,1 -k4,4n -S2G Mus_musculus.GRCm38.75_chr1_random.gtf
```

The -S argument understands suffixes like K for kilobyte, M for megabyte, and G for gigabyte, as well as % for specifying what percent of total memory to use (e.g., 50% with -S 50%).

Another option (only available in GNU sort) is to run sort with the --parallel option. For example, to use four cores to sort *Mus_musculus.GRCm38.75_chr1_random.gtf*:

```
$ sort -k1,1 -k4,4n --parallel 4 Mus_musculus.GRCm38.75_chr1_random.gtf
```

But note that *Mus_musculus.GRCm38.75_chr1_random.gtf* is much too small for either increasing the buffer size or parallelization to make any difference. In fact, because there is a fixed cost to parallelizing operations, parallelizing an operation run on a small file could actually be slower! In general, don't obsess with performance tweaks like these unless your data is truly large enough to warrant them.

So, when is it more efficient to work with sorted output? As we'll see when we work with range data in Chapter 9, working with sorted data can be much faster than working with unsorted data. Many tools have better performance when working on sorted files. For example, BEDTools' bedtools intersect allows the user to indicate whether a file is sorted with -sorted. Using bedtools intersect with a sorted file is both more memory-efficient and faster. Other tools, like tabix (covered in more depth in "Fast Access to Indexed Tab-Delimited Files with BGZF and Tabix" on page 425) require that we presort files before indexing them for fast random-access.

Finding Unique Values in Uniq

Unix's uniq takes lines from a file or standard input stream, and outputs all lines with consecutive duplicates removed. While this is a relatively simple functionality, you will use uniq very frequently in command-line data processing. Let's first see an example of its behavior:

```
$ cat letters.txt
A
A
B
C
B
C
C
C
$ uniq letters.txt
A
B
C
B
C
```

As you can see, uniq *does not* return the unique values *letters.txt*—it only removes consecutive duplicate lines (keeping one). If instead we did want to find all unique lines in a file, we would first sort all lines using sort so that all identical lines are grouped next to each other, and then run uniq. For example:

```
$ sort letters.txt | uniq
A
B
C
```

If we had lowercase letters mixed in this file as well, we could add the option -i to uniq to be case insensitive.

uniq also has a tremendously useful option that's used very often in command-line data processing: -c. This option shows the counts of occurrences next to the unique lines. For example:

```
$ uniq -c letters.txt
   2 A
   1 B
   1 C
   1 B
   3 C
$ sort letters.txt | uniq -c
   2 A
   2 B
   4 C
```

Both sort | uniq and sort | uniq -c are frequently used shell idioms in bioinformatics and worth memorizing. Combined with other Unix tools like grep and cut, sort and uniq can be used to summarize columns of tabular data:

```
$ grep -v "^#" Mus_musculus.GRCm38.75_chr1.gtf | cut -f3 | sort | uniq -c
25901 CDS
7588 UTR
36128 exon
2027 gene
2290 start_codon
2299 stop_codon
4993 transcript
```

If we wanted these counts in order from most frequent to least, we could pipe these results to sort -rn:

```
$ grep -v "^#" Mus_musculus.GRCm38.75_chr1.gtf | cut -f3 | sort | uniq -c | \
    sort -rn
36128 exon
25901 CDS
7588 UTR
4993 transcript
2299 stop_codon
```

```
2290 start_codon
2027 gene
```

Because sort and uniq are line-based, we can create lines from multiple columns to count combinations, like how many of each feature (column 3 in this example GTF) are on each strand (column 7):

```
$ grep -v "^#" Mus_musculus.GRCm38.75_chr1.gtf | cut -f3,7 | sort | uniq -c
12891   CDS         +
13010   CDS         -
3754    UTR         +
3834    UTR         -
18134   exon        +
17994   exon        -
1034    gene        +
993     gene        -
1135    start_codon +
1155    start_codon -
1144    stop_codon  +
1155    stop_codon  -
2482    transcript  +
2511    transcript  -
```

Or, if you want to see the number of features belonging to a particular gene identifier:

```
$ grep "ENSMUSG00000033793" Mus_musculus.GRCm38.75_chr1.gtf | cut -f3 | sort \
  | uniq -c
  13 CDS
   3 UTR
  14 exon
   1 gene
   1 start_codon
   1 stop_codon
   1 transcript
```

These count tables are incredibly useful for summarizing columns of categorical data. Without having to load data into a program like R or Excel, we can quickly calculate summary statistics about our plain-text data files. Later on in Chapter 11, we'll see examples involving more complex alignment data formats like SAM.

uniq can also be used to check for duplicates with the -d option. With the -d option, uniq outputs duplicated lines only. For example, the *mm_gene_names.txt* file (which contains a list of gene names) does not have duplicates:

```
$ uniq -d mm_gene_names.txt
# no output
$ uniq -d mm_gene_names.txt | wc -l
       0
```

A file with duplicates, like the *test.bed* file, has multiple lines returned:

```
uniq -d test.bed | wc -l
   22925
```

Join

The Unix tool `join` is used to join different files together by a common column. This is easiest to understand with simple test data. Let's use our *example.bed* BED file, and *example_lengths.txt*, a file containing the same chromosomes as *example.bed* with their lengths. Both files look like this:

```
$ cat example.bed
chr1    26      39
chr1    32      47
chr3    11      28
chr1    40      49
chr3    16      27
chr1    9       28
chr2    35      54
chr1    10      19
$ cat example_lengths.txt
chr1    58352
chr2    39521
chr3    24859
```

Our goal is to append the chromosome length alongside each feature (note that the result will *not* be a valid BED-formatted file, just a tab-delimited file). To do this, we need to *join* both of these tabular files by their common column, the one containing the chromosome names (the first column in both *example.bed* and *example_lengths.txt*).

To append the chromosome lengths to *example.bed*, we first need to sort both files by the column to be joined on. This is a vital step—Unix's `join` will not work unless both files are sorted by the column to join on. We can appropriately sort both files with `sort`:

```
$ sort -k1,1 example.bed > example_sorted.bed
$ sort -c -k1,1 example_lengths.txt # verifies is already sorted
```

Now, let's use `join` to join these files, appending the chromosome lengths to our *example.bed* file. The basic syntax is `join -1 <file_1_field> -2 <file_2_field> <file_1> <file_2>`, where `<file_1>` and `<file_2>` are the two files to be joined by a column `<file_1_field>` in `<file_1>` and column `<file_2_field>` in `<file_2>`. So, with *example.bed* and *example_lengths.txt* this would be:

```
$ join -1 1 -2 1 example_sorted.bed example_lengths.txt
  > example_with_lengths.txt
$ cat example_with_lengths.txt
chr1 10 19 58352
chr1 26 39 58352
chr1 32 47 58352
chr1 40 49 58352
chr1 9  28 58352
chr2 35 54 39521
```

```
chr3   11   28   24859
chr3   16   27   24859
```

There are many types of joins; we will talk about each kind in more depth in Chapter 13. For now, it's important that we make sure join is working as we expect. Our expectation is that this join should not lead to fewer rows than in our *example.bed* file. We can verify this with wc -l:

```
$ wc -l example_sorted.bed example_with_lengths.txt
      8 example_sorted.bed
      8 example_with_lengths.txt
     16 total
```

We see that we have the same number of lines in our original file and our joined file. However, look what happens if our second file, *example_lengths.txt*, is truncated such that it doesn't have the lengths for chr3:

```
$ head -n2 example_lengths.txt > example_lengths_alt.txt # truncate file
$ join -1 1 -2 1 example_sorted.bed example_lengths_alt.txt
chr1 10 19 58352
chr1 26 39 58352
chr1 32 47 58352
chr1 40 49 58352
chr1 9 28 58352
chr2 35 54 39521
$ join -1 1 -2 1 example_sorted.bed example_lengths_alt.txt | wc -l
      6
```

Because chr3 is absent from *example_lengths_alt.txt*, our join omits rows from *example_sorted.bed* that do not have an entry in the first column of *example_lengths_alt.txt*. In some cases (such as this), we don't want this behavior. GNU join implements the -a option to include unpairable lines—ones that do not have an entry in either file. (This option is not implemented in BSD join.) To use -a, we specify which file is allowed to have unpairable entries:

```
$ join -1 1 -2 1 -a 1 example_sorted.bed example_lengths_alt.txt # GNU join only
chr1   10   19   58352
chr1   26   39   58352
chr1   32   47   58352
chr1   40   49   58352
chr1   9    28   58352
chr2   35   54   39521
chr3   11   28
chr3   16   27
```

Unix's join is just one of many ways to join data, and is most useful for simple quick joins. Joining data by a common column is a common task during data analysis; we'll see how to do this in R and with SQLite in future chapters.

Text Processing with Awk

Throughout this chapter, we've seen how we can use simple Unix tools like `grep`, `cut`, and `sort` to inspect and manipulate plain-text tabular data in the shell. For many trivial bioinformatics tasks, these tools allow us to get the job done quickly and easily (and often very efficiently). Still, some tasks are slightly more complex and require a more expressive and powerful tool. This is where the language and tool Awk excels—extracting data from and manipulating tabular plain-text files. Awk is a tiny, specialized language that allows you to do a variety of text-processing tasks with ease.

We'll introduce the basics of Awk in this section—enough to get you started with using Awk in bioinformatics. While Awk is a fully fledged programming language, it's a lot less expressive and powerful than Python. If you need to implement something complex, it's likely better (and easier) to do so in Python. The key to using Awk effectively is to reserve it for the subset of tasks it's best at: quick data-processing tasks on tabular data. Learning Awk also prepares us to learn `bioawk`, which we'll cover in "Bioawk: An Awk for Biological Formats" on page 163.

Gawk versus Awk

As with many other Unix tools, Awk comes in a few flavors. First, you can still find the original Awk written by Alfred Aho, Peter Weinberger, and Brian Kernighan (whose last names create the name Awk) on some systems. If you use Mac OS X, you'll likely be using the BSD Awk. There's also GNU Awk, known as Gawk, which is based on the original Awk but has many extended features (and an excellent manual; see `man gawk`). In the examples in this section, I've stuck to a common subset of Awk functionality shared by all these Awks. Just take note that there are multiple Awk implementations. If you find Awk useful in your work (which can be a personal preference), it's worthwhile to use Gawk.

To learn Awk, we'll cover two key parts of the Awk language: how Awk processes records, and pattern-action pairs. After understanding these two key parts the rest of the language is quite simple.

First, Awk processes input data a *record* at a time. Each record is composed of *fields*, separate chunks that Awk automatically separates. Because Awk was designed to work with tabular data, each record is a line, and each field is a column's entry for that record. The clever part about Awk is that it automatically assigns the entire record to the variable $0, and field one's value is assigned to $1, field two's value is assigned to $2, field three's value is assigned to $3, and so forth.

Second, we build Awk programs using one or more of the following structures:

```
pattern { action }
```

Each *pattern* is an expression or regular expression pattern. Patterns are a lot like `if` statements in other languages: if the pattern's expression evaluates to true or the regular expression matches, the statements inside *action* are run. In Awk lingo, these are *pattern-action* pairs and we can chain multiple pattern-action pairs together (separated by semicolons). If we omit the pattern, Awk will run the action on all records. If we omit the action but specify a pattern, Awk will print all records that match the pattern. This simple structure makes Awk an excellent choice for quick text-processing tasks. This is a lot to take in, but these two basic concepts—records and fields, and pattern-action pairs—are the foundation of writing text-processing programs with Awk. Let's see some examples.

First, we can simply mimic `cat` by omitting a pattern and printing an entire record with the variable `$0`:

```
$ awk '{ print $0 }' example.bed
chr1    26      39
chr1    32      47
chr3    11      28
chr1    40      49
chr3    16      27
chr1    9       28
chr2    35      54
chr1    10      19
```

`print` prints a string. Optionally, we could omit the `$0`, because `print` called without an argument would print the current record.

Awk can also mimic `cut`:

```
$ awk '{ print $2 "\t" $3 }' example.bed
26      39
32      47
11      28
40      49
16      27
9       28
35      54
10      19
```

Here, we're making use of Awk's string concatenation. Two strings are concatenated if they are placed next to each other with no argument. So for each record, `$2"\t"$3` concatenates the second field, a tab character, and the third field. This is far more typing than `cut -f2,3`, but demonstrates how we can access a certain column's value for the current record with the numbered variables `$1`, `$2`, `$3`, etc.

Let's now look at how we can incorporate simple pattern matching. Suppose we wanted to write a filter that only output lines where the length of the feature (end position - start position) was greater than 18. Awk supports arithmetic with the standard operators +, -, *, /, % (remainder), and ^ (exponentiation). We can subtract within a pattern to calculate the length of a feature, and filter on that expression:

```
$ awk '$3 - $2 > 18' example.bed
chr1    9       28
chr2    35      54
```

See Table 7-2 for reference to Awk comparison and logical operators.

Table 7-2. Awk comparison and logical operations

Comparison	Description
a == b	a is equal to b
a != b	a is not equal to b
a < b	a is less than b
a > b	a is greater than b
a <= b	a is less than or equal to b
a >= b	a is greater than or equal to b
a ~ b	a matches regular expression pattern b
a !~ b	a does not match regular expression pattern b
a && b	logical and a and b
a || b	logical or a and b
!a	not a (logical negation)

We can also chain patterns, by using logical operators && (AND), || (OR), and ! (NOT). For example, if we wanted all lines on chromosome 1 with a length greater than 10:

```
$ awk '$1 ~ /chr1/ && $3 - $2 > 10' example.bed
chr1    26      39
chr1    32      47
chr1    9       28
```

The first pattern, $1 ~ /chr1/, is how we specify a regular expression. Regular expressions are in slashes. Here, we're matching the first field, 1, against the regular expression chr1. The tilde, ~ means match; to not match the regular expression we would use !~ (or !($1 ~ /chr1/)).

We can combine patterns and more complex actions than just printing the entire record. For example, if we wanted to add a column with the length of this feature (end position - start position) for only chromosomes 2 and 3, we could use:

```
$ awk '$1 ~ /chr2|chr3/ { print $0 "\t" $3 - $2 }' example.bed
chr3    11    28    17
chr3    16    27    11
chr2    35    54    19
```

So far, these exercises have illustrated two ways Awk can come in handy:

- For filtering data using rules that can combine regular expressions and arithmetic
- Reformatting the columns of data using arithmetic

These two applications alone make Awk an extremely useful tool in bioinformatics, and a huge time saver. But let's look at some slightly more advanced use cases. We'll start by introducing two special patterns: BEGIN and END.

Like a bad novel, beginning and end are optional in Awk. The BEGIN pattern specifies what to do *before* the first record is read in, and END specifies what to do *after* the last record's processing is complete. BEGIN is useful to initialize and set up variables, and END is useful to print data summaries at the end of file processing. For example, suppose we wanted to calculate the mean feature length in *example.bed*. We would have to take the sum feature lengths, and then divide by the total number of records. We can do this with:

```
$ awk 'BEGIN{ s = 0 }; { s += ($3-$2) }; END{ print "mean: " s/NR };' example.bed
mean: 14
```

There's a special variable we've used here, one that Awk automatically assigns in addition to $0, $1, $2, etc.: NR. NR is the current record number, so on the last record NR is set to the total number of records processed. In this example, we've initialized a variable s to 0 in BEGIN (variables you define do not need a dollar sign). Then, for each record we increment s by the length of the feature. At the end of the records, we print this sum s divided by the number of records NR, giving the mean.

Setting Field, Output Field, and Record Separators

While Awk is designed to work with whitespace-separated tabular data, it's easy to set a different field separator: simply specify which separator to use with the -F argument. For example, we could work with a CSV file in Awk by starting with awk -F",".

It's also possible to set the record (RS), output field (OFS), and output record (ORS) separators. These variables can be set using Awk's -v argument, which sets a variable using the syntax awk -v VAR=val. So, we could convert a three-column CSV to a tab file by just setting the field separator F and output field separator OFS: awk -F"," -v OFS="\t" {print $1,$2,$3}. Setting OFS="\t" saves a few extra characters when outputing tab-delimited results with statements like print "$1 "\t" $2 "\t" $3.

We can use NR to extract ranges of lines, too; for example, if we wanted to extract all lines between 3 and 5 (inclusive):

```
awk 'NR >= 3 && NR <= 5' example.bed
chr3    11      28
chr1    40      49
chr3    16      27
```

Awk makes it easy to convert between bioinformatics files like BED and GTF. For example, we could generate a three-column BED file from *Mus_musculus.GRCm38.75_chr1.gtf* as follows:

```
$ awk '!/^#/ { print $1 "\t" $4-1 "\t" $5 }' Mus_musculus.GRCm38.75_chr1.gtf | \
    head -n 3
1       3054232 3054733
1       3054232 3054733
1       3054232 3054733
```

Note that we subtract 1 from the start position to convert to BED format. This is because BED uses zero-indexing while GTF uses 1-indexing; we'll learn much more about this in Chapter 10. This is a subtle detail, certainly one that's been missed many times. In the midst of analysis, it's easy to miss these small details.

Awk also has a very useful data structure known as an *associative array*. Associative arrays behave like Python's dictionaries or hashes in other languages. We can create an associative array by simply assigning a value to a key. For example, suppose we wanted to count the number of features (third column) belonging to the gene "Lypla1." We could do this by incrementing their values in an associative array:

```
# This example has been split on multiple lines to improve readability
$ awk '/Lypla1/ { feature[$3] += 1 }; \
    END { for (k in feature)             \
    print k "\t" feature[k] }' Mus_musculus.GRCm38.75_chr1.gtf
exon        69
```

```
CDS          56
UTR          24
gene          1
start_codon   5
stop_codon    5
transcript    9
```

This example illustrates that Awk really is a programming language—within our action blocks, we can use standard programming statements like if, for, and while, and Awk has several useful built-in functions (see Table 7-3 for some useful common functions). However, when Awk programs become complex or start to span multiple lines, I usually prefer to switch to Python at that point. You'll have much more functionality at your disposal for complex tasks with Python: Python's standard library, the Python debugger (PDB), and more advanced data structures. However, this is a personal preference—there are certainly programmers who write lengthy Awk processing programs.

Table 7-3. Useful built-in Awk functions

length(s)	Length of a string s.
tolower(s)	Convert string s to lowercase
toupper(s)	Convert string s to uppercase
substr(s, i, j)	Return the substring of s that starts at i and ends at j
split(s, x, d)	Split string s into chunks by delimiter d, place chunks in array x
sub(f, r, s)	Find regular expression f in s and replace it with r (modifying s in place); use gsub for global substitution; returns a positive value if string is found

It's worth noting that there's an entirely Unix way to count features of a particular gene: grep, cut, sort, and uniq -c:

```
$ grep "Lypla1" Mus_musculus.GRCm38.75_chr1.gtf | cut -f 3 | sort | uniq -c
 56 CDS
 24 UTR
 69 exon
  1 gene
  5 start_codon
  5 stop_codon
  9 transcript
```

However, if we needed to also filter on column-specific information (e.g., strand), an approach using just base Unix tools would be quite messy. With Awk, adding an additional filter would be trivial: we'd just use && to add another expression in the pattern.

Bioawk: An Awk for Biological Formats

Imagine extending Awk's powerful processing of tabular data to processing tasks involving common bioinformatics formats like FASTA/FASTQ, GTF/GFF, BED, SAM, and VCF. This is exactly what Bioawk, a program written by Heng Li (author of other excellent bioinformatics tools such as BWA and Samtools) does. You can download, compile, and install Bioawk from source (*http://github.com/lh3/bioawk*), or if you use Mac OS X's Homebrew package manager, Bioawk is also in homebrew-science (so you can install with brew tap homebrew/science; brew install bio awk).

The basic idea of Bioawk is that we specify what bioinformatics format we're working with, and Bioawk will automatically set variables for each field (just as regular Awk sets the columns of a tabular text file to $1, $1, $2, etc.). For Bioawk to set these fields, specify the format of the input file or stream with -c. Let's look at Bioawk's supported input formats and what variables these formats set:

```
$ bioawk -c help
bed:
        1:chrom 2:start 3:end 4:name 5:score 6:strand 7:thickstart
            8:thickend 9:rgb 10:blockcount 11:blocksizes 12:blockstarts
sam:
        1:qname 2:flag 3:rname 4:pos 5:mapq 6:cigar 7:rnext 8:pnext
            9:tlen 10:seq 11:qual
vcf:
        1:chrom 2:pos 3:id 4:ref 5:alt 6:qual 7:filter 8:info
gff:
        1:seqname 2:source 3:feature 4:start 5:end 6:score 7:filter
            8:strand 9:group 10:attribute
fastx:
        1:name 2:seq 3:qual 4:comment
```

As an example of how this works, let's read in *example.bed* and append a column with the length of the feature (end position - start position) for all protein coding genes:

```
$ bioawk -c gff '$3 ~ /gene/ && $2 ~ /protein_coding/ \
    {print $seqname,$end-$start}' Mus_musculus.GRCm38.75_chr1.gtf | head -n 4
1       465597
1       16807
1       5485
1       12533
```

We could add gene names too, but this gets trickier as we need to split key/values out of the group column (column 9). We can do this with Awk's split command and with some cleanup after using sed, but this is stretching what we should attempt with a one-liner. See this chapter's *README.md* file in the book's GitHub repository for a comparison of using Bioawk and Python for this problem.

Bioawk is also quite useful for processing FASTA/FASTQ files. For example, we could use it to turn a FASTQ file into a FASTA file:

```
bioawk -c fastx '{print ">"$name"\n"$seq}' contam.fastq | head -n 4
>DJB775P1:248:D0MDGACXX:7:1202:12362:49613
TGCTTACTCTGCGTTGATACCACTGCTTAGATCGGAAGAGCACACGTCTGAA
>DJB775P1:248:D0MDGACXX:7:1202:12782:49716
CTCTGCGTTGATACCACTGCTTACTCTGCGTTGATACCACTGCTTAGATCGG
```

Note that Bioawk detects whether to parse input as FASTQ or FASTA when we use -c fastx.

Bioawk can also serve as a method of counting the number of FASTQ/FASTA entries:

```
$ bioawk -c fastx 'END{print NR}' contam.fastq
8
```

Or Bioawk's function revcomp() can be used to reverse complement a sequence:

```
$ bioawk -c fastx '{print ">"$name"\n"revcomp($seq)}' contam.fastq | head -n 4
>DJB775P1:248:D0MDGACXX:7:1202:12362:49613
TTCAGACGTGTGCTCTTCCGATCTAAGCAGTGGTATCAACGCAGAGTAAGCA
>DJB775P1:248:D0MDGACXX:7:1202:12782:49716
CCGATCTAAGCAGTGGTATCAACGCAGAGTAAGCAGTGGTATCAACGCAGAG
```

Bioawk is also useful for creating a table of sequence lengths from a FASTA file. For example, to create a table of all chromosome lengths of the *Mus musculus* genome:

```
$ bioawk -c fastx '{print $name,length($seq)}' \
    Mus_musculus.GRCm38.75.dna_rm.toplevel.fa.gz > mm_genome.txt
$ head -n 4 mm_genome.txt
1    195471971
10   130694993
11   122082543
12   120129022
13   120421639
14   124902244
```

Finally, Bioawk has two options that make working with plain tab-delimited files easier: -t and -c hdr. -t is for processing general tab-delimited files; it sets Awk's field separator (FS) and output field separator (OFS) to tabs. The option -c hdr is for unspecific tab-delimited formats with a header as the first line. This option sets field variables, but uses the names given in the header. Suppose we had a simple tab-delimited file containing variant names and genotypes for individuals (in columns):

```
$ head -n 4 genotypes.txt
id      ind_A   ind_B   ind_C   ind_D
S_000   T/T     A/T     A/T     T/T
S_001   G/C     C/C     C/C     C/G
S_002   C/A     A/C     C/C     C/C
```

If we wanted to return all variants for which individuals ind_A and ind_B have identical genotypes (note that this assumes a fixed allele order like ref/alt or major/minor):

```
$ bioawk -c hdr '$ind_A == $ind_B {print $id}' genotypes.txt
S_001
S_003
S_005
S_008
S_009
```

Stream Editing with Sed

In "The Almighty Unix Pipe: Speed and Beauty in One" on page 45, we covered how Unix pipes are fast because they operate on streams of data (rather than data written to disk). Additionally, pipes don't require that we load an entire file in memory at once—instead, we can operate one line at a time. We've used pipes throughout this chapter to transform plain-text data by selecting columns, sorting values, taking the unique values, and so on. Often we need to make trivial edits to a stream (we did this in Example 7-1), usually to prepare it for the next step in a Unix pipeline. The stream editor, or sed, allows you to do exactly that. sed is remarkably powerful, and has capabilities that overlap other Unix tools like grep and awk. As with awk, it's best to keep your sed commands simple at first. We'll cover a small subset of sed's vast functionality that's most useful for day-to-day bioinformatics tasks.

GNU Sed versus BSD Sed

As with many other Unix tools, the BSD and GNU versions of sed differ considerably in behavior. In this section and in general, I recommend you use the GNU version of sed. GNU sed has some additional features, and supports functionality such as escape codes for special characters like tab (\t) we expect in command-line tools.

sed reads data from a file or standard input and can edit a line at a time. Let's look at a very simple example: converting a file (*chroms.txt*) containing a single column of chromosomes in the format "chrom12," "chrom2," and so on to the format "chr12," "chr2," and so on:

```
$ head -n 3 chroms.txt  # before sed
chrom1  3214482 3216968
chrom1  3216025 3216968
chrom1  3216022 3216024
$ sed 's/chrom/chr/' chroms.txt | head -n 3
chr1    3214482 3216968
```

```
chr1     3216025 3216968
chr1     3216022 3216024
```

It's a simple but important change: although *chroms.txt* is a mere 10 lines long, it could be 500 gigabytes of data *and we can edit it without opening the entire file in memory*. This is the beauty of stream editing, sed's bread and butter. Our edited output stream is then easy to redirect to a new file (don't redirect output to the input file; this causes bad things to happen).

Let's dissect how the sed command in the preceding code works. First, this uses sed's *substitute* command, by far the most popular use of sed. sed's substitute takes the first occurrence of the pattern between the first two slashes, and replaces it with the string between the second and third slashes. In other words, the syntax of sed's substitute is s/pattern/replacement/.

By default, sed only replaces the *first occurrence* of a match. Very often we need to replace all occurrences of strings that match our pattern. We can enable this behavior by setting the global flag g after the last slash: s/pattern/replacement/g. If we need matching to be case-insensitive, we can enable this with the flag i (e.g., s/pattern/replacement/i).

By default, sed's substitutions use POSIX Basic Regular Expressions (BRE). As with grep, we can use the -E option to enable POSIX Extended Regular Expressions (ERE). Whether basic or extended, sed's regular expressions give us considerable freedom to match patterns. Perhaps most important is the ability to capture chunks of text that match a pattern, and use these chunks in the replacement (often called grouping and capturing). For example, suppose we wanted to capture the chromosome name, and start and end positions in a string containing a genomic region in the format "chr1:28427874-28425431", and output this as three columns. We could use:

```
$ echo "chr1:28427874-28425431" | \
    sed -E 's/^(chr[^:]+):([0-9]+)-([0-9]+)/\1\t\2\t\3/'
chr1     28427874        28425431
```

That looks quite complex! Let's dissect each part of this. The first component of this regular expression is ^\(chr[^:]+\):. This matches the text that begins at the start of the line (the anchor ^ enforces this), and then captures everything between \(and \). The pattern used for capturing begins with "chr" and matches one or more characters that are not ":", our delimiter. We match until the first ":" through a character class defined by everything that's not ":", [^:]+.

The second and third components of this regular expression are the same: match and capture more than one number. Finally, our replacement is these three captured groups, interspersed with tabs, \t. If you're struggling to understand the details of this regular expression, don't fret—regular expressions are tricky and take time and practice to master.

Explicitly capturing each component of our genomic region is one way to tackle this, and nicely demonstrates sed's ability to capture patterns. But just as there's always more than one way to bash a nail into wood, there are numerous ways to use sed or other Unix tools to parse strings like this. Here are a few more ways to do the same thing:

```
$ echo "chr1:28427874-28425431" | sed 's/[:-]/\t/g' ❶
chr1    28427874        28425431
$ echo "chr1:28427874-28425431" | sed 's/:/\t/' | sed 's/-/\t/' ❷
chr1    28427874        28425431
$ echo "chr1:28427874-28425431" | tr ':-' '\t' ❸
chr1    28427874        28425431
```

❶ Rather than explicitly capturing each chunk of information (the chromosome, start position, and end position), here we just replace both delimiters (: and -) with a tab. Note that we've enabled the global flag g, which is necessary for this approach to work.

❷ Here, we use two sed commands to carry out these edits separately. For complex substitutions, it can be much easier to use two or more calls to sed rather than trying to do this with one regular expression. sed has a feature to chain pattern/ replacements within sed too, using -e. For example, this line is equivalent to: sed -e 's/:/\t/' -e 's/-/\t/'.

❸ Using tr to translate both delimiters to a tab character is also another option. tr translates all occurrences of its first argument to its second (see man tr for more details).

By default, sed prints *every* line, making replacements to matching lines. Sometimes this behavior isn't what we want (and can lead to erroneous results). Imagine the following case: we want to use capturing to capture all transcript names from the last (9th) column of a GTF file. Our pipeline would look like (working with the first three lines to simplify output):

```
# warning: this is incorrect!
$ grep -v "^#" Mus_musculus.GRCm38.75_chr1.gtf | head -n 3 | \
    sed -E 's/.*transcript_id "([^"]+)".*/\1/'

1    pseudogene    gene    3054233    3054733    .    [...]
ENSMUST00000160944
ENSMUST00000160944
```

What happened? Some lines of the last column of *Mus_musculus.GRCm38.75_chr1.gtf* don't contain transcript_id, so sed prints the entire line rather than the captured group. One way to solve this would be to use `grep "tran script_id"` before `sed` to only work with lines containing the string `"tran script_id"`. However, `sed` offers a cleaner way. First, disable `sed` from outputting all lines with `-n`. Then, by appending p after the last slash `sed` will print all lines it's made a replacement on. The following is an illustration of `-n` used with p:

```
$ grep -v "^#" Mus_musculus.GRCm38.75_chr1.gtf | head -n 3 | \
    sed -E -n 's/.*transcript_id "([^"]+)".*/\1/p'
ENSMUST00000160944
ENSMUST00000160944
```

This example uses an important regular expression idiom: capturing text between delimiters (in this case, quotation marks). This is a useful pattern, so let's break it down:

1. First, match zero or more of any character (.*) before the string `"tran script_id"`.

2. Then, match and capture (because there are parentheses around the pattern) one or more characters that *are not* a quote. This is accomplished with [^"]+, the important idiom in this example. In regular extension jargon, the brackets make up a *character class*. Character classes specify what characters the expression is allowed to match. Here, we use a caret (^) inside the brackets to match anything *except* what's inside these brackets (in this case, a quote). The end result is that we match and capture one or more nonquote characters (because there's a trailing +). This approach is *nongreedy*; often beginners make the mistake of taking a *greedy* approach and use .*. Consider:

```
$ echo 'transcript_id "ENSMUST00000160944"; gene_name "Gm16088"'
  > greedy_example.txt

$ sed -E 's/transcript_id "(.*)".*/\1/' greedy_example.txt
ENSMUST00000160944"; gene_name "Gm16088
$ sed -E 's/transcript_id "([^"]+)".*/\1/' greedy_example.txt
ENSMUST00000160944
```

The first example was greedy; it not only captured the transcript identifier inside the quotes, but *everything* until the last quotation mark (which follows the gene name)!

It's also possible to select and print certain ranges of lines with `sed`. In this case, we're not doing pattern matching, so we don't need slashes. To print the first 10 lines of a file (similar to `head -n 10`), we use:

```
$ sed -n '1,10p' Mus_musculus.GRCm38.75_chr1.gtf
```

If we wanted to print lines 20 through 50, we would use:

```
$ sed -n '20,50p' Mus_musculus.GRCm38.75_chr1.gtf
```

Substitutions make up the majority of sed's usage cases, but this is just scratching the surface of sed's capabilities. sed has features that allow you to make any type of edit to a stream of text, but for complex stream processing tasks it can be easier to write a Python script than a long and complicated sed command. Remember the KISS principal: Keep Incredible Sed Simple.

Advanced Shell Tricks

With both the Unix shell refresher in Chapter 3 and the introduction to Unix data tools in this chapter, we're ready to dig into a few more advanced shell tricks. If you're struggling with the shell, don't hesitate to throw a sticky note on this page and come back to this section later.

Subshells

The first trick we'll cover is using Unix subshells. Before explaining this trick, it's helpful to remember the difference between *sequential* commands (connected with && or ;), and *piped* commands (connected with |). Sequential commands are simply run one after the other; the previous command's standard output does *not* get passed to the next program's standard in. In contrast, connecting two programs with pipes means the first program's standard out will be piped into the next program's standard in.

The difference between sequential commands linked with && and ; comes down to exit status, a topic covered in "Exit Status: How to Programmatically Tell Whether Your Command Worked" on page 52. If we run two commands with command1 ; command2, command2 will always run, regardless of whether command1 exits successfully (with a zero exit status). In contrast, if we use command1 && command2, command2 will only run if command1 completed with a zero-exit status. Checking exit status with pipes unfortunately gets tricky (we'll explore this important topic in Chapter 12).

So how do subshells fit into all of this? Subshells allow us to execute sequential commands together in a separate shell process. This is useful primarily to *group* sequential commands together (such that their output is a single stream). This gives us a new way to construct clever one-liners and has practical uses in command-line data processing. Let's look at a toy example first:

```
$ echo "this command"; echo "that command" | sed 's/command/step/'
this command
that step
$ (echo "this command"; echo "that command") | sed 's/command/step/'
this step
that step
```

In the first example, only the second command's standard out is piped into sed. This is because your shell interprets echo "this command" as one command and echo "that command" | sed 's/command/step/' as a second separate command. But grouping both echo commands together using parentheses causes these two commands to be run in a separate subshell, and *both* commands' combined standard output is passed to sed. Combining two sequential commands' standard output into a single stream with a subshell is a useful trick, and one we can apply to shell problems in bioinformatics.

Consider the problem of sorting a GTF file with a metadata header. We can't simply sort the entire file with sort, because this header could get shuffled in with rows of data. Instead, we want to sort everything *except* the header, but still include the header at the top of the final sorted file. We can solve this problem using a subshell to group sequential commands that print the header to standard out and sort all other lines by chromosome and start position, printing all lines to standard out after the header. Admittedly, this is a bit of Unix hackery, but it turns out to be a useful trick.

Let's do this using the gzipped GTF file *Mus_musculus.GRCm38.75_chr1.gtf.gz*, and zgrep (the gzip analog of grep). We'll pipe the output to less first, to inspect our results:

```
$ (zgrep "^#" Mus_musculus.GRCm38.75_chr1.gtf.gz; \
    zgrep -v "^#" Mus_musculus.GRCm38.75_chr1.gtf.gz | \
    sort -k1,1 -k4,4n) | less
```

Because we've used a subshell, all standard output from these sequential commands will be combined into a single stream, which here is piped to less. To write this stream to a file, we could redirect this stream to a file using something like > Mus_mus culus.GRCm38.75_chr1_sorted.gtf. But a better approach would be to use gzip to compress this stream before writing it to disk:

```
$ (zgrep "^#" Mus_musculus.GRCm38.75_chr1.gtf.gz; \
    zgrep -v "^#" Mus_musculus.GRCm38.75_chr1.gtf.gz | \
    sort -k1,1 -k4,4n) | gzip > Mus_musculus.GRCm38.75_chr1_sorted.gtf.gz
```

Subshells give us another way to compose commands in the shell. While these are indeed tricky to understand at first (which is why this section is labelled "advanced"), these do have tangible, useful applications—especially in manipulating large data streams. Note that we could have written a custom Python script to do this same task, but there are several disadvantages with this approach: it would be more work to implement, it would require more effort to have it work with compressed streams, and sorting data in Python would almost certainly be much, much less efficient than using Unix sort.

Named Pipes and Process Substitution

Throughout this chapter, we've used pipes to connect command-line tools to build custom data-processing pipelines. However, some programs won't interface with the Unix pipes we've come to love and depend on. For example, certain bioinformatics tools read in multiple input files and write to multiple output files:

```
$ processing_tool --in1 in1.fq --in2 in2.fq --out1 out2.fq --out2.fq
```

In this case, the imaginary program `processing_tool` requires two separate input files, and produces two separate output files. Because each file needs to be provided separately, we can't pipe the previous processing step's results through `process ing_tool`'s standard in. Likewise, this program creates two separate output files, so it isn't possible to pipe `processing_tool`'s standard output to another program in the processing pipeline. This isn't just an unlikely toy example; many bioinformatics programs that process two paired-end sequencing files together operate like this.

In addition to the inconvenience of not being able to use a Unix pipe to interface `processing_tool` with other programs, there's a more serious problem: we would have to write and read four intermediate files to disk to use this program. If `process ing_tool` were in the middle of a data-processing pipeline, this would lead to a significant computational bottleneck (remember from "The Almighty Unix Pipe: Speed and Beauty in One" on page 45 that reading and writing to disk is very slow).

Fortunately, Unix provides a solution: *named pipes*. A named pipe, also known as a FIFO (First In First Out, a concept in computer science), is a special sort of file. Regular pipes are anonymous—they don't have a name, and only persist while both processes are running. Named pipes behave like files, and are persistent on your filesystem. We can create a named pipe with the program `mkfifo`:

```
$ mkfifo fqin
$ ls -l fqin
prw-r--r--    1 vinceb  staff           0 Aug  5 22:50 fqin
```

You'll notice that this is indeed a special type of file: the p before the file permissions is for pipe. Just like pipes, one process writes data into the pipe, and another process reads data out of the pipe. As a toy example, we can simulate this by using `echo` to redirect some text into a named pipe (running it in the background, so we can have our prompt back), and then `cat` to read the data back out:

```
$ echo "hello, named pipes" > fqin & ❶
[1] 16430
$ cat fqin ❷
[1]  + 16430 done
hello, named pipes
$ rm fqin ❸
```

❶   Write some lines to the named pipe we created earlier with `mkfifo`.

❷ Treating the named pipe just as we would any other file, we can access the data we wrote to it earlier. Any process can access this file and read from this pipe. Like a standard Unix pipe, data that has been read from the pipe is no longer there.

❸ Like a file, we clean up by using `rm` to remove it.

Although the syntax is similar to shell redirection to a file, *we're not actually writing anything to our disk*. Named pipes provide all of the computational benefits of pipes with the flexibility of interfacing with files. In our earlier example, *in1.fq* and *in2.fq* could be named pipes that other processes are writing input to. Additionally, *out1.fq* and *out2.fq* could be named pipes that `processing_tool` is writing to, that other processes read from.

However, creating and removing these file-like named pipes is a bit tedious. Programmers like syntactic shortcuts, so there's a way to use named pipes without having to explicitly create them. This is called *process substitution*, or sometimes known as anonymous named pipes. These allow you to invoke a process, and have its standard output go directly to a named pipe. However, your shell treats this process substitution block like a file, so you can use it in commands as you would a regular file or named pipe. This is a bit confusing at first, but it should be clearer with some examples.

If we were to re-create the previous toy example with process substitution, it would look as follows:

```
$ cat <(echo "hello, process substitution")
hello, process substitution
```

Remember that `cat` takes file arguments. The chunk `<(echo "hello, process sub stition")` runs the `echo` command and pipes the output to an anonymous named pipe. Your shell then replaces this chunk (the `<(...)` part) with the path to this anonymous named pipe. No named pipes need to be explicitly created, but you get the same functionality. Process substitution allows us to connect two (or potentially more) programs, even if one doesn't take input through standard input. This is illustrated in Figure 7-3.

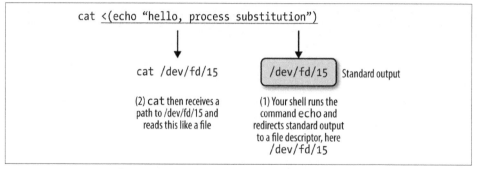

Figure 7-3. Process substitution broken down by what your shell does.

In the `program` example we saw earlier, two inputs were needed (`--in1` and `--in2`). Because this program takes two inputs, and not one from standard input, there's no way to use a classic Unix pipe (`|`) if these input files are the *outputs* of another program. Instead of creating two explicit named pipes with `mkfifo`, we can use process substitution. For the sake of this example, assume that a program called `makein` is creating the input streams for `--in1` and `--in2`:

```
program --in1 <(makein raw1.txt) --in2 <(makein raw2.txt) \
    --out1 out1.txt --out2 out2.txt
```

The last thing to know about process substitution is that it can also be used to capture an *output* stream. This is used often in large data processing to compress output on the fly before writing it to disk. In the preceding `program` example, assume that we wanted to take the output that's being written to files *out1.txt* and *out2.txt* and compress these streams to files *out1.txt.gz* and *out2.txt.gz*. We can do this using the intuitive analog to `<(...)`, `>(...)`. For clarity, I am omitting our previous process substitutions:

```
program --in1 in1.txt --in2 in2.txt \
    --out1 >(gzip > out1.txt.gz) --out2 >(gzip > out2.txt.gz)
```

This creates two anonymous named pipes, and their input is then passed to the `gzip` command. `gzip` then compresses these and writes to standard out, which we redirect to our gzipped files.

The Unix Philosophy Revisited

Throughout this chapter, the Unix philosophy—equipped with the power of the Unix pipe—allowed us to rapidly stitch together tiny programs using a rich set of Unix tools. Not only are Unix piped workflows fast to construct, easy to debug, and versatile, but they're often the most computationally efficient solution, too. It's a testament to the incredible design of Unix that so much of the way we approach modern bioinformatics is driven by the almighty Unix pipe, a piece of technology invented over 40 years ago in "one feverish night" by Ken Thompson (as described by Doug McIlroy).

## CHAPTER 8

A Rapid Introduction to the R Language

In summary, data analysis, like experimentation, must be considered as an open-ended, highly interactive, iterative process, whose actual steps are selected segments of a stubbily branching, tree-like pattern of possible actions.

> — *Data analysis and statistics: an expository overview* J. W. Tukey and M. B. Wilk (1966)

...exploratory data analysis is an attitude, a state of flexibility, a willingness to look for those things that we believe are not there, as well as for those we believe might be there. Except for its emphasis on graphs, its tools are secondary to its purpose.

> —J. W. Tukey in a comment to E. Parzen (1979)

Many biologists are first exposed to the R language by following a cookbook-type approach to conduct a statistical analysis like a *t*-test or an analysis of variance (ANOVA). Although R excels at these and more complicated statistical tasks, R's real power is as a data programming language you can use to explore and understand data in an open-ended, highly interactive, iterative way. Learning R as a data programming language will give you the freedom to experiment and problem solve during data analysis—exactly what we need as bioinformaticians. In particular, we'll focus on the subset of the R language that allows you to conduct *exploratory data analysis (EDA)*. Note, however, that EDA is only one aspect of the R language—R also includes state-of-the-art statistical and machine learning methods.

Popularized by statistician John W. Tukey, EDA is an approach that emphasizes understanding data (and its limitations) through interactive investigation rather than explicit statistical modeling. In his 1977 book *Exploratory Data Analysis*, Tukey described EDA as "detective work" involved in "finding and revealing the clues" in data. As Tukey's quote emphasizes, EDA is more an approach to exploring data than using specific statistical methods. In the face of rapidly changing sequencing technologies, bioinformatics software, and statistical methods, EDA skills are not only widely applicable and comparatively stable—they're also essential to making sure that our analyses are robust to these new data and methods.

Exploratory data analysis plays an integral role throughout an entire bioinformatics project. Exploratory data analysis skills are just as applicable in analyzing intermediate bioinformatics data (e.g., are fewer reads from this sequencing lane aligning?) as they are in making sense of results from statistical analyses (e.g., what's the distribution of these *p*-values, and do they correlate with possible confounders like gene length?). These exploratory analyses need not be complex or exceedingly detailed (many patterns are visible with simple analyses and visualization); it's just about wanting to look into the data and having the skill set to do so.

In many cases, exploratory data analysis—and especially visualization—can reveal patterns in bioinformatics data we might overlook with statistical modeling or hypothesis testing. Bioinformatics data is high dimensional and messy, with each data point being a possible mix of biological signal, measurement noise, bias due to ad hoc bioinformatics filtering criteria or analysis steps, and confounding by both technical and biological covariates. Our brains are the most sophisticated pattern-finding instruments on the planet, and exploratory data analysis is the craft of presenting data in different ways to allow our brains to find patterns—both those that indicate interesting biological signals or suggest potential problems. Compared to our brains, statistical tests are a blunt instrument—and one that's even duller when working with the complex, high-dimensional datasets widespread in bioinformatics.

Although this chapter emphasizes exploratory data analysis, statistical analysis of high-dimensional genomics data is just as important; in fact, this topic is *so* important that you should seek books and articles that cover it in depth. As stated in Chapter 1, no amount of post-experiment data analysis can rescue a poorly designed experiment. Likewise, no amount of terrific exploratory data analysis is substitute for having a good experimental question and applying appropriate statistical methods. Rather, EDA techniques like visualization should play an ongoing role throughout statistical analysis and complement other statistical methods, assessing the output at each step. The objective of this chapter is to teach you the EDA skills that give you the freedom to explore and experiment with your data at any stage of analysis.

Getting Started with R and RStudio

To capitalize on R's interactive capabilities, we need a development environment that promotes interactive coding. The most popular option is RStudio, an open source integrated development environment for R. RStudio supports many features useful for working in R: syntax highlighting, quick access to R's internal help and documentation system, and plots visible alongside code. Additionally, RStudio has an intuitive interface, is easy to use, and is the editor I'd recommend for beginner R users.

Developing code in R is a back-and-forth between writing code in a rerunnable script and exploring data interactively in the R interpreter. To be reproducible, all steps that lead to results you'll use later must be recorded in the R script that accompanies your analysis and interactive work. While R can save a history of the commands you've entered in the interpreter during a session (with the command savehistory()), storing your steps in a well-commented R script makes your life much easier when you need to backtrack to understand what you did or change your analysis. RStudio simplifies developing R scripts this way by allowing you to send lines from your R script to the R interpreter with a simple keyboard shortcut: Command-Enter (for OS X) or Control-Enter (for Windows and Linux).

The Comprehensive R Archive Network (CRAN)

Among R's greatest strengths are the numerous packages that extend its functionality. The R Project hosts many of these packages (over 6,400) on the *Comprehensive R Archive Network*, or *CRAN*, on a variety of servers that mirror the R project's server. You can install these packages directly from within R with the install.packages() function. For example, we can install the ggplot2 package (which we'll use throughout this chapter) as follows:

```
> install.packages("ggplot2")
trying URL 'http://cran.cnr.Berkeley.edu/src/contrib/ggplot2_1.0.0.tar.gz'
Content type 'application/x-gzip' length 2351447 bytes (2.2 Mb)
opened URL
==================================================
downloaded 2.2 Mb

Loading required package: devtools
* installing *source* package 'ggplot2' ...
** package 'ggplot2' successfully unpacked and MD5 sums checked
** R
** data
*** moving datasets to lazyload DB
** inst
** preparing package for lazy loading
** help
*** installing help indices
** building package indices
** installing vignettes
** testing if installed package can be loaded
Loading required package: devtools
* DONE (ggplot2)
```

This downloads and installs ggplot2 from a CRAN repository (for information on setting various options, including how to configure your repository to a nearby mirror, see help(install.packages)).

CRAN isn't the only R package repository—in Chapter 9, we'll use many packages from the Bioconductor project. Bioconductor has its own way of installing packages that we'll learn about in "Installing and Working with Bioconductor Packages" on page 269.

To get started with the examples in this chapter, we first need to install R. You can do this by either downloading R from the R-Project's website (*http://cran.r-project.org/*) or installing it from a ports or package manager like Ubuntu's `apt-get` or OS X's Homebrew. Then, you'll need to install RStudio IDE from RStudio's website (*http://www.rstudio.com/*). Be sure to keep both R and RStudio up to date. I've included some additional resources on getting started with RStudio in this chapter's *README* on GitHub.

R Language Basics

Before getting our hands dirty working with real data in R, we need to learn the basics of the R language. Even if you've poked around in R and seen these concepts before, I would still recommend skimming through this section. Many of the difficulties beginning R users face stem from misunderstanding the language's basic concepts, data structures, and behavior (which can differ quite significantly from other languages like Python and Perl). In this section, we'll learn how to do simple calculations in R, assign values to variables, and call functions. Then, we'll look at R's vectors, vector data types, and vectorization. Vectors and vectorization underpin how we approach many problems in R—a theme we'll continue to see throughout this chapter.

Simple Calculations in R, Calling Functions, and Getting Help in R

Let's begin by looking at some simple calculations in R. Open RStudio (or a terminal R prompt) and try some basic arithmetic calculations in R:

```
> 4 + 3
[1] 7
> 4 - 3
[1] 1
> 4 * 3
[1] 12
> 4 / 3
[1] 1.333333
```

You'll need to familiarize yourself with some R lingo: we say each line contains an *expression* that is *evaluated* by R when you press Enter. Whitespace around the arithmetic operations does not change how R evaluates these expressions. In some cases, you will need to surround parts of an expression with parentheses to indicate the order of evaluation. For example:

```
> 3 + 4/2
[1] 5
> (3 + 4)/2
[1] 3.5
```

Other mathematical operations can be performed by using *functions*. Functions take in zero or more *arguments*, do some work, and output a *return value*. A very important fact about R's functions is that they copy their arguments, and do *not* modify their arguments in place (there are some technical exceptions to this rule we'll ignore). In contrast, Python's functions can modify their arguments in place. Functions are the bread and butter of working in R, so it's necessary to understand and be able to work with functions, function arguments, and return values.

R has numerous mathematical functions (see Table 8-1 for some commonly used ones). For example, we *call* the function sqrt() on a (nonnegative) numeric argument:

```
> sqrt(3.5)
[1] 1.870829
```

Table 8-1. Common mathematic functions

Function Name	Description	Example
exp(x)	Exponential function	exp(1), exp(2)
log(x, base=exp(1)), log10(), log2()	Natural, base 10, and base 2 logarithms	log(2), log10(100), log2(16)
sqrt(x)	Square root	sqrt(2)
sin(x), cos(x), tan(x), etc.	Trigonometric functions (see help(sin) for more)	sin(pi)
abs(x)	Absolute value	abs(-3)
factorial(x)	Factorial	factorial(5)
choose(n, k)	Binomial coefficient	choose(5, 3)

Significant Digits, print(), and Options in R

By default, R will print seven significant digits (which is what it did when we executed sqrt(3.5)). While seven significant digits is the default, it's easy to print more digits of a value by using the function print(). For example:

```
> print(sqrt(3.5), digits=10)
[1] 1.870828693
```

Behind the scenes, R uses `print()` to format the R objects you see printed as output. (note that you won't see `print()` explicitly called).

Alternatively, you can change the default number of significant digits R uses by changing the *global option* in R. You can view the current default value of an option by calling `getOption()` with the option name as an argument; for example, you can retrieve the number of significant digits printed as follows:

```
> getOption('digits')
[1] 7
```

A new option value can be set using the function `options()`:

```
> options(digits=9)
```

`options()` contains numerous user-customizable global options. See `help(options)` for more information.

Here, 1.870829 is the *return value* of this function. This value can either be assigned to a variable (which we'll see later), or passed directly to other functions as an argument. For example, we could pass this return value into another function, `round()`, to round our square root:

```
> round(sqrt(3.5))
[1] 2
```

The `round()` function rounds `sqrt(3.5)` to 2, keeping zero decimal places, because the `round()` function's second argument (`digits`) has a default value of zero. You can learn about a function's arguments and their default values through `round()`'s documentation, which you can access with `help(round)` (see "Getting Help in R" on page 180 for more depth on R's help and documentation system). We can change the number of digits `round()` uses by specifying the value of this second `digits` argument either of two ways:

```
> round(sqrt(3.5), digits=3)
[1] 1.871
> round(sqrt(3.5), 3)
[1] 1.871
```

Getting Help in R

As would be expected from a sophisticated scientific and statistical computing language, R has oodles of functions—far more than any reasonable human can expect to learn and remember. Consequently, you'll need to master two of the most practical R skills:

- Knowing how to look up a function's documentation to recall its arguments and how it works

- Being able to discover new useful functions

Each of R's functions (and other objects such as constants like pi, classes, and packages) has integrated documentation accessible within R. R's documentation includes descriptions and details about the function, all arguments of a function, and useful usage examples. You access access R's built-in documentation with the help() function or its syntactic shortcut, ?:

```
> help(log)
> ?log
```

In RStudio, this opens a special help window containing log()'s documentation. In terminal R, this documentation is handled by your default pager (probably the program less). Operators such as + and ^ need to be quoted (e.g., help('+')). R also has documentation for general topics available; see, for example, help(Quotes).

R's help() function is useful when you already know the name of the function you need documentation for. Unfortunately, we often only have a fuzzier idea of what we need help with (e.g., what was the function in R that calculates cross tabulate vectors?). For tasks like this, we can search R's help system with the function help.search(), or its shortcut ??:

```
> help.search("cross tabulate")
> ??"cross tabulate"
```

In this case, help.search() would help you find the function table(), which is useful in creating counts and cross tabulations from vectors.

Also, R has the neat feature that all examples in an R help file can be executed with the function example(). For example:

```
> example(log)

log> log(exp(3))
[1] 3
[...]
```

Finally, R also has functions for listing all functions in a package (e.g., library(help="base")) and finding functions by name (e.g., apropos(norm)), which are often useful in remembering a function's name.

First, values are matched to arguments by name. Technically, R also allows partial matching of argument names but I would discourage this, as it decreases code readability. Second, values are matched based on position within the argument list. For functions that have many arguments with default values such as foo(x, a=3, b=4, c=5), it's easier to set an argument later in the function by specifying it by name. For example, compare calling foo() with c=5 by using foo(2, 3, 4, 5) with foo(2, c=5).

Variables and Assignment

To store a value for future use, we *assign* it to a *variable* (also known as a symbol in R jargon) using the <- assignment operator:

```
> x <- 3.1
```

Once the variable is assigned, we can retrieve its value by evaluating it on a line:

```
> x
3.1
```

Variables can also be used in expressions and passed to functions as arguments. R will substitute the variable's value during evaluation and return the results. These results can then be assigned to other variables:

```
> (x + 2)/2
[1] 2.55
> exp(x)
[1] 22.1979513
> y <- exp(2*x) - 1
> y
[1] 491.749041
```

 RStudio Assignment Operator Shortcut

In RStudio, you can create the <- assignment operator in one keystroke using Option - (that's a dash) on OS X or Alt - on Windows/Linux.

It's also possible to use = for assignment, but <- is more conventional (and is what we'll use throughout the chapter).

When we assign a value in our R session, we're assigning it to an *environment* known as the *global environment*. The objects we create by assigning values to variables are kept in environments. We can see objects we've created in the global environment with the function ls():

```
> ls()
[1] "x"
```

Here, we see the variable name x, which we assigned the value 3.1 earlier. When R needs to lookup a variable name, it looks in the search path. Calling the function search() returns where R looks when searching for the value of a variable—which includes the global environment (*.GlobalEnv*) and attached packages.

Vectors, Vectorization, and Indexing

Arguably the most important feature of the R language is its vectors. A *vector* is a container of contiguous data. Unlike most languages, R does not have a type for a single value (known as a *scalar*) such as 3.1 or "AGCTACGACT." Rather, these values are stored in a vector of length 1. We can verify that values like 3.1 are vectors of length 1 by calling the function length() (which returns the length of a vector) on them:

```
> length(3.1)
[1] 1
```

To create longer vectors, we combine values with the function c():

```
> x <- c(56, 95.3, 0.4)
> x
[1] 56.0 95.3  0.4
> y <- c(3.2, 1.1, 0.2)
> y
[1] 3.2 1.1 0.2
```

R's vectors are the basis of one of R's most important features: *vectorization*. Vectorization allows us to loop over vectors elementwise, without the need to write an explicit loop. For example, R's arithmetic operators are all vectorized:

```
> x + y
[1] 59.2 96.4  0.6
> x - y
[1] 52.8 94.2  0.2
> x/y
[1] 17.50000 86.63636  2.00000
```

Integer sequences crop up all over computing and statistics, so R has a few ways of creating these vectors. We'll use these later in this section:

```
> seq(3, 5)
[1] 3 4 5
> 1:5
[1] 1 2 3 4 5
```

There's one important subtle behavior of vectorized operations applied to two vectors simultaneously: if one vector is longer than the other, R will *recycle* the values in the shorter vector. This is an intentional behavior, so R won't warn you when this happens (unless the recycled shorter vector's length isn't a multiple of the longer vector's length). Recycling is what allows you to add a single value to all elements of a vector; the shorter vector (the single value in this case) is recycled to all elements of the longer vector:

```
> x
[1] 56.0 95.3  0.4
> x - 3
[1] 53.0 92.3 -2.6
```

```
> x / 2
[1] 28.00 47.65  0.20
```

R does warn if it recycles the shorter vector and there are remainder elements left. Consider the following examples:

```
> c(1, 2) + c(0, 0, 0, 0) ❶
[1] 1 2 1 2
> c(1, 2) + c(0, 0, 0) ❷
[1] 1 2 1
Warning message:
In c(1, 2) + c(0, 0, 0) :
  longer object length is not a multiple of shorter object length
```

❶ R adds a shorter vector c(1, 2) to a longer vector c(0, 0, 0, 0) by recycling the shorter values. The longer vector is all zeros so this is easier to see.

❷ When the shorter vector's length isn't a multiple of the longer vector's length, there will be a remainder element. R warns about this.

In addition to operators like + and *, many of R's mathematical functions (e.g., sqrt(), round(), log(), etc.) are all vectorized:

```
> sqrt(x)
[1] 7.4833148 9.7621719 0.6324555
> round(sqrt(x), 3)
[1] 7.483 9.762 0.632
> log(x)/2 + 1 # note how we can combined vectorized operations
[1] 3.0126758 3.2785149 0.5418546
```

This vectorized approach is not only more clear and readable, it's also often computationally faster. Unlike other languages, R allows us to completely forgo explicitly looping over vectors with a for loop. Later on, we'll see other methods used for more explicit looping.

We can access specific elements of a vector through *indexing*. An index is an integer that specifies which element in the vector to retrieve. We can use indexing to get or set values to certain elements from a vector:

```
> x <- c(56, 95.3, 0.4)
> x[2] ❶
[1] 95.3
> x[1]
[1] 56
> x[4] ❷
[1] NA
> x[3] <- 0.5 ❸
> x
[1] 56.0 95.3  0.5
```

❶ R's vectors are 1-indexed, meaning that the index 1 corresponds to the first element in a list (in contrast to 0-indexed languages like Python). Here, the value 95.3 is the 2nd item, and is accessed with x[2].

❷ Trying to access an element that doesn't exist in the vector leads R to return NA, the "not available" missing value.

❸ We can change specific vector elements by combining indexing and assignment.

Vectors can also have names, which you can set while combining values with c():

```
> b <- c(a=3.4, b=5.4, c=0.4)
> b
  a   b   c
3.4 5.4 0.4
```

The names of a vector can be both accessed and set with names():

```
> names(b)
[1] "a" "b" "c"
> names(b) <- c("x", "y", "z") # change these names
> b
  x   y   z
3.4 5.4 0.4
```

And just as we can access elements by their positional index, we can also access them by their name:

```
> b['x']
  x
3.4
> b['z']
  z
0.4
```

It is also possible to extract more than one element simultaneously from a vector using indexing. This is more commonly known as *subsetting*, and it's a primary reason why the R language is so terrific for manipulating data. Indices like 3 in x[3] are just vectors themselves, so it's natural to allow these vectors to have more than one element. R will return each element at the positions in the indexing vector:

```
> x[c(2, 3)]
[1] 95.3  0.4
```

Vectorized indexing provides some incredibly powerful ways to manipulate data, as we can use indexes to slice out sections of vector, reorder elements, and repeat values. For example, we can use the methods we used to create contiguous integer sequences we learned in "Vectors, Vectorization, and Indexing" on page 183 to create indexing vectors, as we often want to extract contiguous slices of a vector:

```
> z <- c(3.4, 2.2, 0.4, -0.4, 1.2)
> z[3:5] # extract third, fourth, and fifth elements
[1]  0.4 -0.4  1.2
```

Out-of-Range Indexing

Be aware that R does not issue a warning if you try to access an element in a position that's greater than the number of elements—instead, R will return a missing value (NA; more on this later). For example:

```
> z[c(2, 1, 10)]
[1] 2.2 3.4  NA
```

Similarly, missing values in indexes leads to an NA too.

It's also possible to *exclude* certain elements from lists using *negative* indexes:

```
> z[c(-4, -5)]   # exclude fourth and fifth elements
[1] 3.4 2.2 0.4
```

Negative Indexes and the Colon Operator

One important subtle gotcha occurs when trying to combine negative indexes with the colon operator. For example, if you wanted to exclude the second through fourth elements of a vector x, you might be tempted to use x[-2:4]. However, if you enter -2:4 in the R interpreter, you'll see that it creates a sequence from -2 to 4—*not* -2, -3, and -4. To remedy this, wrap the sequence in parentheses:

```
> -(2:4)
[1] -2 -3 -4
```

Indices are also often used to reorder elements. For example, we could rearrange the elements of this vector z with:

```
> z[c(3, 2, 4, 5, 1)]
[1]  0.4  2.2 -0.4  1.2  3.4
```

Or we could reverse the elements of this vector by creating the sequence of integers from 5 down to 1 using 5:1:

```
> z[5:1]
[1]  1.2 -0.4  0.4  2.2  3.4
```

Similarly, we can use other R functions to create indexes for us. For example, the function order() returns a vector of indexes that indicate the (ascending) order of the elements. This can be used to reorder a vector into increasing or decreasing order:

```
> order(z)
[1] 4 3 5 2 1
> z[order(z)]
```

```
[1] -0.4  0.4  1.2  2.2  3.4
> order(z, decreasing=TRUE)
[1] 1 2 5 3 4
> z[order(z, decreasing=TRUE)]
[1]  3.4  2.2  1.2  0.4 -0.4
```

Again, R's vector index rule is simple: R will return the element at the ith position for each i in the indexing vector. This also allows us to repeat certain values in vectors:

```
> z[c(2, 2, 1, 4, 5, 4, 3)]
[1]  2.2  2.2  3.4 -0.4  1.2 -0.4  0.4
```

Again, often we use functions to generate indexing vectors for us. For example, one way to resample a vector (with replacement) is to randomly sample its indexes using the sample() function:

```
> set.seed(0)   # we set the random number seed so this example is reproducible
> i <- sample(length(z), replace=TRUE)
> i
[1] 5 2 2 3 5
> z[i]
[1] 1.2 2.2 2.2 0.4 1.2
```

Just as we use certain functions to generate indexing vectors, we can use *comparison operators* like ==, !=, <, <=, >, and >= (see Table 8-2) to build *logical vectors* of TRUE and FALSE values indicating the result of the comparison test for each element in the vector. R's comparison operators are also vectorized (and will be recycled according to R's rule). Here are some examples:

```
> v <- c(2.3, 6, -3, 3.8, 2, -1.1)
> v == 6
[1] FALSE  TRUE FALSE FALSE FALSE FALSE
> v <= -3
[1] FALSE FALSE  TRUE FALSE FALSE FALSE
> abs(v) > 5
[1] FALSE  TRUE FALSE FALSE FALSE FALSE
```

Logical vectors are useful because they too can be used as indexing vectors—R returns the elements with corresponding TRUE values in the indexing vector (see Example 8-1).

Example 8-1. Indexing vectors with logical vectors

```
> v[c(TRUE, TRUE, FALSE, TRUE, FALSE, FALSE)]
[1] 2.3 6.0 3.8
```

But it's tedious to construct these logical vectors of TRUE and FALSE by hand; if we wanted to select out particular elements using an integer, indexes would involve much less typing. But as you might have guessed, the power of using logical vectors in subsetting comes from creating logical vectors using comparison operators. For example, to subset v such that only elements greater than 2 are kept, we'd use:

```
> v[v > 2]
[1] 2.3 6.0 3.8
```

Note that there's no magic or special evaluation going on here: we are simply creating a logical vector using v > 2 and then using this logical vector to select out certain elements of v. Lastly, we can use comparison operators (that return logical vectors) with vectorized logical operations (also known as *Boolean algebra* in other disciplines) such as & (AND), | (OR), and ! (NOT). For example, to find all elements of v greater than 2 and less than 4, we'd construct a logical vector and use this to index the vector:

```
> v > 2 & v < 4
[1]  TRUE FALSE FALSE  TRUE FALSE FALSE
> v[v > 2 & v < 4]
[1] 2.3 3.8
```

Table 8-2. R's comparison and logical operators

Operator	Description
>	Greater than
<	Less than
>=	Greater than or equal to
<=	Less than or equal to
==	Equal to
!	Not equal to
&	Elementwise logical AND
|	Elementwise logical OR
!	Elementwise logical NOT
&&	Logical AND (first element only, for `if` statements)
||	Logical OR (first element only, for `if` statements)

We'll keep returning to this type of subsetting, as it's the basis of some incredibly powerful data manipulation capabilities in R.

Vector types

The last topic about vectors to cover before diving into working with real data is R's vector types. Unlike Python's lists or Perl's arrays, R's vectors *must* contain elements of the same *type*. In the context of working with statistical data, this makes sense: if we add two vectors x and y together with a vectorized approach like x + y, we want to know ahead of time whether all values in both vectors have the same type and can indeed be added together. It's important to be familiar with R's types, as it's common as an R beginner to run into type-related issues (especially when loading in data that's messy).

R supports the following vector types (see also Table 8-3):

Numeric

Numeric vectors (also known as *double* vectors) are used for real-valued numbers like 4.094, –12.4, or 23.0. By default, all numbers entered into R (e.g., c(5, 3, 8)) create numeric vectors, regardless of whether they're an integer.

Integer

R also has an integer vector type for values like –4, 39, and 23. Because R defaults to giving integer values like –4, 39, and 23 the type numeric, you can explicitly tell R to treat a value as an integer by appending a capital L after the value (e.g., -4L, 39L, and 23L). (Note that the seemingly more sensible *i* or *I* aren't used to avoid confusion with complex numbers with an imaginary component like 4 + 3i).

Character

Character vectors are used to represent text data, also known as *strings*. Either single or double quotes can be used to specify a character vector (e.g., c("AGTC GAT", "AGCTGGA")). R's character vectors recognize special characters common to other programming languages such as newline (\n) and tab (\t).

Logical

Logical values are used to represent Boolean values, which in R are TRUE and FALSE. T and F are assigned the values TRUE and FALSE, and while you might be tempted to use these shortcuts, *do not*. Unlike TRUE and FALSE, T and F can be redefined in code. Defining T <- 0 will surely cause problems.

Table 8-3. R's vector types

Type	Example	Creation function	Test function	Coercion function
Numeric	c(23.1, 42, -1)	numeric()	is.numeric()	as.numeric()
Integer	c(1L, -3L, 4L)	integer()	is.integer()	as.integer()

Type	Example	Creation function	Test function	Coercion function
Character	c("a", "c")	character()	is.character()	as.character()
Logical	c(TRUE, FALSE)	logical()	is.logical()	as.logical()

In addition to double, integer, character, and logical vectors, R has two other vector types: *complex* to represent complex numbers (those with an imaginary component), and *raw*, to encode raw bytes. These types have limited application in bioinformatics, so we'll skip discussing them.

R's Special Values

R has four special values (NA, NULL, Inf/-Inf, and NaN) that you're likely to encounter in your work:

NA
: NA is R's built-in value to represent missing data. Any operation on an NA will return an NA (e.g., 2 + NA returns NA). There are numerous functions to handle NAs in data; see na.exclude() and complete.cases(). You can find which elements of a vector are NA with the function is.na().

NULL
: NULL represents not having a value (which is different than having a value that's missing). It's analogous to Python's None value. You can test if a value is NULL with the function is.null().

-Inf, Inf
: These are just as they sound, negative infinite and positive infinite values. You can test whether a value is finite with the function is.finite() (and its complement is.infinite()).

NaN
: NaN stands for "not a number," which can occur in some computations that don't return numbers, i.e., 0/0 or Inf + -Inf. You can test for these with is.nan().

Again, the most important thing to remember about R's vectors and data types is that vectors are of *homogenous* type. R enforces this restriction through *coercion*, which like recycling is a subtle behavior that's important to remember about R.

Type Coercion in R

Because all elements in a vector must have homogeneous data type, R will silently coerce elements so that they have the same type.

R's coercion rules are quite simple; R coerces vectors to a type that leads to no information loss about the original value. For example, if you were to try to create a vector containing a logical value and a numeric value, R would coerce the logical TRUE and FALSE values to 1 and 0, as these represent TRUE and FALSE without a loss of information:

```
> c(4.3, TRUE, 2.1, FALSE)
[1] 4.3 1.0 2.1 0.0
```

Similarly, if you were to try to create a vector containing integers and numerics, R would coerce this to a numeric vector, because integers can be represented as numeric values without any loss of information:

```
> c(-9L, 3.4, 1.2, 3L)
[1] -9.0  3.4  1.2  3.0
```

Lastly, if a string were included in a vector containing integers, logicals, or numerics, R would convert everything to a character vector, as this leads to the least amount of information loss:

```
> c(43L, TRUE, 3.2413341, "a string")
[1] "43"        "TRUE"      "3.2413341" "a string"
```

We can see any object's type (e.g., a vector's type) using the function typeof():

```
> q <- c(2, 3.5, -1.1, 3.8)
> typeof(q)
[1] "double"
```

Factors and classes in R

Another kind of vector you'll encounter are *factors*. Factors store categorical variables, such as a treatment group (e.g., "high," "medium," "low," "control"), strand (forward or reverse), or chromosome ("chr1," "chr2," etc.). Factors crop up all over R, and occasionally cause headaches for new R users (we'll discuss why in "Loading Data into R" on page 194).

Suppose we had a character vector named chr_hits containing the *Drosophila melanogaster* chromosomes where we find a particular sequence aligns. We can create a factor from this vector using the function factor():

```
> chr_hits <- c("chr2", "chr2", "chr3", "chrX", "chr2", "chr3", "chr3")
> hits <- factor(chr_hits)
> hits
```

```
[1] chr2 chr2 chr3 chrX chr2 chr3 chr3
Levels: chr2 chr3 chrX
```

Printing the `hits` object shows the original sequence (i.e., chr2, chr2, chr3, etc.) as well as all of this factor's *levels*. The levels are the possible values a factor can contain (these are fixed and must be unique). We can view a factor's levels by using the function `levels()`:

```
> levels(hits)
[1] "chr2" "chr3" "chrX"
```

Biologically speaking, our set of levels isn't complete. *Drosophila melanogaster* has two other chromosomes: chrY and chr4. Although our data doesn't include these chromosomes, they are valid categories and should be included as levels in the factor. When creating our factor, we could use the argument `levels` to include all relevant levels:

```
> hits <- factor(chr_hits, levels=c("chrX", "chrY", "chr2", "chr3", "chr4"))
> hits
[1] chr2 chr2 chr3 chrX chr2 chr3 chr3
Levels: chrX chrY chr2 chr3 chr4
```

If we've already created a factor, we can add or rename existing levels with the function `levels()`. This is similar to how we assigned vector names using `names(obj) <- `. When setting names, we use a named character vector to provide a mapping between the original names and the new names:

```
> levels(hits) <- list(chrX="chrX", chrY="chrY", chr2="chr2",
                       chr3="chr3", chr4="chr4")
> hits
[1] chr2 chr2 chr3 chrX chr2 chr3 chr3
Levels: chrX chrY chr2 chr3 chr4
```

We can count up how many of each level there are in a factor using the function `table()`:

```
> table(hits)
hits
chrX chrY chr2 chr3 chr4
   1    0    3    3    0
```

Factors are a good segue into briefly discussing *classes* in R. An object's class endows objects with higher-level properties that can affect how functions treat that R object. We won't get into the technical details of creating classes or R's object orientation system in this section (see a text like Hadley Wickham's *Advanced R* for these details). But it's important to have a working familiarity with the idea that R's objects have a class and this can change how certain functions treat R objects.

To discern the difference between an object's class and its type, notice that factors are just integer vectors under the hood:

```
> typeof(hits)
[1] "integer"
> as.integer(hits)
[1] 3 3 4 1 3 4 4
```

Functions like `table()` are generic—they are designed to work with objects of all kinds of classes. Generic functions are also designed to do the right thing depending on the class of the object they're called on (in programming lingo, we say that the function is *polymorphic*). For example, `table()` treats a factor differently than it would treat an integer vector. As another example, consider how the function `summary()` (which summarizes an object such as vector) behaves when it's called on a vector of numeric values versus a factor:

```
> nums <- c(0.97, -0.7, 0.44, 0.25, -1.38, 0.08)
> summary(nums)
    Min. 1st Qu.  Median    Mean 3rd Qu.    Max.
-1.38000 -0.50500  0.16500 -0.05667  0.39250  0.97000
> summary(hits)
chrX chrY chr2 chr3 chr4
   1    0    3    3    0
```

When called on numeric values, `summary()` returns a numeric summary with the quartiles and the mean. This numeric summary wouldn't be meaningful for the categorical data stored in a factor, so instead `summary()` returns the level counts like `table()` did. This is function polymorphism, and occurs because `nums` has class "numeric" and `hits` has class "factor":

```
> class(nums)
[1] "numeric"
> class(hits)
[1] "factor"
```

These classes are a part of R's *S3 object-oriented system*. R actually has three object orientation systems (S3, S4, and reference classes). Don't worry too much about the specifics; we'll encounter classes in this chapter (and also in Chapter 9), but it won't require an in-depth knowledge of R's OO systems. Just be aware that in addition to a type, objects have a class that changes how certain functions treat that object.

Working with and Visualizing Data in R

With a knowledge of basic R language essentials from the previous section, we're ready to start working with real data. We'll work a few different datasets in this chapter. All files to load these datasets into R are available in this chapter's directory on GitHub.

The dataset we'll use for learning data manipulation and visualization skills is from the 2006 paper "The Influence of Recombination on Human Genetic Diversity" (*http://journals.plos.org/plosgenetics/article?id=10.1371/journal.pgen.0020148*)   by

Spencer et al. I've chosen this dataset (*Dataset_S1.txt* on GitHub) for the following reasons:

- It's an excellent piece of genomic research with interesting biological findings.
- The article is open access and thus freely accessible the public.
- The raw data is also freely available and is tidy, allowing us to start exploring it immediately.
- All scripts used in the authors' analyses are freely available (making this a great example of reproducible research).

In addition, this type of genomic dataset is also characteristic of the data generated from lower-level bioinformatics workflows that are then analyzed in R.

Dataset_S1.txt contains estimates of population genetics statistics such as nucleotide diversity (e.g., the columns Pi and Theta), recombination (column *Recombination*), and sequence divergence as estimated by percent identity between human and chimpanzee genomes (column Divergence). Other columns contain information about the sequencing depth (depth), and GC content (percent.GC). We'll only work with a few columns in our examples; see the description of *Dataset_S1.txt* in this paper's supplementary information for more detail. *Dataset_S1.txt* includes these estimates for 1kb windows in human chromosome 20.

Loading Data into R

The first step of any R data analysis project is loading data into R. For some datasets (e.g., *Dataset_S1.txt*), this is quite easy—the data is tidy enough that R's functions for loading in data work on the first try. In some cases, you may need to use Unix tools (Chapter 7) to reformat the data into a tab-delimited or CSV plain-text format, or do some Unix sleuthing to find strange characters in the file (see "Decoding Plain-Text Data: hexdump" on page 145). In other cases, you'll need to identify and remove improper values from data within R or coerce columns.

Before loading in this file, we need to discuss R's *working directory*. When R is running, the process runs in a specific working directory. It's important to mind this working directory while loading data into R, as which directory R is running in will affect how you specify relative paths to data files. By default, command-line R will use the directory you start the R program with; RStudio uses your home directory (this is a customizable option). You can use getwd() to get R's current working directory and setwd() to set the working directory:

```
> getwd()
[1] "/Users/vinceb"
> setwd("~/bds-files/chapter-08-r") # path to this chapter's
                                     # directory in the Github repository.
```

For all examples in this chapter, I'll assume R's working directory is this chapter's directory in the book's GitHub repository. We'll come back to working directories again when we discuss R scripts in "Working with R Scripts" on page 254.

Next, it's wise to first inspect a file from the command line before loading it into R. This will give you a sense of the column delimiter being used, whether there are comment lines (e.g., lines that don't contain data and begin with a character like #), and if the first line is a header containing column names. Either head or less work well for this:

```
$ head -n 3 Dataset_S1.txt
start,end,total SNPs,total Bases,depth,unique SNPs,dhSNPs, [...]
55001,56000,0,1894,3.41,0,0, [...]
56001,57000,5,6683,6.68,2,2, [...]
```

From this, we see *Dataset_S1.txt* is a comma-separated value file with a header. If you explore this file in more detail with less, you'll notice this data is tidy and organized. Each column represents a single variable (e.g., window start position, window end position, percent GC, etc.) and each row represents an observation (in this case, the values for a particular 1kb window). Loading tidy plain-text data like *Dataset_S1.txt* requires little effort, so you can quickly get started working with it. Your R scripts should organize data in a similar tidy fashion (and in a format like tab-delimited or CSV), as it's much easier to work with tidy data using Unix tools and R.

Large Genomics Data into R: colClasses, Compression, and More

It's quite common to encounter genomics datasets that are difficult to load into R because they're large files. This is either because it takes too long to load the entire dataset into R, or your machine simply doesn't have enough memory. In many cases, the best strategy is to reduce the size of your data somehow: summarizing data in earlier processing steps, omitting unnecessary columns, splitting your data into chunks (e.g., working with a chromosome at a time), or working on a random subset of your data. Many bioinformatics analyses do not require working on an entire genomic dataset at once, so these strategies can work quite well. These approaches are also the only way to work with data that is truly too large to fit in your machine's memory (apart from getting a machine with more memory).

If your data will fit in your machine's memory, it's still possible that loading the data into R may be quite slow. There are a few tricks to make the read.csv() and read.delim() functions load large data files more quickly. First, we could explicitly set the class of each column through the colClasses argument. This saves R time (usually making the R's data reading functions twice as fast), as it takes R time to figure out what type of class a column has and convert between classes. If your dataset has columns you don't need in your analysis (and unnecessarily take up memory), you can set their value in the colClasses vector to "NULL" (in quotes) to force R to skip them.

Additionally, specifying how many rows there are in your data by setting `nrow` in `read.delim()` can lead to some performance gains. It's OK to somewhat overestimate this number; you can get a quick estimate using `wc -l`. If these solutions are still too slow, you can install the `data.table` package and use its `fread()` function, which is the fastest alternative to `read.*` functions (though be warned: `fread()` returns a `data.table`, not a `data.frame`, which behaves differently; see the manual).

If your data is larger than the available memory on your machine, you'll need to use a strategy that keeps the bulk of your data out of memory, but still allows for easy access from R. A good solution for moderately large data is to use SQLite and query out subsets for computation using the R package `RSQLite` (we'll cover SQLite and other strategies for data too large to fit in memory in Chapter 13).

Finally, as we saw in Chapter 6, many Unix data tools have versions that work on gzipped files: `zless`, `zcat` (`gzcat` on BSD-derived systems like Max OS X), and others. Likewise, R's data-reading functions can also read gzipped files directly—there's no need to uncompress gzipped files first. This saves disk space and there can be some slight performance gains in reading in gzipped files, as there are fewer bytes to read off of (slow) hard disks.

With our working directory properly set, let's load *Dataset_S1.txt* into R. The R functions `read.csv()` and `read.delim()` are used for reading CSV and tab-delimited files, respectively. Both are thin wrappers around the function `read.table()` with the proper arguments set for CSV and tab-delimited files—see `help(read.table)` for more information. To load the CSV file *Dataset_S1.txt*, we'd use `read.csv()` with `"Dataset_S1.txt"` as the `file` argument (the first argument). To avoid repeatedly typing a long name when we work with our data, we'll assign this to a variable named d:

```
> d <- read.csv("Dataset_S1.txt")
```

Note that the functions `read.csv()` and `read.delim()` have the argument `header` set to `TRUE` by default. This is important because `Dataset_S1`'s first line contains column names rather than the first row of data. Some data files don't include a header, so you'd need to set `header=FALSE`. If your file lacks a column name header, it's a good idea to assign column names as you read in the data using the `col.names` argument. For example, to load a fake tab-delimited file named *noheader.bed* that contains three columns named `chrom`, `start`, and `end`, we'd use:

```
> bd <- read.delim("noheader.bed", header=FALSE,
                    col.names=c("chrom", "start", "end"))
```

R's `read.csv()` and `read.delim()` functions have numerous arguments, many of which will need to be adjusted for certain files you'll come across in bioinformatics. See Table 8-4 for a list of some commonly used arguments, and consult help(read.csv) for full documentation. Before we move on to working with the data in *Dataset_S1.txt*, we need to discuss one common stumbling block when loading data into R: factors. As we saw in "Factors and classes in R" on page 191, factors are R's way of encoding categorical data in a vector. By default, R's read.delim() and read.csv() functions will coerce a column of strings to a factor, rather than treat it as a character vector. It's important to know that R does this so you can disable this coercion when you need a column as a character vector. To do this, set the argument stringsAsFactors=FALSE (or use asis; see help(read.table) for more information). In "Factors and classes in R" on page 191, we saw how factors are quite useful, despite the headaches they often cause new R users.

Table 8-4. Commonly used read.csv() and read.delim() arguments

Argument	Description	Additional comments
header	A TRUE/FALSE value indicating whether the first row contains column names rather than data	
sep	A character value indicating what delimits columns; using the empty string " " treats all whitespace as a separator	
stringsAs Factors	Setting this argument as FALSE prevents R from coercing character vectors to factors for all columns; see argument asis in help(read.delim) to prevent coercion to factor on specific columns	This is an important argument to be aware of, because R's default behavior of coercing character vector columns to factors is a common stumbling block for beginners.
col.names	A character vector to be used as column names	
row.names	A vector to use for row names, or a single integer or column name indicating which column to use as row names	
na.strings	A character vector of possible missing values to convert to NA	Files using inconsistent missing values (e.g., a mix of "NA," "Na," "na") can be corrected using na.strings=c("NA", "Na", "na").
colClasses	A character vector of column classes; "NULL" indicates a column should be skipped and NA indicates R should infer the type	colClasses can drastically decrease the time it takes to load a file into R, and is a useful argument when working with large bioinformatics files.
com ment.char	Lines beginning with this argument are ignored; the empty string (i.e., " ") disables this feature	This argument is useful in ignoring metadata lines in bioinformatics files.

Getting Data into Shape

Quite often, data we load in to R will be in the wrong *shape* for what we want to do with it. Tabular data can come in two different formats: *long* and *wide*. With wide data, each measured variable has its own column (Table 8-5).

Table 8-5. A gene expression counts table by tissue in wide format

Gene	Meristem	Root	Flower
gene_1	582	91	495
gene_2	305	3505	33

With long data, one column is used to store what type of variable was measured and another column is used to store the measurement (Table 8-6).

Table 8-6. A gene expression counts table by tissue in long format

Gene	Tissue	Expression
gene_1	meristem	582
gene_2	meristem	305
gene_1	root	91
gene_2	root	3503
gene_1	flower	495
gene_2	flower	33

In many cases, data is recorded by humans in wide format, but we need data in long format when working with and plotting statistical modeling functions. Hadley Wickham's `reshape2` package provides functions to reshape data: the function `melt()` turns wide data into long data, and `cast()` turns long data into wide data. There are numerous resources for learning more about the `reshape2` package:

- reshape2 CRAN page (*http://bit.ly/reshape-2*)
- Hadley Wickham's reshape page (*http://had.co.nz/reshape/*)

Exploring and Transforming Dataframes

The *Dataset_S1.txt* data we've loaded into R with `read.csv()` is stored as a *dataframe*. Dataframes are R's workhorse data structure for storing tabular data. Dataframes consist of *columns* (which represent variables in your dataset), and *rows* (which represent observations). In this short section, we'll learn the basics of accessing the dimensions, rows, columns, and row and column names of a dataframe. We'll also see how to transform columns of a dataframe and add additional columns.

Each of the columns of a dataframe are vectors just like those introduced in "Vectors, Vectorization, and Indexing" on page 183. Consequently, each element in a dataframe column has the same type. But a dataframe can contain many columns of all different types; storing columns of heterogeneous types of vectors is what dataframes are designed to do.

First, let's take a look at the dataframe we've loaded in with the function `head()`. By default, `head()` prints the first six lines of a dataframe, but we'll limit this to three using the n argument:

```
> head(d, n=3)
  start   end total.SNPs total.Bases depth unique.SNPs dhSNPs reference.Bases
1 55001 56000          0        1894  3.41           0      0             556
2 56001 57000          5        6683  6.68           2      2            1000
3 57001 58000          1        9063  9.06           1      0            1000
  Theta     Pi Heterozygosity   X.GC Recombination Divergence Constraint SNPs
1 0.000  0.000          0.000 54.8096   0.009601574 0.003006012          0    0
2 8.007 10.354          7.481 42.4424   0.009601574 0.018018020          0    0
3 3.510  1.986          1.103 37.2372   0.009601574 0.007007007          0    0
```

(Note that R has wrapped the columns of this dataset.)

Other things we might want to know about this dataframe are its dimensions, which we can access using `nrow()` (number of rows), `ncol()` (number of columns), and `dim()` (returns both):

```
> nrow(d)
[1] 59140
> ncol(d)
[1] 16
> dim(d)
[1] 59140    16
```

We can also print the columns of this dataframe using `col.names()` (there's also a `row.names()` function):

```
> colnames(d)
 [1] "start"         "end"           "total.SNPs"    "total.Bases"
 [5] "depth"         "unique.SNPs"   "dhSNPs"        "reference.Bases"
 [9] "Theta"         "Pi"            "Heterozygosity" "X.GC"
[13] "Recombination" "Divergence"    "Constraint"    "SNPs"
```

Note that R's read.csv() function has automatically renamed some of these columns for us: spaces have been converted to periods and the percent sign in %GC has been changed to an "X." "X.GC" isn't a very descriptive column name, so let's change this. Much like we've set the names of a vector using names() <-, we can set column names with col.names() <-. Here, we only want to change the 12th column name, so we'd use:

```
> colnames(d)[12] # original name
[1] "X.GC"
> colnames(d)[12] <- "percent.GC"
> colnames(d)[12] # after change
[1] "percent.GC"
```

Creating Dataframes from Scratch

R's data loading functions read.csv() and read.delim() read in data from a file and return the results as a data.frame. Sometimes you'll need to create a dataframe from scratch from a set of vectors. You can do this with the function data.frame(), which creates a dataframe from vector arguments (recycling the shorter vectors when necessary). One nice feature of data.frame() is that if you provide vectors as named arguments, data.frame() will use these names as column names. For example, if we simulated data from a simple linear model using sample() and rnorm(), we could store it in a dataframe with:

```
> x <- sample(1:50, 300, replace=TRUE)
> y <- 3.2*x + rnorm(300, 0, 40)
> d_sim <- data.frame(y=y, x=x)
```

As with R's read.csv() and read.delim() functions, data.frame() will convert character vectors into factors. You can disable this behavior by setting stringsAsFactors=FALSE.

Similarly, we could set row names using row.names() <- (note that row names must be unique).

The most common way to access a single column of a dataframe is with the dollar sign operator. For example, we could access the column depth in d using:

```
> d$depth
 [1]  3.41  6.68  9.06 10.26  8.06  7.05 [...]
```

This returns the depth column as a vector, which we can then pass to R functions like mean() or summary() to get an idea of what depth looks like across this dataset:

```
> mean(d$depth)
[1] 8.183938
> summary(d$depth)
```

```
    Min. 1st Qu.  Median   Mean 3rd Qu.    Max.
   1.000   6.970   8.170  8.184   9.400  21.910
```

The dollar sign operator is a syntactic shortcut for a more general bracket operator used to access rows, columns, and cells of a dataframe. Using the bracket operator is similar to accessing the elements of a vector (e.g., vec[2]), except as two-dimensional data structures, dataframes use two indexes separated by a comma: df[row, col]. Just as with indexing vectors, these indexes can be vectors to select multiple rows and columns simultaneously. Omitting the row index retrieves all rows, and omitting the column index retrieves all columns. This will be clearer with some examples. To access the first two columns (and all rows), we'd use:

```
> d[ , 1:2]
  start   end
1 55001 56000
2 56001 57000
3 57001 58000
[...]
```

It's important to remember the comma (and note that whitespace does not matter here; it's just to increase readability). Selecting two columns like this returns a dataframe. Equivalently, we could use the column names (much like we could use names to access elements of a vector):

```
> d[, c("start", "end")]
  start   end
1 55001 56000
2 56001 57000
3 57001 58000
[...]
```

If we only wanted the first row of start and end positions, we'd use:

```
> d[1, c("start", "end")]
  start   end
1 55001 56000
```

Similarly, if we wanted the first row of data for all columns, we'd omit the column index:

```
> d[1, ]
  start   end total.SNPs total.Bases depth unique.SNPs dhSNPs reference.Bases
1 55001 56000          0        1894  3.41           0      0             556
  Theta Pi Heterozygosity percent.GC Recombination  Divergence Constraint SNPs
1     0  0             0    54.8096   0.009601574 0.003006012          0    0
```

Single cells can be accessed by specifying a single row and a single column:

```
>  d[2, 3]
[1] 5
```

However, in practice we don't usually need to access single rows or cells of a dataframe during data analysis (see Fragile Code and Accessing Rows and Columns in Dataframes).

Fragile Code and Accessing Rows and Columns in Dataframes

It's a good idea to avoid referring to specific dataframe rows in your analysis code. This would produce code fragile to row permutations or new rows that may be generated by rerunning a previous analysis step. In every case in which you might need to refer to a specific row, it's avoidable by using subsetting (see "Exploring Data Through Slicing and Dicing: Subsetting Dataframes" on page 203).

Similarly, it's a good idea to refer to columns by their column name, *not* their position. While columns may be less likely to change across dataset versions than rows, it still happens. Column names are more specific than positions, and also lead to more readable code.

When accessing a single column from a dataframe, R's default behavior is to return this as a vector—*not* a dataframe with one column. Sometimes this can cause problems if downstream code expects to work with a dataframe. To disable this behavior, we set the argument drop to FALSE in the bracket operator:

```
> d[, "start", drop=FALSE]
  start
1 55001
2 56001
3 57001
[...]
```

Now, let's add an additional column to our dataframe that indicates whether a window is in the centromere region. The positions of the chromosome 20 centromere (based on Giemsa banding) are 25,800,000 to 29,700,000 (see this chapter's *README* on GitHub to see how these coordinates were found). We can append to our d dataframe a column called cent that has TRUE/FALSE values indicating whether the current window is fully within a centromeric region using comparison and logical operations:

```
> d$cent <- d$start >= 25800000 & d$end <= 29700000
```

Note the single ampersand (&), which is the vectorized version of logical *AND*. & operates on each element of the two vectors created by dd$start >= 25800000 and dd$end <= 29700000 and returns TRUE when both are >true. How many windows fall into this centromeric region? There are a few ways to tally the TRUE values. First, we could use table():

```
> table(d$cent)
FALSE   TRUE
58455    685
```

Another approach uses the fact that sum( ) will coerce logical values to integers (so TRUE has value 1 and FALSE has value 0). To count how many windows fall in this centromeric region, we could use:

```
> sum(d$cent)
[1] 685
```

Lastly, note that according to the supplementary material of this paper, the diversity estimates (columns Theta, Pi, and Heterozygosity) are all scaled up by 10x in the dataset (see supplementary Text S1 for more details). We'll use the nucleotide diversity column Pi later in this chapter in plots, and it would be useful to have this scaled as per basepair nucleotide diversity (so as to make the scale more intuitive). We can create a new rescaled column called diversity with:

```
> d$diversity <- d$Pi / (10*1000)  # rescale, removing 10x and making per bp
> summary(d$diversity )
      Min.   1st Qu.    Median      Mean   3rd Qu.      Max.
0.0000000 0.0005577 0.0010420 0.0012390 0.0016880 0.0265300
```

Average nucleotide diversity per basepair in this data is around 0.001 (0.12%), roughly what we'd expect from other estimates of human diversity (Hernandez et al., 2012, Perry et al., 2012).

Exploring Data Through Slicing and Dicing: Subsetting Dataframes

The most powerful feature of dataframes is the ability to slice out specific rows by applying the same vector subsetting techniques we saw in "Vectors, Vectorization, and Indexing" on page 183 to columns. Combined with R's comparison and logical operators, this leads to an incredibly powerful method to query out rows in a dataframe. Understanding and mastering dataframe subsetting is one of the most important R skills to develop, as it gives you considerable power to interrogate and explore your data. We'll learn these skills in this section by applying them to the *Dataset_S1.txt* dataset to explore some features of this data.

Let's start by looking at the total number of SNPs per window. From summary( ), we see that this varies quite considerably across all windows on chromosome 20:

```
> summary(d$total.SNPs)
   Min. 1st Qu.  Median    Mean 3rd Qu.    Max.
  0.000   3.000   7.000   8.906  12.000  93.000
```

Notice how right-skewed this data is: the third quartile is 12 SNPs, but the maximum is 93 SNPs. Often we want to investigate such outliers more closely. Let's use data subsetting to select out some rows that have 85 or more SNPs (this arbitrary threshold is just for exploratory data analysis, so it doesn't matter much). We can create a logical vector containing whether each observation (row) has 85 or more SNPs using the following:

```
> d$total.SNPs >= 85
[1] FALSE FALSE FALSE FALSE FALSE FALSE [...]
```

Recall from "Vectors, Vectorization, and Indexing" on page 183 that in addition to integer indexing vectors, we can use logical vectors (Example 8-1). Likewise, we can use the logical vector we created earlier to extract the *rows* of our dataframe. R will keep only the rows that have a TRUE value:

```
> d[d$total.SNPs >= 85, ]
          start       end total.SNPs total.Bases depth unique.SNPs dhSNPs
2567    2621001   2622000         93       11337 11.34          13     10
12968  13023001  13024000         88       11784 11.78          11      1
43165  47356001  47357000         87       12505 12.50           9      7
       reference.Bases  Theta     Pi Heterozygosity percent.GC Recombination
2567              1000 43.420 50.926         81.589    43.9439   0.000706536
12968             1000 33.413 19.030         74.838    28.8288   0.000082600
43165             1000 29.621 27.108         69.573    46.7467   0.000500577
       Divergence Constraint SNPs  cent
2567   0.01701702          0    1 FALSE
12968  0.01401401          0    1 FALSE
43165  0.02002002          0    7 FALSE
```

This subset of the data gives a view of windows with 85 or greater SNPs. With the start and end positions of these windows, we can see if any potential confounders stand out. If you're curious, explore these windows using the UCSC Genome Browser (*http://genome.ucsc.edu*); remember to use the NCBI34/hg16 version of the human genome).

We can build more elaborate queries by chaining comparison operators. For example, suppose we wanted to see all windows where Pi (nucleotide diversity) is greater than 16 and percent GC is greater than 80. Equivalently, we could work with the rescaled diversity column but subsetting with larger numbers like 16 is easier and less error prone than with numbers like 0.0016. We'd use:

```
> d[d$Pi > 16 & d$percent.GC > 80, ]
          start       end total.SNPs total.Bases depth unique.SNPs dhSNPs
58550  63097001  63098000          5         947  2.39           2      1
58641  63188001  63189000          2        1623  3.21           2      0
58642  63189001  63190000          5        1395  1.89           3      2
       reference.Bases  Theta     Pi Heterozygosity percent.GC Recombination
58550             397 37.544 41.172         52.784    82.0821   0.000781326
58641             506 16.436 16.436         12.327    82.3824   0.000347382
58642             738 35.052 41.099         35.842    80.5806   0.000347382
       Divergence Constraint SNPs  cent
58550  0.03826531        226    1 FALSE
58641  0.01678657        148    0 FALSE
58642  0.01793722          0    0 FALSE
```

In these examples, we're extracting all columns by omitting the column argument in the bracket operator (e.g., col in df[row, col]). If we only care about a few particular columns, we could specify them by their position or their name:

```
> d[d$Pi > 16 & d$percent.GC > 80, c("start", "end", "depth", "Pi")]
          start       end depth      Pi
58550 63097001 63098000  2.39 41.172
58641 63188001 63189000  3.21 16.436
58642 63189001 63190000  1.89 41.099
```

Similarly, you could reorder columns by providing the column names or column indexes in a different order. For example, if you wanted to swap the order of depth and Pi, use:

```
> d[d$Pi > 16 & d$percent.GC > 80, c("start", "end", "Pi", "depth")]
          start       end      Pi depth
58550 63097001 63098000 41.172  2.39
58641 63188001 63189000 16.436  3.21
58642 63189001 63190000 41.099  1.89
```

Remember, columns of a dataframe are just vectors. If you only need the data from one column, just subset it as you would a vector:

```
> d$percent.GC[d$Pi > 16]
[1] 39.1391 38.0380 36.8368 36.7367 43.0430 41.1411 [...]
```

This returns all of the percent GC values for observations where that observation has a Pi value greater than 16. Note that there's no need to use a comma in the bracket because d$percent is a vector, *not* a two-dimensional dataframe.

Subsetting columns can be a useful way to summarize data across two different conditions. For example, we might be curious if the average depth in a window (the depth column) differs between very high GC content windows (greater than 80%) and all other windows:

```
> summary(d$depth[d$percent.GC >= 80])
   Min. 1st Qu.  Median    Mean 3rd Qu.    Max.
   1.05    1.89    2.14    2.24    2.78    3.37
> summary(d$depth[d$percent.GC < 80])
   Min. 1st Qu.  Median    Mean 3rd Qu.    Max.
  1.000   6.970   8.170   8.185   9.400  21.910
```

This is a fairly large difference, but it's important to consider how many windows this includes. Indeed, there are only nine windows that have a GC content over 80%:

```
> sum(d$percent.GC >= 80)
[1] 9
```

As another example, consider looking at Pi by windows that fall in the centromere and those that do not. Because d$cent is a logical vector, we can subset with it directly (and take its complement by using the negation operator, !):

```
> summary(d$Pi[d$cent])
   Min. 1st Qu.  Median    Mean 3rd Qu.    Max.
   0.00    7.95   16.08   20.41   27.36  194.40
> summary(d$Pi[!d$cent])
   Min. 1st Qu.  Median    Mean 3rd Qu.    Max.
  0.000   5.557  10.370  12.290  16.790 265.300
```

Indeed, the centromere does appear to have higher nucleotide diversity than other regions in this data. Extracting specific observations using subsetting, and summarizing particular columns is a quick way to explore relationships within the data. Later on in "Working with the Split-Apply-Combine Pattern" on page 239, we'll learn ways to group observations and create per-group summaries.

In addition to using logical vectors to subset dataframes, it's also possible to subset rows by referring to their integer positions. The function which() takes a vector of logical values and returns the positions of all TRUE values. For example:

```
> d$Pi > 3
[1] FALSE  TRUE FALSE  TRUE  TRUE  TRUE [...]
> which(d$Pi > 3)
[1]  2  4  5  6  7 10 [...]
```

Thus, d[$Pi > 3,] is identical to d[which(d$Pi > 3),]; subsetting operations can be expressed using either method. In general, you should omit which() when subsetting dataframes and use logical vectors, as it leads to simpler and more readable code. Under other circumstances, which() is necessary—for example, if we wanted to select the four first TRUE values in a vector:

```
> which(d$Pi > 10)[1:4]
[1]  2 16 21 23
```

which() also has two related functions that return the index of the first minimum or maximum element of a vector: which.min() and which.max(). For example:

```
> d[which.min(d$total.Bases),]
        start       end total.SNPs total.Bases depth [...]
25689 25785001 25786000          0         110  1.24 [...]

> d[which.max(d$depth),]
       start      end total.SNPs total.Bases depth [...]
8718 8773001 8774000         58       21914 21.91 [...]
```

Sometimes subsetting expressions inside brackets can be quite redundant (because each column must be specified like dPi, ddepth, etc). A useful convenience function (intended primarily for interactive use) is the R function subset(). subset() takes two arguments: the dataframe to operate on, and then conditions to include a row. With subset(), d[d$Pi > 16 & d$percent.GC > 80,] can be expressed as:

```
$ subset(d, Pi > 16 & percent.GC > 80)
        start       end total.SNPs total.Bases depth [...]
58550 63097001 63098000          5         947  2.39 [...]
```

```
58641 63188001 63189000        2       1623  3.21 [...]
58642 63189001 63190000        5       1395  1.89 [...]
```

Optionally, a third argument can be supplied to specify which columns (and in what order) to include:

```
> subset(d, Pi > 16 & percent.GC > 80,
    c(start, end, Pi, percent.GC, depth))
        start      end     Pi percent.GC depth
58550 63097001 63098000 41.172    82.0821  2.39
58641 63188001 63189000 16.436    82.3824  3.21
58642 63189001 63190000 41.099    80.5806  1.89
```

Note that we (somewhat magically) don't need to quote column names. This is because subset() follows special evaluation rules, and for this reason, subset() is best used only for interactive work.

Exploring Data Visually with ggplot2 I: Scatterplots and Densities

> Instead of spending time making your graph look pretty, [ggplot2 allows you to] focus on creating a graph that best reveals the messages in your data.
>
> — *ggplot2: Elegant Graphics for Data Analysis* Hadley Wickham

Exploratory data analysis emphasizes visualization as the best tool to understand and explore our data—both to learn what the data says and what its limitations are. We'll learn visualization in R through Hadley Wickham's powerful ggplot2 package, which is just one of a few ways to plot data in R; R also has a built-in graphics system (known as base graphics) and the lattice package. Every R user should be acquainted with base graphics (at some point, you'll encounter a base graphics plot you need to modify), but we're going to skip base graphics in this chapter to focus entirely on learning visualization with ggplot2. The reason for this is simple: you'll be able to create more informative plots with less time and effort invested with ggplot2 than possible with base graphics.

As with other parts of this chapter, this discussion of ggplot2 will be short and deficient in some areas. This introduction is meant to get you on your feet so you can start exploring your data, rather than be an exhaustive reference. The best up-to-date reference for ggplot2 is the ggplot2 online documentation (*http://docs.ggplot2.org/*). As you shift from beginning ggplot2 to an intermediate user, I highly recommend the books *ggplot2: Elegant Graphics for Data Analysis* by Hadley Wickham (Springer, 2010) and *R Graphics Cookbook* by Winston Chang (O'Reilly, 2012) for more detail.

First, we need to load the ggplot2 package with R's library() function. If you don't have ggplot2 installed, you'll need to do that first with install.packages():

```
> install.packages("ggplot2")
> library(ggplot2)
```

ggplot2 is quite different from R's base graphics because it's built on top of a grammar inspired by Leland Wilkinson's *Grammar of Graphics* (Springer, 2005). This grammar provides an underlying logic to creating graphics, which considerably simplifies creating complex plots. Each ggplot2 plot is built by adding layers to a plot that map the *aesthetic properties* of *geometric objects* to data. Layers can also apply statistical transformations to data and change the scales of axes and colors. This may sound abstract now, but you'll become familiar with gplot2's grammar through examples.

Let's look at how we'd use ggplot2 to create a scatterplot of nucleotide diversity along the chromosome in the diversity column in our d dataframe. Because our data is window-based, we'll first add a column called position to our dataframe that's the midpoint between each window:

```
> d$position <- (d$end + d$start) / 2
> ggplot(d) + geom_point(aes(x=position, y=diversity))
```

This creates Figure 8-1.

There are two components of this ggplot2 graphic: the call to the function ggplot(), and the call to the function geom_point(). First, we use ggplot(d) to supply this plot with our d dataframe. ggplot2 works exclusively with dataframes, so you'll need to get your data tidy and into a dataframe before visualizing it with ggplot2.

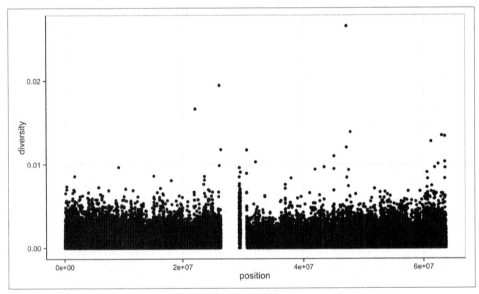

Figure 8-1. ggplot2 scatterplot nucleotide diversity by position for human chromosome 20

Second, with our data specified, we then add layers to our plot (remember: `ggplot2` is layer-based). To add a layer, we use the same + operator that we use for addition in R. Each layer updates our plot by adding geometric objects such as the points in a scatterplot, or the lines in a line plot.

We add `geom_point()` as a layer because we want points to create a scatterplot. `geom_point()` is a type of geometric object (or *geom* in `ggplot2` lingo). `ggplot2` has many geoms (e.g., `geom_line()`, `geom_bar()`, `geom_density()`, `geom_boxplot()`, etc.), which we'll see throughout this chapter. Geometric objects have many aesthetic attributes (e.g., *x* and *y* positions, color, shape, size, etc.). Different geometric objects will have different aesthetic attributes (`ggplot2` documentation refers to these as *aesthetics*). The beauty of `ggplot2`s grammar is that it allows you to map geometric objects' aesthetics to columns in your dataframe. In our diversity by position scatterplot, we mapped the *x* position aesthetic to the `position` column, and the *y* position to the `diversity` column. We specify the mapping of aesthetic attributes to columns in our dataframe using the function `aes()`.

Axis Labels, Plot Titles, and Scales

As my high school mathematics teacher Mr. Williams drilled into my head, no plot is complete without proper axis labels and a title. While `ggplot2` chooses smart labels based on your column names, you might want to change this down the road. `ggplot2` makes specifying labels easy: simply use the `xlab()`, `ylab()`, and `ggtitle()` functions to specify the *x*-axis label, *y*-axis label, and plot title. For example, we could change our *x*- and *y*-axis labels when plotting the diversity data with `p + xlab("chromosome position (basepairs)") + ylab("nucleotide diversity")`. To avoid clutter in examples in this book, I'll just use the default labels.

You can also set the limits for continuous axes using the function `scale_x_continuous(limits=c(start, end))` where `start` and `end` are the start and end of the axes (and `scale_y_continuous()` for the y axis). Similarly, you can change an axis to a log10-scale using the functions `scale_x_log10()` and `scale_y_log10()`. `ggplot2` has numerous other scale options for discrete scales, other axes transforms, and color scales; see *http://docs.ggplot2.org* for more detail.

Aesthetic mappings can also be specified in the call to `ggplot()`—geoms will then use this mapping. Example 8-2 creates the exact same scatterplot as Figure 8-1.

Example 8-2. Including aes() in ggplot()

```
> ggplot(d, aes(x=position, y=diversity)) + geom_point()
```

Notice the missing diversity estimates in the middle of this plot. What's going on in this region? ggplot2's strength is that it makes answering these types of questions with exploratory data analysis techniques effortless. We simply need to map a possible confounder or explanatory variable to another aesthetic and see if this reveals any unexpected patterns. In this case, let's map the color aesthetic of our point geometric objects to the column cent, which indicates whether the window falls in the centromeric region of this chromosome (see Example 8-3).

Example 8-3. A simple diversity scatterplot with ggplot2

```
> ggplot(d) + geom_point(aes(x=position, y=diversity, color=cent))
```

As you can see from Figure 8-2, the region with missing diversity estimates is around the centromere. This is intentional; centromeric and heterochromatic regions were excluded from this study.

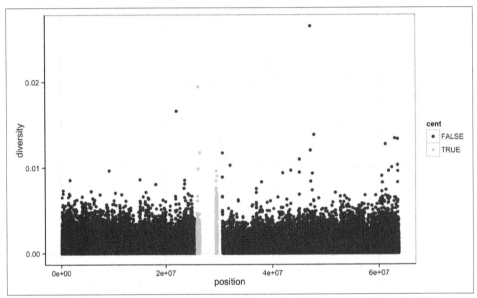

Figure 8-2. ggplot2 scatterplot nucleotide diversity by position coloring by whether windows are in the centromeric region

Throughout this chapter, I've used a slightly different ggplot theme than the default to optimize graphics for print and screen. All of the code to produce the plots exactly as they appear in this chapter is in this chapter's directory on GitHub.

A particularly nice feature of `ggplot2` is that it has well-chosen default behaviors, which allow you to quickly create or adjust plots without having to consider technical details like color palettes or a legend (though these details are customizable). For example, in mapping the `color` aesthetic to the `cent` column, `ggplot2` considered `cent`'s type (logical) when choosing a color palette. Discrete color palettes are automatically used with columns of logical or factor data mapped to the `color` aesthetic; continuous color palettes are used for numeric data. We'll see further examples of mapping data to the `color` aesthetic later on in this chapter.

As Tukey's quote at the beginning of this chapter explains, exploratory analysis is an interactive, iterative process. Our first plot gives a quick first glance at what the data say, but we need to keep exploring to learn more. One problem with this plot is the degree of *overplotting* (data oversaturating a plot so as to obscure the information of other data points). We can't get a sense of the distribution of diversity from this figure —everything is saturated from about 0.05 and below.

One way to alleviate overplotting is to make points somewhat transparent (the transparency level is known as the *alpha*). Let's make points almost entirely transparent so only regions with multiple overlapping points appear dark (this produces a plot like Figure 8-3):

```
> ggplot(d) + geom_point(aes(x=position, y=diversity), alpha=0.01)
```

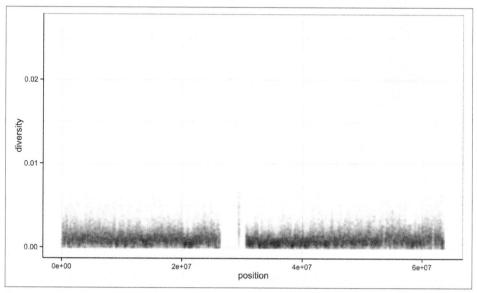

Figure 8-3. Using transparency to address overplotting

There's a subtlety here that illustrates a very important `ggplot2` concept: we set `alpha=0.01` *outside* of the aesthetic mapping function `aes()`. This is because we're not mapping the `alpha` aesthetic to a column of data in our dataframe, but rather giving it a fixed value for all data points.

Other than highlighting the lack of diversity estimates in centromeric windows, the position axes isn't revealing any positional patterns in diversity. Part of the problem is still overplotting (which occurs often when visualizing genomic data). But the more severe issue is that these windows span 63 megabases, and it's difficult to detect regional patterns with data at this scale.

Let's now look at the density of diversity across all positions. We'll use a different geometric object, `geom_density()`, which is slightly different than `geom_point()` in that it takes the data and calculates a density from it for us (see Figure 8-4):

```
> ggplot(d) + geom_density(aes(x=diversity), fill="black")
```

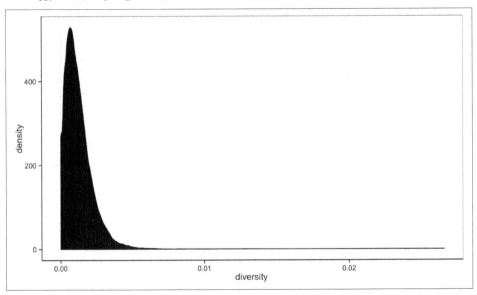

Figure 8-4. Density plot of diversity

By default, `ggplot2` uses lines to draw densities. Setting `fill="black"` fills the density so it's clearer to see (try running this same command without this `fill` argument).

We can also map the `color` aesthetic of `geom_density()` to a discrete-valued column in our dataframe, just as we did with `geom_point()` in Example 8-3. `geom_density()` will create separate density plots, grouping data by the column mapped to the `color` aesthetic and using colors to indicate the different densities. To see both overlapping densities, we use `alpha` to set the transparency to half (see Figure 8-5):

```
> ggplot(d) + geom_density(aes(x=diversity, fill=cent), alpha=0.4)
```

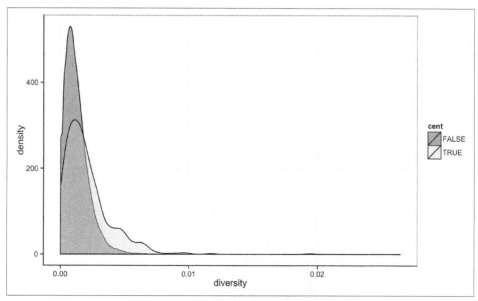

Figure 8-5. Densities of diversity, colored by whether a window is in the centromere or not

Immediately we're able to see a trend that wasn't clear by using a scatterplot: diversity is skewed to more extreme values in centromeric regions. Try plotting this same figure without mapping the color aesthetic to cent—you'll see there's no indication of bimodality. Again (because this point is worth repeating), mapping columns to additional aesthetic attributes can reveal patterns and information in the data that may not be apparent in simple plots. We'll see this again and again throughout this chapter.

Exploring Data Visually with ggplot2 II: Smoothing

Let's look at the *Dataset_S1.txt* data using another useful ggplot2 feature: smoothing. We'll use ggplot2 in particular to investigate potential confounders in genomic data. There are numerous potential confounders in genomic data (e.g., sequencing read depth; GC content; mapability, or whether a region is capable of having reads correctly align to it; batch effects; etc.). Often with large and high-dimension datasets, visualization is the easiest and best way to spot these potential issues.

In the previous section, we saw how overplotting can obscure potential relationships between two variables in a scatterplot. The number of observations in whole genome datasets almost ensures that overplotting will be a problem during visualization. Earlier, we used transparency to give us a sense of the most dense regions. Another strategy is to use ggplot2's geom_smooth() to add a smoothing line to plots and look for an unexpected trend. This geom requires x and y aesthetics, so it can fit the smoothed curve to the data. Because we often superimpose a smooth curve on a scatterplot created from the same x and y mappings, we can specify the aesthetic in ggplot() as we did in Example 8-2. Let's use a scatterplot and smoothing curve to look at the relationship between the sequencing depth (the depth column) and the total number of SNPs in a window (the total.SNPs column; see Figure 8-6):

```
> ggplot(d, aes(x=depth, y=total.SNPs)) + geom_point() + geom_smooth()
```

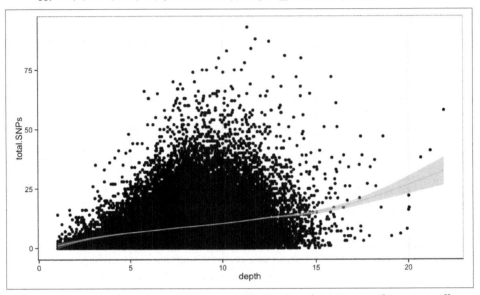

Figure 8-6. A scatterplot (demonstrating overplotting) and GAM smoothing curve illustrating how total number of SNPs in a window depends on sequencing depth

By default, ggplot2 uses generalized additive models (GAM) to fit this smoothed curve for datasets with more than 1,000 rows (which ours has). You can manually specify the smoothing method through the method argument of geom_smooth() (see help(stat_smooth) for the method options). Also, ggplot2 adds confidence intervals around the smoothing curve; this can be disabled by using geom_smooth(se=FALSE).

Visualizing the data this way reveals a well-known relationship between depth and SNPs: higher sequencing depth increases the power to detect and call SNPs, so in general more SNPs will be found in higher-depth regions. However, this isn't the entire story—the relationship among these variables is made more complex by GC content. Both higher and lower GC content regions have been shown to decrease read coverage, likely through less efficient PCR in these regions (Aird et al., 2011). We can get a sense of the effect GC content has on depth in our own data through a similar scatterplot and smoothing curve plot:

```
> ggplot(d, aes(x=percent.GC, y=depth)) + geom_point() + geom_smooth()
```

The trajectory of the smoothing curve (in Figure 8-7) indicates that GC content does indeed have an effect on sequencing depth in this data. There's less support in the data that low GC content leads to lower depth, as there are few windows that have a GC content below 25%. However, there's clearly a sharp downward trend in depth for GC contents above 60%.

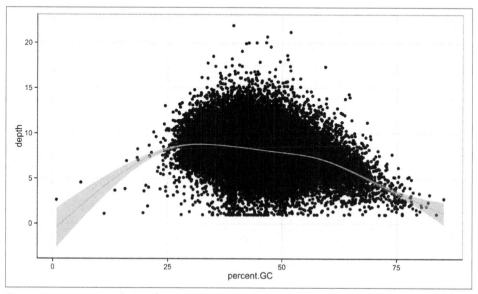

Figure 8-7. A scatterplot and GAM smoothing curve show a relationship between extreme GC content windows and sequencing depth

Binning Data with cut() and Bar Plots with ggplot2

Another way we can extract information from complex datasets is by reducing the resolution of the data through *binning* (or *discretization*). Binning takes continuous numeric values and places them into a discrete number of ranged bins. At first, it might sound counterintuitive to reduce the resolution of the data through binning to learn more about the data. The benefit is that discrete bins facilitate *conditioning* on a variable. Conditioning is an incredibly powerful way to reveal patterns in data. This idea stems from William S. Cleveland's concept of a coplot (a portmanteau of conditioning plot); I encourage you to read Cleveland's *Visualizing Data* for more information if this topic is of interest. We'll create plots like Cleveland's coplots when we come to gplot2's facets in "Using ggplot2 Facets" on page 224.

In R, we bin data through the cut() function (see Example 8-4):

Example 8-4. Using cut() to bin GC content

```
> d$GC.binned <- cut(d$percent.GC, 5)
> d$GC.binned
[1] (51.6,68.5] (34.7,51.6] (34.7,51.6] (34.7,51.6] (34.7,51.6]
[...]
Levels: (0.716,17.7] (17.7,34.7] (34.7,51.6] (51.6,68.5] (68.5,85.6]
```

When cut()'s second argument breaks is a single number, cut() divides the data into that number of equally sized bins. The returned object is a factor, which we introduced in "Factors and classes in R" on page 191. The levels of the factor returned from cut() will always have labels like (34.7,51.6], which indicate the particular bin that value falls in. We can count how many items fall into a bin using table():

```
> table(d$GC.binned)

(0.716,17.7]  (17.7,34.7]  (34.7,51.6]  (51.6,68.5]  (68.5,85.6]
          6         4976        45784         8122          252
```

We can also use prespecified ranges when binning data with cut() by setting breaks to a vector. For example, we could cut the percent GC values with breaks at 0, 25, 50, 75, and 100:

```
> cut(d$percent.GC, c(0, 25, 50, 75, 100))
[1] (50,75] (25,50] (25,50] (25,50] (25,50] (25,50]
[...]
Levels: (0,25] (25,50] (50,75] (75,100]
```

An important gotcha to remember about cut() is that if you manually specify breaks that don't fully enclose all values, values outside the range of breaks will be given the value NA. You can check if your manually specified breaks have created NA values using any(is.na(cut(x, breaks))) (for your particular vector x and breaks).

Bar plots are the natural visualization tool to use when looking at count data like the number of occurrences in bins created with cut(). ggplot2's geom_bar() can help us visualize the number of windows that fall into each GC bin we've created previously:

```
> ggplot(d) + geom_bar(aes(x=GC.binned))
```

When geom_bar()'s x aesthetic is a factor (e.g., d$binned.GC), ggplot2 will create a bar plot of counts (see Figure 8-8, left). When geom_bar()'s x aesthetic is mapped to a continuous column (e.g., percent.GC) geom_bar() will automatically bin the data itself, creating a histogram (see Figure 8-8, right):

```
> ggplot(d) + geom_bar(aes(x=percent.GC))
```

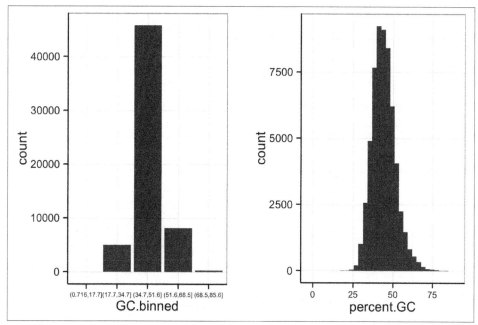

Figure 8-8. ggplot2's geom_bar() used to visualize the GC.binned column we created with cut(), and the numeric percent.GC column using its own binning scheme

The bins created from cut() are useful in grouping data (a concept we often use in data analysis). For example, we can use the GC.binned column to group data by the 10 GC content bins to see how GC content has an impact on other variables. To do this, we map aesthetics like color, fill, or linetype to our GC.binned column. Again, looking at sequencing depth and GC content:

```
> ggplot(d) + geom_density(aes(x=depth, linetype=GC.binned), alpha=0.5)
```

The same story of depth and GC content comes out in Figure 8-9: both the lowest GC content windows and the highest GC content windows have lower depth. Try to create this sample plot, except *don't* group by GC.binned—the entire story disappears. Also, try plotting the densities of other variables like Pi and Total.SNPs, grouping by GC.binned again.

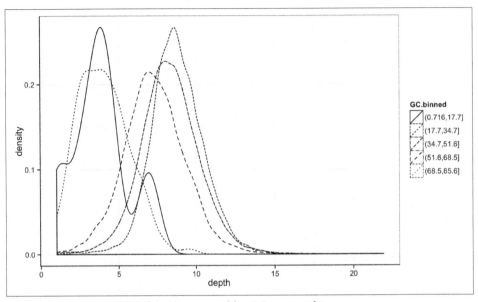

Figure 8-9. Density plots of depth, grouped by GC content bin

Finding the Right Bin Width

Notice in Figure 8-8 how different bin widths can drastically change the way we view and understand the data. Try creating a histogram of Pi with varying binwidths using: `ggplot(d)` + `geom_bar(aes(x=Pi)`, `binwidth=1)` + `scale_x_continu ous(limits=c(0.01, 80))`.

Using `scale_x_continuous()` just ignores all windows with 0 Pi and zooms into the figure. Try binwidths of 0.05, 0.5, 1, 5, and 10. Smaller bin widths can fit the data better (revealing more subtle details about the distribution), but there's a trade-off. As bin widths become narrower, each bin will contain fewer data points and consequently be more noisy (and *undersmoothed*). Using wider bins smooth over this noise. However, bins that are too wide result in *oversmoothing*, which can hide details about the data. This trade-off is a case of the more general *bias-variance trade-off* in statistics; see the Wikipedia pages on the bias–variance trade-off (*http://bit.ly/ bias-var-to*) and histograms (*http://en.wikipedia.org/wiki/Histo gram*) for more information on these topics. In your own data, be sure to explore a variety of bin widths.

We can learn an incredible amount about our data by grouping by possible confounding factors and creating simple summaries like densities by group. Because it's a powerful exploratory data analysis skill, we'll look at other ways to group and summarize data in "Working with the Split-Apply-Combine Pattern" on page 239.

Merging and Combining Data: Matching Vectors and Merging Dataframes

We're now going to switch topics to merging and combining data so we can create example data for the next sections. Bioinformatics analyses involve connecting many numerous datasets: sequencing data, genomic features (e.g., gene annotation), functional genomics data, population genetic data, and so on. As data piles up in repositories, the ability to connect different datasets together to tell a cohesive story will become an increasingly more important analysis skill. In this section, we'll look at some canonical ways to combine datasets together in R. For more advanced joins (or data approaching the limits of what R can store in memory), using a database (which we learn about in Chapter 13) may be a better fit.

The simplest operation to match two vectors is checking whether some of a vector's values are in another vector using R's %in% operator. x %in% y returns a logical vector indicating which of the values of x are in y. As a simple example:

```
> c(3, 4, -1) %in% c(1, 3, 4, 8)
[1]  TRUE  TRUE FALSE
```

We often use %in% to select rows from a dataframe by specifying the levels a factor column can take. We'll use the dataset *chrX_rmsk.txt*, the repeats on human chromosome X found by Repeat Masker, to illustrate this. Unlike *Dataset_S1.txt*, this data is on human reference genome version hg17 (because these same Repeat Masker files are used in a later example that replicates findings that also use hg17). Let's load in and look at this file:

```
> reps <- read.delim("chrX_rmsk.txt.gz", header=TRUE)
> head(reps, 3)
  bin swScore milliDiv milliDel milliIns genoName genoStart genoEnd   genoLeft
1 585     342        0        0        0     chrX         0      38 -154824226
2 585     392      109        0        0     chrX        41     105 -154824159
3 585     302      240       31       20     chrX       105     203 -154824061
  strand    repName      repClass    repFamily repStart repEnd repLeft id
1      + (CCCTAA)n Simple_repeat Simple_repeat        3     40       0  1
2      +    LTR12C           LTR         ERV1     1090   1153    -425  2
3      +    LTR30           LTR         ERV1      544    642     -80  3
```

repClass is an example of a factor column—try class(d$repClass) and levels(d $repClass) to verify for yourself. Suppose we wanted to select out rows for a few common repeat classes: DNA, LTR, LINE, SINE, and Simple_repeat. Even with just five repeat classes, it would be error prone and tedious to construct a statement to select these values using logical operators: reps$repClass == "SINE" | reps$repClass

`== "LINE" | reps$repClass == "LTR" |` and so on. Instead, we can create a vector common_repclass and use %in%:

```
> common_repclass <- c("SINE", "LINE", "LTR", "DNA", "Simple_repeat")
> reps[reps$repClass %in% common_repclass, ]
  bin swScore milliDiv milliDel milliIns genoName genoStart genoEnd   genoLeft
1 585     342        0        0        0     chrX         0      38 -154824226
2 585     392      109        0        0     chrX        41     105 -154824159
3 585     302      240       31       20     chrX       105     203 -154824061
[...]
  strand    repName      repClass      repFamily repStart repEnd repLeft id
1      + (CCCTAA)n Simple_repeat Simple_repeat        3     40       0  1
2      +     LTR12C           LTR          ERV1     1090   1153    -425  2
3      +     LTR30           LTR          ERV1      544    642     -80  3
[...]
```

It's worth noting that we can also create vectors like common_repclass programmatically. For example, we could always just directly calculate the five most common repeat classes using:

```
> sort(table(reps$repClass), decreasing=TRUE)[1:5]

         SINE          LINE           LTR           DNA Simple_repeat
        45709         30965         14854         11347          9163
```

```
> top5_repclass <- names(sort(table(reps$repClass), decreasing=TRUE)[1:5])
> top5_repclass
[1] "LINE"         "SINE"         "LTR"              "Simple_repeat"
[5] "DNA"
```

The %in% operator is a simplified version of another function, match(). x %in% y returns TRUE/FALSE for each value in x depending on whether it's in y. In contrast, match(x, y) returns the first occurrence of each of x's values in y. If match() can't find one of x's elements in y, it returns its nomatch argument (which by default has the value NA).

It's important to remember the directionality of match(x, y): the first argument x is what you're searching for (the proverbial needle) in the second argument y (the haystack). The positions returned are always the first occurrences in y (if an occurrence was found). Here's a simple example:

```
> match(c("A", "C", "E", "A"), c("A", "B", "A", "E"))
[1]  1 NA  4  1
```

Study this example carefully—to use match() safely, it's important to understand its subtle behavior. First, although there are two "A" values in the second argument, the position of the first one is returned. Second, "C" does not occur in the second argument, so match() returns NA. Lastly, the vector returned will always have the same length as the first argument and contains positions in the second argument.

Because match() returns where it finds a particular value, match()'s output can be used to join two dataframes together by a shared column. We'll see this in action by stepping through a merge of two datasets. I've intentionally chosen data that is a bit tricky to merge; the obstacles in this example are ones you're likely to encounter with real data. I'll review various guidelines you should follow when applying the lessons of this section. Our reward for merging these datasets is that we'll use the result to replicate an important finding in human recombination biology.

For this example, we'll merge two datasets to explore recombination rates around a degenerate sequence motif that occurs in repeats. This motif has been shown to be enriched in recombination hotspots (see Myers et al., 2005; Myers et al., 2008) and is common in some repeat classes. The first dataset (*motif_recombrates.txt* in the Git-Hub directory) contains estimates of the recombination rate for all windows within 40kb of each motif (for two motif variants). The second dataset (*motif_repeats.txt*) contains which repeat each motif occurs in. Our goal is to merge these two datasets so that we can look at the local effect of recombination of each motif on specific repeat backgrounds.

Creating These Example Datasets

Both of these datasets were created using the GenomicRanges tools we will learn about in Chapter 9, from tracks downloaded directly from the UCSC Genome Browser. With the appropriate tools and bioinformatics data skills, it takes surprisingly few steps to replicate part of this important scientific finding (though the original paper did much more than this—see Myers et al., 2008). For the code to reproduce the data used in this example, see the *motif-example/* directory in this chapter's directory on GitHub.

Let's start by loading in both files and peeking at them with head():

```
> mtfs <- read.delim("motif_recombrates.txt", header=TRUE)
> head(mtfs, 3)
   chr motif_start motif_end     dist recomb_start recomb_end   recom
1 chrX    35471312  35471325  39323.0     35430651   35433340  0.0015
2 chrX    35471312  35471325  36977.0     35433339   35435344  0.0015
3 chrX    35471312  35471325  34797.5     35435343   35437699  0.0015
          motif
1 CCTCCCTGACCAC
2 CCTCCCTGACCAC
3 CCTCCCTGACCAC

> rpts <- read.delim("motif_repeats.txt", header=TRUE)
> head(rpts, 3)
   chr    start      end name motif_start
1 chrX 63005829 63006173   L2    63005830
```

```
2 chrX  67746983  67747478   L2     67747232
3 chrX 118646988 118647529   L2    118647199
```

The first guideline of combining data is always to carefully consider the structure of
both datasets. In mtfs, each motif is represented across multiple rows. Each row gives
the distance between a focal motif and a window over which recombination rate was
estimated (in centiMorgans). For example, in the first three rows of mtfs, we see
recombination rate estimates across three windows, at 39,323, 36,977, and 34,797
bases away from the motif at position chrX:35471312-35471325. The dataframe rpts
contains the positions of THE1 or L2 repeats that completely overlap motifs, and the
start positions of the motifs they overlap.

Our goal is to merge the column name in the rpts dataframe into the mtfs column, so
we know which repeat each motif is contained in (if any). The link between these two
datasets are the positions of each motif, identified by the chromosome and motif start
position columns chr and motif_start. When two or more columns are used as a
link between datasets, concatenating these columns into a single *key* string column
can simplify merging. We can merge these two columns into a string using the function paste(), which takes vectors and combines them with a separating string specified by sep:

```
> mtfs$pos <- paste(mtfs$chr, mtfs$motif_start, sep="-")
> rpts$pos <- paste(rpts$chr, rpts$motif_start, sep="-")

> head(mtfs, 2) # results
    chr motif_start motif_end  dist recomb_start recomb_end  recom        motif
1 chrX    35471312  35471325 39323     35430651   35433340 0.0015 CCTCCCTGACCAC
2 chrX    35471312  35471325 36977     35433339   35435344 0.0015 CCTCCCTGACCAC
             pos
1 chrX-35471312
2 chrX-35471312
> head(rpts, 2)
   chr    start      end name motif_start          pos
1 chrX 63005829 63006173   L2    63005830 chrX-63005830
2 chrX 67746983 67747478   L2    67747232 chrX-67747232
```

Now, this pos column functions as a common key between the two datasets.

The second guideline in merging data is to validate that your keys overlap in the way
you think they do before merging. One way to do this is to use table() and %in% to
see how many motifs in mtfs have a corresponding entry in rpts:

```
> table(mtfs$pos %in% rpts$pos)

FALSE  TRUE
10832  9218
```

This means there are 9,218 motifs in mtfs with a corresponding entry in rpts and
10,832 without. Biologically speaking, this means 10,832 motifs in our dataset don't

---

overlap either the THE1 or L2 repeats. By the way I've set up this data, all repeats in rpts have a corresponding motif in mtfs, but you can see for yourself using table(rpts$pos %in% mtfs$pos). Remember: directionality matters—you don't go looking for a haystack in a needle!

Now, we use match() to find where each of the mtfs$pos keys occur in the rpts$pos. We'll create this indexing vector first before doing the merge:

```
> i <- match(mtfs$pos, rpts$pos)
```

All motif positions without a corresponding entry in rpts are NA; our number of NAs is exactly the number of mts$pos elements not in rpts$pos:

```
> table(is.na(i))

FALSE   TRUE
 9218  10832
```

Finally, using this indexing vector we can select out the appropriate elements of rpts $name and merge these into mtfs:

```
> mtfs$repeat_name <- rpts$name[i]
```

Often in practice you might skip assigning match()'s results to i and use this directly:

```
> mtfs$repeat_name <- rpts$name[match(mtfs$pos, rpts$pos)]
```

The third and final guideline of merging data: validate, validate, validate. As this example shows, merging data is tricky; it's easy to make mistakes. In our case, good external validation is easy: we can look at some rows where mtfs$repeat_name isn't NA and check with the UCSC Genome Browser that these positions do indeed overlap these repeats (you'll need to visit UCSC Genome Browser and do this yourself):

```
> head(mtfs[!is.na(mtfs$repeat_name), ], 3)
      chr motif_start motif_end    dist recomb_start recomb_end  recom
102 chrX    63005830  63005843 37772.0     62965644    62970485 1.4664
103 chrX    63005830  63005843 34673.0     62970484    62971843 0.0448
104 chrX    63005830  63005843 30084.5     62971842    62979662 0.0448
             motif           pos repeat_name
102 CCTCCCTGACCAC chrX-63005830          L2
103 CCTCCCTGACCAC chrX-63005830          L2
104 CCTCCCTGACCAC chrX-63005830          L2
```

Our result is that we've combined the rpts$name vector directly into our mtfs dataframe (technically, this type of join is called a *left outer join*). Not all motifs have entries in rpts, so some values in mfs$repeat_name are NA. We could easily remove these NAs with:

```
> mtfs_inner <- mtfs[!is.na(mtfs $repeat_name), ]
> nrow(mtfs_inner)
[1] 9218
```

In this case, only motifs in `mtfs` contained in a repeat in `rpts` are kept (technically, this type of join is called an *inner join*). Inner joins are the most common way to merge data. We'll talk more about the different types of joins in Chapter 13.

We've learned `match()` first because it's a general, extensible way to merge data in R. `match()` reveals some of the gritty details involved in merging data necessary to avoid pitfalls. However, R does have a more user-friendly merging function: `merge()`. Merge can directly merge two datasets:

```
> recm <- merge(mtfs, rpts, by.x="pos", by.y="pos")
> head(recm, 2)
            pos chr.x motif_start.x motif_end    dist recomb_start recomb_end
1 chr1-101890123  chr1     101890123 101890136 34154.0    101855215  101856736
2 chr1-101890123  chr1     101890123 101890136 35717.5    101853608  101855216
   recom         motif repeat_name chr.y     start       end  name
1 0.0700 CCTCCCTAGCCAC        THE1B  chr1 101890032 101890381 THE1B
2 0.0722 CCTCCCTAGCCAC        THE1B  chr1 101890032 101890381 THE1B
  motif_start.y
1     101890123
2     101890123
> nrow(recm)
[1] 9218
```

`merge()` takes two dataframes, x and y, and joins them by the columns supplied by `by.x` and `by.y`. If they aren't supplied, `merge()` will try to infer what these columns are, but it's much safer to supply them explicitly. By default, `merge()` behaves like our `match()` example after we removed the NA values in `repeat_name` (technically, `merge()` uses a variant of an inner join known as a natural join). But `merge()` can also perform joins similar to our first `match()` example (left outer joins), through the argument `all.x=TRUE`:

```
> recm <- merge(mtfs, rpts, by.x="pos", by.y="pos", all.x=TRUE)
```

Similarly, `merge()` can also perform joins that keep all rows of the second argument (a join known as a *right outer join*) through the argument `all.y=TRUE`. If you want to keep all rows in both datasets, you can specify `all=TRUE`. See `help(merge)` for more details on how `merge()` works. To continue our recombination motif example, we'll use `mtfs` because unlike the `recm` dataframe created by `merge()`, `mtfs` doesn't have any duplicated columns.

Using ggplot2 Facets

After merging our two datasets in the last example, we're ready to explore this data using visualization. One useful visualization technique we'll introduce in this section is *facets*. Facets allow us to visualize grouped data by creating a series of separate adjacent plots for each group. Let's first glimpse at the relationship between recombination rate and distance to a motif using the `mtfs` dataframe created in the previous section. We'll construct this graphic in steps:

```
> p <- ggplot(mtfs, aes(x=dist, y=recom)) + geom_point(size=1)

> p <- p + geom_smooth(method="loess", se=FALSE, span=1/10)
> print(p)
```

This creates Figure 8-10. Note that I've turned off `geom_smooth()`'s standard error estimates, adjusted the smoothing with `span`, and set the smoothing method to `"loess"`. Try playing with these settings yourself to become familiar with how each changes the visualization and what we learn from the data. From this data, we only see a slight bump in the smoothing curve where the motifs reside. However, this data is a convolution of two different motif sequences on many different genomic backgrounds. In other words, there's a large amount of heterogeneity we're not accounting for, and this could be washing out our signal. Let's use faceting to pick apart this data.

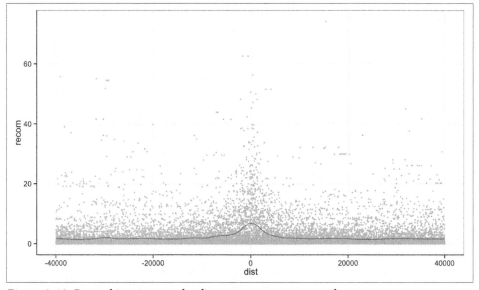

Figure 8-10. Recombination rate by distance to sequence motif

First, if you've explored the `mtfs` dataframe, you'll notice that the `mtfs$motif` column contains two variations of the sequence motif:

```
> unique(mtfs$motif)
[1] CCTCCCTGACCAC CCTCCCTAGCCAC
Levels: CCTCCCTAGCCAC CCTCCCTGACCAC
```

We might wonder if these motifs have any noticeably different effects on local recombination. One way to compare these is by grouping and coloring the loess curves by motif sequence. I've omitted this plot to save space, but try this on your own:

```
> ggplot(mtfs, aes(x=dist, y=recom)) + geom_point(size=1) +
    geom_smooth(aes(color=motif), method="loess", se=FALSE, span=1/10)
```

Alternatively, we can split these motifs apart visually with facets using ggplot2's facet_wrap() (shown in Figure 8-11):

```
> p <- ggplot(mtfs, aes(x=dist, y=recom)) + geom_point(size=1, color="grey")
> p <- p + geom_smooth(method='loess', se=FALSE, span=1/10)
> p <- p + facet_wrap(~ motif)
> print(p)
```

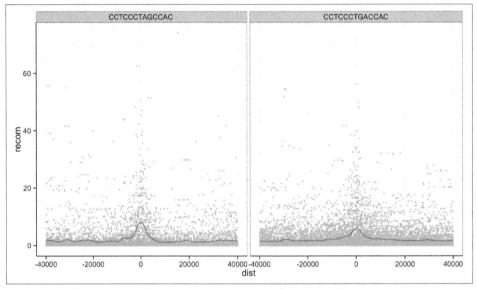

Figure 8-11. Faceting plots by motif sequence using facet_wrap()

ggplot2 has two facet methods: facet_wrap() and facet_grid(). facet_wrap() (used earlier) takes a factor column and creates a panel for each level and wraps around horizontally. facet_grid() allows finer control of facets by allowing you to specify the columns to use for vertical and horizontal facets. For example:

```
> p <- ggplot(mtfs, aes(x=dist, y=recom)) + geom_point(size=1, color="grey")
> p <- p + geom_smooth(method='loess', se=FALSE, span=1/16)
> p <- p + facet_grid(repeat_name ~ motif)
> print(p)
```

Figure 8-12 shows some of the same patterns seen in Figure 1 of Myers et al., 2008. Motif CCTCCCTAGCCAC on a THE1B repeat background has a strong effect, as does CCTCCCTGACCAC on a L2 repeat background. You can get a sense of the data that goes into this plot with table(mtfs$repeat_name, mtfs$motif, useNA="ifany").

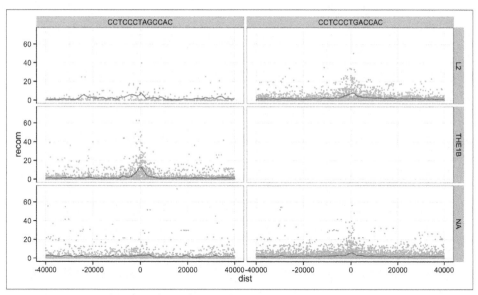

Figure 8-12. Using facet_grid() to facet by both repeat background and motif sequence; the empty panel indicates there is no data for this particular motif/repeat combination

The tilde (~) used with `facet_wrap()` and `facet_grid()` is how we specify model formula in R. If you've used R to fit linear models, you've encountered ~ before. We can ignore the specifics when using it in `facet_wrap()` (but see `help(formula)` if you're curious).

One important feature of `facet_wrap()` and `facet_grid()` is that by default, x- and y-scales will be the same across all panels. This is a good default because people have a natural tendency to compare adjacent graphics as if they're on the same scale. However, forcing facets to have fixed scales can obscure patterns that occur on different scales. Both `facet_grid()` and `facet_wrap()` have `scales` arguments that by default are `"fixed"`. You can set scales to be free with `scales="free_x"` and `scales="free_y"` (to free the x- and y-scales, respectively), or `scales="free"` to free both axes. For example (see Figure 8-13):

```
> p <- ggplot(mtfs, aes(x=dist, y=recom)) + geom_point(size=1, color="grey")
> p <- p + geom_smooth(method='loess', se=FALSE, span=1/10)
> p <- p + facet_wrap( ~ motif, scales="free_y")
> print(p)
```

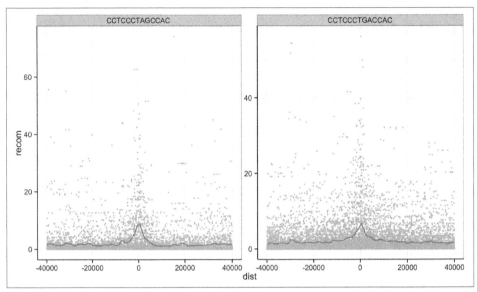

Figure 8-13. Recombination rates around two sequence motifs, with the y-scales free

Try using facets to look at this data when grouped by chromosome with `facet_wrap( ~ chr)`.

More R Data Structures: Lists

Thus far, we've used two R data structures for everything: vectors and dataframes. In this section, we'll learn about another R data structure as important as these two: the *list*. Recall the following points about R's vectors:

- R vectors require all elements to have the same data type (that is, vectors are *homogenous*)
- They only support the six data types discussed earlier (integer, double, character, logical, complex, and raw)

In contrast, R's lists are more versatile:

- Lists can contain elements of different types (they are *heterogeneous*)
- Elements can be *any* object in R (vectors with different types, other lists, environments, dataframes, matrices, functions, etc.)
- Because lists can store other lists, they allow for storing data in a recursive way (in contrast, vectors cannot contain other vectors)

The versatility of lists make them indispensable in programming and data analysis with R. You've actually been using lists already without knowing it—dataframes are built using R's lists. This makes sense, because the columns of a dataframe are vectors and each column can have a different type. The R data structure used to store heterogeneous elements is the list. See for yourself that dataframes are truly lists; try `is.list(mtfs)`.

We create lists with the `list()` function. Like creating dataframes with the `data.frame()` function or combining vectors with `c()`, `list()` will interpret named arguments as the names of the elements. For example, suppose we wanted to store a specific genomic position using a list:

```
> adh <- list(chr="2L", start=14615555L, end=14618902L, name="Adh")
> adh
$chr
[1] "2L"
$start
[1] 14615555
$end
[1] 14618902
$name
[1] "Adh"
```

Had we tried to store these three values in a vector, vector coercion would coerce them into a character vector. Lists allow heterogeneous typed elements, so the character vectors "chr2L" and "Adh", and integer vectors 14,615,555 and 14,618,902 can exist in the same list without being coerced.

As with R's vectors, we can extract subsets of a list or change values of specific elements using indexing. However, accessing elements from an R list is slightly different than with vectors. Because R's lists can contain objects with different types, a subset containing multiple list elements could contain objects with different types. Consequently, the only way to return a subset of more than one list element is with another list. As a result, there are two indexing operators for lists: one for accessing a subset of multiple elements as a list (the single bracket; e.g., `adh[1:2]`) and one for accessing an element within a list (the double bracket; e.g., `adh[[3]]`).

For example, if we were to access the first two elements of the list x we created before, we would use single bracket indexing:

```
> adh[1:2]
$chr
[1] "2L"
$start
[1] 14615555
```

Peeking into R's Structures with str()

Because R's lists can be nested and can contain any type of data, list-based data structures can grow to be quite complex. In some cases, it can be difficult to understand the overall structure of some lists. The function `str()` is a convenient R function for inspecting complex data structures. `str()` prints a line for each contained data structure, complete with its type, length (or dimensions), and the first few elements it contains. We can see an example by creating an artificially complex nested list (for the sake of this example) and looking at its structure:

```
> z <- list(a=list(rn1=rnorm(20), rn2=rnorm(20)), b=rnorm(10))
> str(z)
List of 2
 $ a:List of 2
  ..$ rn1: num [1:20] -2.8126 1.0328 -0.6777 0.0821 0.7532 ...
  ..$ rn2: num [1:20] 1.09 1.27 1.31 2.03 -1.05 ...
 $ b: num [1:10] 0.571 0.929 1.494 1.123 1.713 ...
```

For deeply nested lists, you can simplify `str()`'s output by specifying the maximum depth of nested structured to return with `str()`'s second argument, `max.level`. By default, `max.level` is NA, which returns all nested structures.

Note that this subset of the original list is returned *as a list*. We can verify this using the function `is.list()`. Additionally, note that single brackets return a list even if a single element is selected (for consistency):

```
> is.list(adh[1:2])
[1] TRUE
> is.list(adh[1])
[1] TRUE
```

Because the single bracket operator always returns subjects of a list as a list, R has the double bracket operator to extract an object from a list position (either by position or name):

```
> adh[[2]]
[1] 14615555
> adh[['start']]
[1] 14615555
```

Unlike the single bracket operator, the double bracket operator will return the value from inside a list position (e.g., not as a list). Because accessing list elements by name is quite common, R has a syntactic shortcut:

```
> adh$chr
[1] "2L"
```

You should be familiar with this syntax already—we used the same syntax to extract columns from a dataframe. This isn't a coincidence: dataframes are built from lists, and each dataframe column is a vector stored as a list element.

We can create new elements or change existing elements in a list using the familiar `<-`. Assigning a list element the value NULL removes it from the list. Some examples are shown here:

```
> adh$id <- "FBgn0000055"
> adh$chr <- "chr2L"
> adh
$chr
[1] "chr2L"
$start
[1] 14615555
$end
[1] 14618902
$name
[1] "Adh"
$id
[1] "FBgn0000055"

> adh$id <- NULL # remove the FlyBase ID
> adh
$chr
[1] "chr2L"
$start
[1] 14615555
$end
[1] 14618902
$name
[1] "Adh"
```

Similar to vectors, list names can be accessed with names() or changed using
names(x) <-. We'll use lists extensively in the next few sections of this book, so it's
important that you're familiar with these basic operations.

Writing and Applying Functions to Lists with lapply() and sapply()

Understanding R's data structures and how subsetting works are fundamental to having the freedom in R to explore data any way you like. In this section, we'll cover
another cornerstone of R: how to write and apply functions to data. This approach of
applying functions to data rather than writing explicit loops follows from a
functional-programming style that R inherited from one of its language influences,
Scheme. Specifically, our focus will be on applying functions to R's lists using
lapply() and sapply(), but the same ideas extend to other R data structures through
similar "apply" functions. Solving common data analysis problems with apply functions is tricky at first, but mastering it will serve you well in R.

Using lapply()

Let's work through a simple example on artificial data first (we'll see how to apply
these concepts to real data in the next section). Suppose you have a list of numeric
values (here, generated at random with rnorm()):

```
> ll <- list(a=rnorm(6, mean=1), b=rnorm(6, mean=4), c=rnorm(6, mean=6))
> ll
```

```
$a
[1]  2.2629543  0.6737666  2.3297993  2.2724293  1.4146414 -0.5399500
$b
[1] 3.071433 3.705280 3.994233 6.404653 4.763593 3.200991
$c
[1] 4.852343 5.710538 5.700785 5.588489 6.252223 5.108079
```

How might we calculate the mean of each vector stored in this list? If you're familiar with for loops in other languages, you may approach this problem using R's for loops (a topic we save for "Control Flow: if, for, and while" on page 253):

```
# create an empty numeric vector for the means
ll_means <- numeric(length(ll))

# loop over each list element and calculate mean
for (i in seq_along(ll)) {
    ll_means[i] <- mean(ll[[i]])
}
```

However, this is not idiomatic R; a better approach is to use an apply function that applies another function to each list element. For example, to calculate the mean of each list element, we'd want to apply the function mean() to each element. To do so, we can use the function lapply() (the *l* is for list, as lapply() returns the result as a list):

```
> lapply(ll, mean)
$a
[1] 0.5103648
$b
[1] 0.09681026
$c
[1] -0.2847329
```

lapply() has several advantages: it creates the output list for us, uses fewer lines of code, leads to clearer code, and is in some cases faster than using a for loop. While using lapply() rather than loops admittedly takes time getting used to, it's worth the effort.

lapply() in Parallel

A great feature of using the `lapply()` approach is that parallelizing `lapply()`s is simple. R's `parallel` package has a parallelized drop-in version of `lapply()` called `mclapply()` ("mc" stands for multi-core). We can use `mclapply()` just as we would use `lapply()`:

```
> library(parallel)
> results <- mclapply(my_samples, slowFunction)
```

This would run the function `slowFunction` on each of the elements of `my_samples` in parallel. By default, `mclapply()` will use as many cores as are set in your options (or two cores if this option is not set). You can explicitly set the number of cores to use by setting this option:

```
> options(cores=3)
> getOption('cores')
[1] 3
```

Even though for some tasks parallelization is absolutely indispensable, it's not a substitute for writing idiomatic, efficient R code. Often, efficient R code will lead to sizable performance gains without requiring parallelization.

A few remarks: first, don't call the function you pass to `lapply()`—for example, don't do `lapply(d_split, mean(x))` or `lapply(d_split, mean())`. Remember, you're *passing* `lapply()` the function you want it to apply to each list element. It's `lapply()`'s job to call the `mean()` function. Note that `lapply()` calls the supplied function using each list element as the function's *first argument.*

Second, in some cases you will need to specify additional arguments to the function you're passing to `lapply()`. For example, suppose an element of `ll` was a vector that contained an NA. In this case, `mean()` would return NA unless we ignore NA values by calling `mean()` with the argument `na.rm=TRUE`. To supply this argument to `mean()`, we could use:

```
> ll$a[3] <- NA    # replace an element with a missing value
> lapply(ll, mean)
$a
[1] NA
$b
[1] 4.19003
$c
[1] 5.53541

> lapply(ll, mean, na.rm=TRUE)
$a
[1] 1.216768
$b
[1] 4.19003
```

```
$c
[1] 5.53541
```

In general, it's a good idea to pass arguments by name—both to help readers understand what's going on and to prevent issues in argument matching.

Writing functions

There's another way to specify additional arguments in functions that illustrates the flexibility of R's functions: write a function that *wraps* another function. R makes writing functions very easy, both because you should be writing lots of them to organize your code and applying functions is such a common operation in R. For example, we could write a simple version of R's mean() function with na.rm=TRUE:

```
> meanRemoveNA <- function(x) mean(x, na.rm=TRUE)
> lapply(ll, meanRemoveNA)
$a
[1] 1.216768
$b
[1] 4.19003
$c
[1] 5.53541
```

The syntax for meanRemoveNA() is a common shortened version of the general syntax for R functions:

```
fun_name <- function(args) {
    # body, containing R expressions
    return(value)
}
```

Function definitions consist of *arguments*, a *body*, and a *return value*. Functions that contain only one line in their body can omit the braces (as the meanRemoveNA() function does). Similarly, using return() to specify the return value is optional; R's functions will automatically return the last evaluated expression in the body. These syntactic shortcuts are commonly used in R, as we often need to quickly write functions to apply to data.

Alternatively, we could forgo creating a function named meanRemoveNA() in our global environment altogether and instead use an *anonymous function* (named so because anonymous functions are functions without a name). Anonymous functions are useful when we only need a function once for a specific task. For example, instead of writing meanRemoveNA(), we could use:

```
> lapply(ll, function(x) mean(x, na.rm=TRUE))
$a
[1] 1.216768
$b
[1] 4.19003
```

```
$c
[1] 5.53541
```

In other cases, we might need to create polished functions that we'll use repeatedly throughout our code. In theses cases, it pays off to carefully document your functions and add some extra features. For example, the following is a version of meanRemo veNA() that will warn the user when it encounters and automatically removes missing values. This more verbose behavior can be disabled by setting the argument warn=FALSE. Note how in constructing this meanRemoveNAVerbose() function, we specify that the argument warn is TRUE by default in the arguments:

```
meanRemoveNAVerbose <- function(x, warn=TRUE) {
  # A function that removes missing values when calculating the mean
  # and warns us about it.
    if (any(is.na(x)) && warn) {
      warning("removing some missing values!")
    }
  mean(x, na.rm=TRUE)
}
```

Don't try to type out long function definitions like meanRemoveNAVerbose() directly in the R interpreter: functions over one line should be kept in a file and sent to the interpreter. RStudio conveniently allows you to send a whole function definition at once with Command-Option-*f* on a Mac and Control-Alt-*f* on Windows.

Function Scope

One of the benefits of using functions in your code is that they organize code into separate units. One way functions separate code is through *scoping*, which is how R's function finds the value of variables. The limited scope of R's functions prevents mistakes due to name collisions; for example:

```
> x <- 3
> fun <- function(y) {
    x <- 2.3
    x + y
  }
> fun(x)
[1] 5.3
> x
[1] 3
```

Note that although we've assigned x a new value in our fun() function, this does not affect the value of x defined earlier in our global environment. The technical name for R scoping rules is *lexical scoping*. Fully understanding how lexical scoping works is outside of the scope of this introductory chapter, but there are many good resource on the subject. Hadley Wickham's terrific book *Advanced R* (*http://bit.ly/lex-scoping*) is a good place to start.

There are several benefits to using functions in code. Functions can easily turn complex code into neatly contained, reusable, easy-to-debug blocks of code. Functions also make code easier to read; it's much easier to figure out what a function named normalizeRNASeqCounts() does than looking at its code (though you should still document your functions). In general, if you find yourself copying and pasting code, this is almost surely a sign you're repeating code that should be turned into a function.

Digression: Debugging R Code

Functions are great because they help organize code into reusable containers. But this can make debugging code more difficult, as without the right tools it can be hard to poke around in misbehaving functions. Fortunately, R has numerous debugging tools. Often it's hard to convince new programmers to take the time to play around and learn these debugging tools. If you're doubtful about taking the time to learn R's debugging tools, I can promise you: you will have bugs in your code, and debugging tools help you find and fix these frustrating bugs faster. With that advice, let's take a quick look.

One of the best ways to debug a function is to pause execution at a *breakpoint* in code and poke around. With execution paused, you can step through code line by line, look at variables' values, and inspect what functions called other functions (known as the *call stack*). The function browser() allows us to do this—let's place a call to browser() in a fake function foo():

```
foo <- function(x) {
  a <- 2
  browser()
  y <- x + a
  return(y)
}
```

Load this function into R, and then run:

```
> foo(1)
Called from: foo(1)
Browse[1]>
```

We use one-letter commands to control stepping through code with browser(). The mostly frequently used are n (execute the next line), c (continue running the code), and Q (exit without continuing to run code). You can see help(browser) for other commands. Within browser(), we can view variables' values:

```
Browse[1]> ls() # list all variables in local scope
[1] "a" "x"
Browse[1]> a
[1] 2
```

If we step to the two next lines, our function assigns a value to y, which we can inspect:

```
Browse[1]> n
debug at #4: y <- x + a
Browse[2]> n
debug at #5: return(y)
Browse[2]> y
[1] 3
```

Then, we can continue with c, and foo(1) runs normally and returns 3:

```
Browse[2]> c
[1] 3
```

Another useful debugging trick is to set options(error=recover). Setting this option will drop you into an interactive debugging session anytime an error is encountered. For example, if you were to have a buggy function bar():

```
> bar <- function(x) x + "1"
> bar(2)
Error in x + "1" : non-numeric argument to binary operator
```

There's not much we can do here. But setting options(error=recover) allows us to select which function (only one in this simple case) we'd like to enter to inspect:

```
> options(error=recover)
> bar(2)
Error in x + "1" : non-numeric argument to binary operator

Enter a frame number, or 0 to exit

1: bar(2)

Selection: 1
Selection: 1
Called from: top level
Browse[1]> # now at browser() prompt
```

To turn this off, enter options(error=NULL).

We've only scratched the surface of R's debugging capabilities. See the help pages for browser(), debug(), traceback(), and recover() for more detail.

More list apply functions: sapply() and mapply()

In the same family as lapply() are the functions sapply() and mapply(). The sap ply() function is similar to lapply(), except that it simplifies the results into a vector, array, or matrix (see "Other Apply Functions for Other R Data Structures" on page 238 for a quick note about these other data structures). For example, if we were to replace our earlier lapply() call with sapply():

```
> sapply(ll, function(x) mean(x, na.rm=TRUE))
       a        b        c
1.216768 4.190030 5.535410
```

sapply() can simplify more complex data structures than this simple list, but occasionally sapply() simplifies something in a strange way, leading to more headaches than it's worth. In "Working with the Split-Apply-Combine Pattern" on page 239, we'll see other ways to combine data from lists into more interpretable structures.

The last apply function we'll discuss is mapply(). mapply() is a multivariate version of sapply(): the function you pass to mapply() can take in and use multiple arguments. Suppose you had two lists of genotypes and you wanted to see how many alleles are shared by calling intersect() pairwise on both lists:

```
> ind_1 <- list(loci_1=c("T", "T"), loci_2=c("T", "G"), loci_3=c("C", "G"))
> ind_2 <- list(loci_1=c("A", "A"), loci_2=c("G", "G"), loci_3=c("C", "G"))
> mapply(function(a, b) length(intersect(a, b)), ind_1, ind_2)
loci_1 loci_2 loci_3
     0      1      2
```

Unlike lapply() and sapply(), mapply()'s first argument is the function you want to apply. mapply() then takes as many vectors as there are needed arguments in the function applied to the data. Here, each loci in these two lists is processed pairwise using an anonymous function across the two lists ind_1 and ind_2. intersect() is one of R's set functions (see help(intersect) for some useful others).

Like sapply(), mapply() tries to simplify the result as much as possible. Unfortunately, sometimes this will wreak havoc on your resulting data. To prevent this, specify SIMPLIFY=FALSE. Also, mapply(fun, x, y, SIMPLIFY=FALSE) is equivalent to using the function Map() like Map(fun, x, y), which saves some typing.

Other Apply Functions for Other R Data Structures

There are two other R data structures we won't cover in this chapter: *arrays* and *matrices*. Both are simply R vectors with *dimensions*. Arrays can be any number of dimensions; matrices are arrays with two dimensions (like the matrices from linear algebra). Because arrays and matrices are simply vectors, they follow the same coercion rules and are of homogeneous type. We use dataframes rather than matrices because most data we encounter has a mix of different column types. However, if you need to implement lower-level statistical or mathematical functionality, you'll likely need to work with R's arrays or matrices. While these topics are out of the scope of this introductory chapter, it's worth mentioning that these data structures have their own useful apply functions—for example, see apply() and sweep().

Working with the Split-Apply-Combine Pattern

Grouping data is a powerful method in exploratory data analysis. With data grouped by a factor, per-group summaries can reveal interesting patterns we may have missed by exploring ungrouped data. We've already used grouping implicitly with ggplot2, through coloring or faceting plots by a factor column (such as motif sequence, repeat, or the GC bins we created with cut()). For example, in Figure 8-9, binning GC content and plotting depth densities per GC bin revealed that both low GC contents and high GC content windows have lower sequencing depth. In this section, we'll learn a common data analysis pattern used to group data, apply a function to each group, and then combine the results. This pattern is *split-apply-combine*, a widely used strategy in data analysis (see Hadley Wickham's paper "The Split-Apply-Combine Strategy for Data Analysis" (*http://www.jstatsoft.org/v40/i01/paper*) for a nice detailed introduction). At first, we'll use standard R functions to implement the split-apply-combine pattern, but later in "Exploring Dataframes with dplyr" on page 243 we'll apply this same strategy using the dplyr package.

Let's get started with a simple example of split-apply-combine: finding the mean depth for the three GC bins we created in Example 8-4 for the d dataframe. This will give us some numeric summaries of the pattern we saw in Figure 8-9.

The first step is to *split* our data. Splitting data combines observations into groups based on the levels of the grouping factor. We split a dataframe or vector using split(x, f), where x is a dataframe/vector and f is a factor. In this example, we'll split the d$depth column into a list based on the factor column d$GC.binned:

```
> d_split <- split(d$depth, d$GC.binned)
> str(d_split)
List of 5
 $ (0.716,17.7]: num [1:6] 4.57 1.12 6.95 2.66 3.69 3.87
 $ (17.7,34.7] : num [1:4976] 8 8.38 9.02 10.31 12.09 ...
 $ (34.7,51.6] : num [1:45784] 6.68 9.06 10.26 8.06 7.05 ...
 $ (51.6,68.5] : num [1:8122] 3.41 7 6.63 7.15 6.97 4.77 5.18 ...
 $ (68.5,85.6] : num [1:252] 8.04 1.96 3.71 1.97 4.82 4.22 3.76 ...
```

Be sure to understand what's happened here: split() returns a list with each element containing all observations for a particular level of the factor used in grouping. The elements in the list returned from split() correspond to the levels of the grouping factor d$GC.binned. You can verify that the list returned by split() contains elements corresponding to the levels of d$GC.binned using names(d_split), levels(d$GC.binned), length(d_split), and nlevels(d$GC.binned).

With our data split into groups, we can then *apply* a function to each group using the lapply() function we learned earlier. Continuing our example, let's find the mean depth of each GC bin by applying the function mean() to d_split:

```
> lapply(d_split, mean)
$`(0.716,17.7]`
[1] 3.81
$`(17.7,34.7]`
[1] 8.788244
$`(34.7,51.6]`
[1] 8.296699
$`(51.6,68.5]`
[1] 7.309941
$`(68.5,85.6]`
[1] 4.037698
```

Finally, the last step is to *combine* this data together somehow (because it's currently split). In this case, the data in this list is already understandable without combining it back together (though this won't always be the case). But we can simplify our split-apply results by converting it to a vector. One way to do this is to call `unlist()`:

```
> unlist(lapply(d_split, mean))
(0.716,17.7]  (17.7,34.7]  (34.7,51.6]  (51.6,68.5]  (68.5,85.6]
   3.810000     8.788244     8.296699     7.309941     4.037698
```

`unlist()` returns a vector with the highest type it can (following R's coercion rules; see `help(unlist)` for more information). Equivalently, we could just replace our call to `lapply()` with `sapply()`:

```
> sapply(d_split, mean)
(0.716,17.7]  (17.7,34.7]  (34.7,51.6]  (51.6,68.5]  (68.5,85.6]
   3.810000     8.788244     8.296699     7.309941     4.037698
```

Now, let's look at an example that involves a slightly trickier combine step: applying the `summary()` function to each group. We'll run both the split and apply steps in one expression:

```
> dpth_summ <- lapply(split(d$depth, d$GC.binned), summary)
> dpth_summ
$`(0.716,17.7]`
   Min. 1st Qu.  Median    Mean 3rd Qu.    Max.
  1.120   2.918   3.780   3.810   4.395   6.950
$`(17.7,34.7]`
   Min. 1st Qu.  Median    Mean 3rd Qu.    Max.
  1.000   7.740   8.715   8.788   9.800  17.780

 [...]
```

`dpth_summ` is a list of depth summary tables for each GC bin. The routine way to combine a list of vectors is by binding each element together into a matrix or data-frame using either `cbind()` (column bind) or `rbind()` (row bind). For example:

```
> rbind(dpth_summ[[1]], dpth_summ[[2]])
      Min. 1st Qu. Median  Mean 3rd Qu.   Max.
[1,] 1.12   2.918  3.780 3.810   4.395   6.95
[2,] 1.00   7.740  8.715 8.788   9.800  17.78
```

```
> cbind(dpth_summ[[1]], dpth_summ[[2]])
           [,1]    [,2]
Min.      1.120   1.000
1st Qu.   2.918   7.740
Median    3.780   8.715
Mean      3.810   8.788
3rd Qu.   4.395   9.800
Max.      6.950  17.780
```

However, this approach won't scale if we needed to bind together many list elements. No one wants to type out rbind(x[[1]], x[[2]], x[[3]], … for a thousand list entries. Fortunately, R's do.call() function takes a function and a list as arguments, and calls the function using the list as the function's arguments (see the following tip). We can use do.call() with rbind() to merge the list our split-apply steps produces into a matrix:

```
> do.call(rbind, lapply(split(d$depth, d$GC.binned), summary))
              Min. 1st Qu. Median  Mean 3rd Qu.   Max.
(0.716,17.7] 1.12   2.918  3.780 3.810   4.395   6.95
(17.7,34.7]  1.00   7.740  8.715 8.788   9.800  17.78
(34.7,51.6]  1.00   7.100  8.260 8.297   9.470  21.91
(51.6,68.5]  1.00   6.030  7.250 7.310   8.540  21.18
(68.5,85.6]  1.00   2.730  3.960 4.038   5.152   9.71
```

Combining this data such that the quantiles and means are columns is the natural way to represent it. Replacing rbind with cbind in do.call() would swap the rows and columns.

There are a few other useful tricks to know about the split-apply-combine pattern built from split(), lapply(), and do.call() with rbind() that we don't have the space to cover in detail here, but are worth mentioning. First, it's possible to group by more than one factor—just provide split() with a list of factors. split() will split the data by all combinations of these factors. Second, you can unsplit a list back into its original vectors using the function unsplit(). unsplit() takes a list and the same factor (or list of factors) used as the second argument of split() to reconstruct the new list back into its original form (see help(split) for more information). Third, although we split single columns of a dataframe (which are just vectors), split() will happily split dataframes. Splitting entire dataframes is necessary when your apply step requires more than one column. For example, if you wanted to fit separate linear models for each set of observations in a group, you could write a function that takes each dataframe passed lapply() and fits a model using its column with lm().

Understanding do.call()

If do.call() seems confusing, this is because it is at first. But understanding do.call() will provide you with an essential tool in problem solving in R. They key point about do.call() is that it *constructs and executes* a function call. Function calls have two parts: the name of the function you're calling, and the arguments supplied to the function. For example, in the function call func(arg1, arg2, arg3), func is the name of the function and arg1, arg2, arg3 are the arguments. All do.call() does is allow you to construct and call a function using the function name and *a list* of arguments. For example, calling func(arg1, arg2, arg3) is the same as do.call(func, list(arg1, arg2, arg3)). If the list passed to do.call() has named elements, do.call() will match these named elements with named arguments in the function call. For example, one could build a call to rnorm() using:

```
> do.call(rnorm, list(n=4, mean=3.3, sd=4))
[1] 8.351817 1.995067 8.619197 8.389717
```

do.call() may seem like a complex way to construct and execute a function call, but it's the most sensible way to handle situations where the arguments we want to pass to a function are already in a list. This usually occurs during data processing when we need to combine a list into a single data structure by using functions like cbind() or rbind() that take any number of arguments (e.g., their first argument is . . .).

Lastly, R has some convenience functions that wrap the split(), lapply(), and combine steps. For example, the functions tapply() and aggregate() can be used to create per-group summaries too:

```
> tapply(d$depth, d$GC.binned, mean)
 (0.716,17.7]   (17.7,34.7]   (34.7,51.6]   (51.6,68.5]   (68.5,85.6]
    3.810000      8.788244      8.296699      7.309941      4.037698

> aggregate(d$depth, list(gc=d$GC.binned), mean)
            gc        x
1 (0.716,17.7] 3.810000
2  (17.7,34.7] 8.788244
3  (34.7,51.6] 8.296699
4  (51.6,68.5] 7.309941
5  (68.5,85.6] 4.037698
```

Both tapply() and aggregate() have the same split-apply-combine pattern at their core, but vary slightly in the way they present their output. If you're interested in similar functions in R, see the help pages for aggregate(), tapply(), and by().

You may be wondering why we slogged through all of the split(), lapply(), and do.call() material given how much simpler it is to call tapply() or aggregate(). The answer is twofold. First, R's base functions like split(), lapply(), and do.call() give you some raw power and flexibility in how you use the split-apply-combine pattern. In working with genomics datasets (and with Bioconductor packages), we often need this flexibility. Second, Hadley Wickham's package dplyr (which we see in the next section) is both simpler and more powerful than R's built-in split-apply-combine functions like tapply() and aggregate().

The take-home point of this section: the split-apply-combine pattern is an essential part of data analysis. As Hadley Wickham's article points out, this strategy is similar to Google's map-reduce framework and SQL's GROUP BY and AGGREGATE functions (which we cover in "SQLite Aggregate Functions" on page 442).

Exploring Dataframes with dplyr

Every data analysis you conduct will likely involve manipulating dataframes at some point. Quickly extracting, transforming, and summarizing information from dataframes is an essential R skill. The split-apply-combine pattern implemented from R base functions like split() and lapply() is versatile, but not always the fastest or simplest approach. R's split-apply-combine convenience functions like tapply() and aggregate() simplify the split-apply-combine, but their output still often requires some cleanup before the next analysis step. This is where Hadley Wickham's dplyr package comes in: dplyr consolidates and simplifies many of the common operations we perform on dataframes. Also, dplyr is very fast; much of its key functionality is written in C++ for speed.

dplyr has five basic functions for manipulating dataframes: arrange(), filter(), mutate(), select(), and summarize(). None of these functions perform tasks you can't accomplish with R's base functions. But dplyr's advantage is in the added consistency, speed, and versatility of its data manipulation interface. dplyr's design drastically simplifies routine data manipulation and analysis tasks, allowing you to more easily and effectively explore your data.

Because it's common to work with dataframes with more rows and columns than fit in your screen, dplyr uses a simple class called tbl_df that wraps dataframes so that they don't fill your screen when you print them (similar to using head()). Let's convert our d dataframe into a tbl_df object with the tbl_df() function:

```
> install.packages("dplyr") # install dplyr if it's not already installed
> library(dplyr)
> d_df <- tbl_df(d)
> d_df
Source: local data frame [59,140 x 20]
```

```
    start    end total.SNPs total.Bases depth unique.SNPs dhSNPs reference.Bases
1 55001 56000          0        1894  3.41           0      0            556
2 56001 57000          5        6683  6.68           2      2           1000
3 57001 58000          1        9063  9.06           1      0           1000
[...]
Variables not shown: Theta (dbl), Pi (dbl), Heterozygosity (dbl), percent.GC
   (dbl), Recombination (dbl), Divergence (dbl), Constraint (int), SNPs (int),
   cent (lgl), diversity (dbl), position (dbl), GC.binned (fctr)
```

Let's start by selecting some columns from d_df using dplyr's `select()` function:

```
> select(d_df, start, end, Pi, Recombination, depth)
Source: local data frame [59,140 x 5]

    start    end      Pi Recombination depth
1 55001 56000   0.000   0.009601574  3.41
2 56001 57000  10.354   0.009601574  6.68
3 57001 58000   1.986   0.009601574  9.06
[...]
```

This is equivalent to d[, c("start", "end", "Pi", "Recombination", "depth")], but dplyr uses special evaluation rules that allow you to omit quoting column names in `select()` (and the returned object is a tbl_df). `select()` also understands ranges of consecutive columns like `select(d_df, start:total.Bases)`. Additionally, you can drop columns from a dataframe by prepending a negative sign in front of the column name (and this works with ranges too):

```
> select(d_df, -(start:cent))
Source: local data frame [59,140 x 3]

    position     GC.binned diversity
1   55500.5 (51.6,68.5] 0.0000000
2   56500.5 (34.7,51.6] 0.0010354
3   57500.5 (34.7,51.6] 0.0001986
[...]
```

Similarly, we can select specific rows as we did using dataframe subsetting in "Exploring Data Through Slicing and Dicing: Subsetting Dataframes" on page 203 using the dplyr function `filter()`. `filter()` is similar to subsetting dataframes using expressions like d[d$Pi > 16 & d$percent.GC > 80,], though you can use multiple statements (separated by commas) instead of chaining them with &:

```
> filter(d_df, Pi > 16, percent.GC > 80)
Source: local data frame [3 x 20]

       start      end total.SNPs total.Bases depth unique.SNPs dhSNPs
1 63097001 63098000          5         947  2.39           2      1
2 63188001 63189000          2        1623  3.21           2      0
3 63189001 63190000          5        1395  1.89           3      2
Variables not shown: reference.Bases (int), Theta (dbl), Pi (dbl),
   Heterozygosity (dbl), percent.GC (dbl), Recombination (dbl), Divergence
```

(dbl), Constraint (int), SNPs (int), cent (lgl), position (dbl), GC.binned
(fctr), diversity (dbl)

To connect statements with logical OR, you need to use the standard logical operator
| we learned about in "Vectors, Vectorization, and Indexing" on page 183.

dplyr also simplifies sorting by columns with the function arrange(), which behaves
like d[order(d$percent.GC),]:

```
> arrange(d_df, depth)
Source: local data frame [59,140 x 20]

      start      end total.SNPs total.Bases depth unique.SNPs dhSNPs
1  1234001  1235000          0         444     1           0      0
2  1584001  1585000          0         716     1           0      0
3  2799001  2800000          0         277     1           0      0
[...]
```

You can sort a column in descending order using arrange() by wrapping its name in
the function desc(). Also, additional columns can be specified to break ties:

```
> arrange(d_df, desc(total.SNPs), desc(depth))
Source: local data frame [59,140 x 20]

       start       end total.SNPs total.Bases depth unique.SNPs dhSNPs
1   2621001   2622000         93       11337 11.34          13     10
2  13023001  13024000         88       11784 11.78          11      1
3  47356001  47357000         87       12505 12.50           9      7
[...]
```

Using dplyr's mutate() function, we can add new columns to our dataframe: For
example, we added a rescaled version of the Pi column as d$diversity—let's drop d
$diversity using select() and then recalculate it:

```
> d_df <- select(d_df, -diversity) # remove our earlier diversity column
> d_df <- mutate(d_df, diversity = Pi/(10*1000))
> d_df
Source: local data frame [59,140 x 20]

    start    end total.SNPs total.Bases depth unique.SNPs dhSNPs reference.Bases
1  55001  56000          0        1894  3.41           0      0             556
2  56001  57000          5        6683  6.68           2      2            1000
3  57001  58000          1        9063  9.06           1      0            1000
[...]
..  ...    ...        ...         ...   ...         ...    ...            ...
Variables not shown: Theta (dbl), Pi (dbl), Heterozygosity (dbl), percent.GC
  (dbl), Recombination (dbl), Divergence (dbl), Constraint (int), SNPs (int),
  cent (lgl), position (dbl), GC.binned (fctr), diversity (dbl)
```

mutate() creates new columns by transforming existing columns. You can refer to
existing columns directly by name, and not have to use notation like d$Pi.

So far we've been using dplyr to get our dataframes into shape by selecting columns, filtering and arranging rows, and creating new columns. In daily work, you'll need to use these and other dplyr functions to manipulate and explore your data. While we could assign output after each step to an intermediate variable, it's easier (and more memory efficient) to chain dplyr operations. One way to do this is to nest functions (e.g., `filter(select(hs_df, seqname, start, end, strand), strand == "+")`). However, reading a series of data manipulation steps from the inside of a function outward is a bit unnatural. To make it easier to read and create data-processing pipelines, dplyr uses `%>%` (known as *pipe*) from the magrittr package (*http://github.com/smbache/magrittr*). With these pipes, the lefthand side is passed as the first argument of the righthand side function, so `d_df %>% filter(percent.GC > 40)` becomes `filter(d_df, percent.GC > 40`. Using pipes in dplyr allows us to clearly express complex data manipulation operations:

```
> d_df %>% mutate(GC.scaled = scale(percent.GC)) %>%
          filter(GC.scaled > 4, depth > 4) %>%
          select(start, end, depth, GC.scaled, percent.GC) %>%
          arrange(desc(depth))

Source: local data frame [18 x 5]
      start       end depth GC.scaled percent.GC
1  62535001 62536000  7.66  4.040263    73.9740
2  63065001 63066000  6.20  4.229954    75.3754
3  62492001 62493000  5.25  4.243503    75.4755
[...]
```

Pipes are a recent innovation, but one that's been quickly adopted by the R community. You can learn more about magrittr's pipes in `help('%>%')`.

dplyr's raw power comes from the way it handles grouping and summarizing data. For these examples, let's use the mtfs dataframe (loaded into R in "Merging and Combining Data: Matching Vectors and Merging Dataframes" on page 219), as it has some nice factor columns we can group by:

```
> mtfs_df <- tbl_df(mtfs)
```

Now let's group by the chromosome column chr. We can group by one or more columns by calling `group_by()` with their names as arguments:

```
> mtfs_df %>% group_by(chr)
Source: local data frame [20,050 x 10]
Groups: chr

    chr motif_start motif_end    dist recomb_start recomb_end   recom
1  chrX    35471312  35471325 39323.0     35430651   35433340  0.0015
2  chrX    35471312  35471325 36977.0     35433339   35435344  0.0015
[...]
```

Note that dplyr's output now includes a line indicating which column(s) the dataset is grouped by. But this hasn't changed the data; now dplyr's functions will be applied *per group* rather than on all data (where applicable). The most common use case is to create summaries as we did with tapply() and aggregate() using the summarize() function:

```
> mtfs_df %>%
    group_by(chr) %>%
    summarize(max_recom = max(recom), mean_recom = mean(recom), num=n())

Source: local data frame [23 x 4]
    chr max_recom mean_recom  num
1  chr1    41.5648   2.217759 2095
2 chr10    42.4129   2.162635 1029
3 chr11    36.1703   2.774918  560
[...]
```

dplyr's summarize() handles passing the relevant column to each function and automatically creates columns with the supplied argument names. Because we've grouped this data by chromosome, summarize() computes per-group summaries. Try this same expression without group_by().

dplyr provides some convenience functions that are useful in creating summaries. Earlier, we saw that n() returns the number of observations in each group. Similarly, n_distinct() returns the unique number of observations in each group, and first(), last() and nth() return the first, last, and n[th] observations, respectively. These latter three functions are mostly useful on data that has been sorted with arrange() (because specific rows are arbitrary in unsorted data).

We can chain additional operations on these grouped and summarized results; for example, if we wanted to sort by the newly created summary column max_recom:

```
> mtfs_df %>%
    group_by(chr) %>%
    summarize(max_recom = max(recom), mean_recom = mean(recom), num=n()) %>%
    arrange(desc(max_recom))

Source: local data frame [23 x 4]
    chr max_recom mean_recom  num
1  chrX    74.0966   2.686840  693
2  chr8    62.6081   1.913325 1727
3  chr3    56.2775   1.889585 1409
4 chr16    54.9638   2.436250  535
[...]
```

dplyr has a few other functions we won't cover in depth: distinct() (which returns only unique values), and sampling functions like sample_n() and sample_frac() (which sample observations). Finally, one of the best features of dplyr is that all of these same methods also work with *database connections*. For example, you can manipulate a SQLite database (the subject of Chapter 13) with all of the same verbs we've used here. See dplyr's databases vignette (*http://bit.ly/dplyr_data*) for more information on this.

Working with Strings

In bioinformatics, we often need to extract data from strings. R has several functions to manipulate strings that are handy when working with bioinformatics data in R. Note, however, that for most bioinformatics text-processing tasks, R is *not* the preferred language to use for a few reasons. First, R works with all data stored in memory; many bioinformatics text-processing tasks are best tackled with the stream-based approaches (discussed in Chapters 3 and 7), which explicitly avoid loading all data in memory at once. Second, R's string processing functions are admittedly a bit clunky compared to Python's. Even after using these functions for years, I still have to constantly refer to their documentation pages.

Despite these limitations, there are many cases when working with strings in R is the best solution. If we've already loaded data into R to explore, it's usually easier to use R's string processing functions than to write and process data through a separate Python script. In terms of performance, we've already incurred the costs of reading data from disk, so it's unlikely there will be performance gains from using Python or another language. With these considerations in mind, let's jump into R's string processing functions.

First, remember that all strings in R are actually character vectors. Recall that this means a single string like "AGCTAG" has a length of 1 (try length("AGCTAG")). If you want to retrieve the number of characters of each element of a character vector, use nchar(). Like many of R's functions, nchar() is vectorized:

```
> nchar(c("AGCTAG", "ATA", "GATCTGAG", ""))
[1] 6 3 8 0
```

We can search for patterns in character vectors using either grep() or regexpr(). These functions differ slightly in their behavior, making both useful under different circumstances. The function grep(pattern, x) returns the positions of all elements in x that match pattern:

```
> re_sites <- c("CTGCAG", "CGATCG", "CAGCTG", "CCCACA")
> grep("CAG", re_sites)
[1] 1 3
```

By default, grep() uses POSIX extended regular expressions, so we could use more sophisticated patterns:

```
> grep("CT[CG]", re_sites)
[1] 1 3
```

grep() and R's other regular-expression handling functions (which we'll see later) support Perl Compatible Regular Expressions (PCRE) with the argument perl=TRUE, and fixed string matching (e.g., not interpreting special characters) with fixed=TRUE. If a regular expression you're writing isn't working, it may be using features only available in the more modern PCRE dialect; try enabling PCRE with perl=TRUE.

Because grep() is returning indices that match the pattern, grep() is useful as a pattern-matching equivalent of match(). For example, we could use grep() to pull out chromosome 6 entries from a vector of chromosome names with sloppy, inconsistent naming:

```
> chrs <- c("chrom6", "chr2", "chr6", "chr4", "chr1", "chr16", " chrom8")
> grep("[^\\d]6", chrs, perl=TRUE)
[1] 1 3
> chrs[grep("[^\\d]6", chrs, perl=TRUE)]
[1] "chrom6" "chr6"
```

There are some subtle details in this example worth discussing. First, we can't use a simpler regular expression like chrs[grep("6", chrs)] because this would match entries like "chr16". We prevent this by writing a restrictive pattern that is any nonnumeric character ([^\\d]) followed by a 6. Note that we need an additional backslash to escape the backslash in \d. Finally, \d is a special symbol available in Perl Compatible Regular Expressions, so we need to specify perl=TRUE. See help(regex) for more information on R's regular expressions.

The Double Backslash

The double backslash is a *very* important part of writing regular expressions in R. Backslashes don't represent themselves in R strings (i.e., they are used as escape characters as in "\"quote\" string" or the newline character \n). To actually include a blackslash in a string, we need to escape the backslash's special meaning with another backslash.

Unlike grep(), regexpr(pattern, x) returns where in each element of x it matched pattern. If an element doesn't match the pattern, regexpr() returns –1. For example:

```
> regexpr("[^\\d]6", chrs, perl=TRUE)
[1]  5 -1  3 -1 -1 -1 -1
attr(,"match.length")
[1]  2 -1  2 -1 -1 -1 -1
attr(,"useBytes")
[1] TRUE
```

regexpr() also returns the length of the matching string using *attributes*. We haven't discussed attributes, but they're essentially a way to store meta-information alongside objects. You can access attributes with the function `attributes()`. For more information on `regexpr`'s output, see `help(regexpr)`.

Clearly, the vector chromosome named `chrs` is quite messy and needs tidying. One way to do this is to find the informative part of each name with `regexpr()` and extract this part of the string using `substr()`. `substr(x, start, stop)` takes a string `x` and returns the characters between `start` and `stop`:

```
> pos <- regexpr("\\d+", chrs, perl=TRUE)
> pos
[1] 6 4 4 4 4 7
attr(,"match.length")
[1] 1 1 1 1 1 2 1
attr(,"useBytes")
[1] TRUE
> substr(chrs, pos, pos + attributes(pos)$match.length)
[1] "6"  "2"  "6"  "4"  "1"  "16" "8"
```

While this solution introduced the helpful `substr()` function, it is fragile code and we can improve upon it. The most serious flaw to this approach is that it isn't robust to all valid chromosome names. We've written code that solves our immediate problem, but may not be robust to data this code may encounter in the future. If our code were rerun on an updated input with chromosomes "chrY" or "chrMt," our regular expression would fail to match these. While it may seem like a far-fetched case to worry about, consider the following points:

- These types of errors can bite you and are time consuming to debug.
- Our code should anticipate biologically realistic input data (like sex and mitochondrial chromosomes).

We can implement a cleaner, more robust solution with the `sub()` function. `sub()` allows us to substitute strings for other strings. Before we continue with our example, let's learn about `sub()` through a simple example that doesn't use regular expressions. `sub(pattern, replacement, x)` replaces the *first* occurrence of `pattern` with `replacement` for each element in character vector `x`. Like `regexpr()` and `grep()`, `sub()` supports `perl=TRUE` and `fixed=TRUE`:

```
> sub(pattern="Watson", replacement="Watson, Franklin,",
      x="Watson and Crick discovered DNA's structure.")
[1] "Watson, Franklin, and Crick discovered DNA's structure."
```

Here, we've replaced the string "Watson" with the string "Watson, Franklin," using `sub()`. Fixed text substitution like this works well for some problems, but to tidy our chr vector we want to capture the informative part of chromosome name (e.g., 1, 2, X, Y, or M) and substitute it into a consistent naming scheme (e.g., chr1, chr2, chrX, or chrM). We do this with regular expression *capturing groups*. If you're completely unfamiliar with these, do study how they work in other scripting languages you use;

capturing groups are extremely useful in bioinformatics string processing tasks. In this case, the parts of our regular expression in parentheses are captured and can be used later by referencing which group they are. Let's look at a few simple examples and then use sub() to help tidy our chromosomes:

```
> sub("gene=(\\w+)", "\\1", "gene=LEAFY", perl=TRUE) ❶
[1] "LEAFY"

> sub(">[^ ]+ *(.*)", "\\1", ">1 length=301354135 type=dna") ❷
[1] "length=301354135 type=dna"

> sub(".*(\\d+|X|Y|M)", "chr\\1", "chr19", perl=TRUE) ❸
[1] "chr9"

> sub(" *[chrom]+(\\d+|X|Y|M) *", "chr\\1", c("chr19", "chrY"), perl=TRUE) ❹
[1] "chr19" "chrY"
```

❶ This line extracts a gene name from a string formatted as gene=<name>. We anchor our expression with gene=, and then use the word character \\w. This matches upper- and lowercase letters, digits, and underscores. Note that this regular expression assumes our gene name will only include alphanumeric characters and underscores; depending on the data, this regular expression may need to be changed.

❷ This expression extracts all text *after* the first space. This uses a common idiom: specify a character class of characters *not* to match. In this case, [^]+ specifies match all characters that aren't spaces. Then, match zero or more spaces, and capture one or more of any character (.*).

❸ Here, we show a common problem with regular expressions and sub(). Our intention was to extract the informative part of a chromosome name like "chr19" (in this case, "19"). However, our regular expression was too *greedy*. Because the part .* matches zero more of *any* character, this matches through to the "1" of "chr19." The "9" is still matched by the rest of the regular expression, captured, and inserted into the replacement string. Note that this error is especially dangerous because it silently makes your data incorrect.

❹ This expression can be used to clean up the messy chromosome name data. Both "chr" and "chrom" (as well as other combinations of these characters) are matched. The informative part of each chromosome name is captured, and replaced into the string "chr" to give each entry a consistent name. We're assuming that we only have numeric, X, Y, and mitochondrion (M) chromosomes in this example.

Parsing inconsistent naming is always a daily struggle for bioinformaticians. Inconsistent naming isn't usually a problem with genome data resources like Ensembl or the UCSC Genome Browser; these resources are well curated and consistent. Data input by humans is often the cause of problems, and unfortunately, no regular expression can handle all errors a human can make when inputting data. Our best strategy is to try to write general parsers and explicitly test that parsed values make sense (see the following tip).

Friendly Functions for Loud Code

The Golden Rule of Bioinformatics is to not trust your data (or tools). We can be proactive about this in code by using functions like stopifnot(), stop() warning(), and message() that stop execution or let the user know of issues that occur when running code. The function stopifnot() errors out if any of its arguments don't evaluate to TRUE. warning() and message() don't stop execution, but pass warnings and messages to users. Occasionally, it's useful to turn R warnings into errors so that they stop execution. We can enable this behavior with options(warn=2) (and set options(warn=0) to return to the default).

Another useful function is paste(), which constructs strings by "pasting" together the parts. paste() takes any number of arguments and concatenates them together using the separating string specified by the sep argument (which is a space by default). Like many of R's functions, paste() is vectorized:

```
> paste("chr", c(1:22, "X", "Y"), sep="")
 [1] "chr1"  "chr2"  "chr3"  "chr4"  "chr5"  "chr6"  "chr7"  "chr8"  "chr9"
[10] "chr10" "chr11" "chr12" "chr13" "chr14" "chr15" "chr16" "chr17" "chr18"
[19] "chr19" "chr20" "chr21" "chr22" "chrX"  "chrY"
```

Here, paste() pasted together the first vector (chr) and second vector (the autosome and sex chromosome names), recycling the shorter vector chr. paste() can also paste all these results together into a single string (see paste()'s argument collapse).

Extracting Multiple Values from a String

For some strings, like "chr10:158395-172881," we might want to extract several chunks. Processing the same string many times to extract different parts is not efficient. A better solution is to combine sub() and strsplit():

```
> region <- "chr10:158395-172881"
> chunks <- sub("(chr[\\d+MYX]+):(\\d+)-(\\d+)",
                "\\1;;\\2;;\\3",
                region, perl=TRUE)
> strsplit(chunks, ";;")
[[1]]
[1] "chr10"  "158395" "172881"
```

The final function essential to string processing in R is strsplit(x, split), which splits string x by split. Like R's other string processing functions, strsplit() supports optional perl and fixed arguments. For example, if we had a string like gene=LEAFY;locus=2159208;gene_model=AT5G61850.1 and we wished to extract each part, we'd need to split by ";":

```
> leafy <- "gene=LEAFY;locus=2159208;gene_model=AT5G61850.1"
> strsplit(leafy, ";")
[[1]]
[1] "gene=LEAFY"        "locus=2159208"        "gene_model=AT5G61850.1"
```

Also, like all of R's other string functions, strsplit() is vectorized, so it can process entire character vectors at once. Because the number of split chunks can vary, strsplit() always returns results in a list.

Developing Workflows with R Scripts

In the last part of this section, we'll focus on some topics that will help you develop the data analysis techniques we've learned so far into reusable workflows stored in scripts. We'll look at control flow, R scripts, workflows for working with many files, and exporting data.

Control Flow: if, for, and while

You might have noticed that we've come this far in the chapter without using any control flow statements common in other languages, such as if, for, or while. Many data analysis tasks in R don't require modifying control flow, and we can avoid using loops by using R's apply functions like lapply(), sapply(), and mapply(). Still, there are some circumstances where we need these classic control flow and looping statements. The basic syntax of if, for, and while are:

```
if (x == some_value) {
        # do some stuff in here
} else {
        # else is optional
}

for (element in some_vector) {
        # iteration happens here
}

while (something_is_true) {
        # do some stuff
}
```

You can break out of for and while loops with a break statement, and advance loops to the next iteration with next. If you do find you need loops in your R code, read the additional notes about loops and pre-allocation in this chapter's *README* on Git-Hub.

Iterating over Vectors

In for loops, it's common to create a vector of indexes like:

```
for (i in 1:length(vec)) {
  # do something
}
```

However, there's a subtle gotcha here—if the vector vec has no elements, it's length is 0, and 1:0 would return the sequence 1, 0. The behavior we want is for the loop to not be evaluated at all. R provides the function seq_along() to handle this situation safely:

```
> vec <- rnorm(3)
> seq_along(vec)
[1] 1 2 3
> seq_along(numeric(0)) # numeric(0) returns an empty
                        # numeric vector
integer(0)
```

seq_len(length.out) is a similar function, which returns a sequence length.out elements long.

R also has a vectorized version of if: the ifelse function. Rather than control program flow, ifelse(test, yes, no) returns the yes value for all TRUE cases of test, and no for all FALSE cases. For example:

```
> x <- c(-3, 1, -5, 2)
> ifelse(x < 0, -1, 1)
[1] -1  1 -1  1
```

Working with R Scripts

Although we've learned R interactively though examples in this chapter, in practice your analyses should be kept in scripts that can be run many times throughout development. Scripts can be organized into project directories (see Chapter 2) and checked into Git repositories (see Chapter 5). There's also a host of excellent R tools to help in creating well-documented, reproducible projects in R; see the following sidebar for examples.

Reproducibility with Knitr and Rmarkdown

For our work to be reproducible (and to make our lives easier if we need to revisit code in the future), it's essential that our code is saved, version controlled, and well documented. Although in-code comments are a good form of documentation, R has two related packages that go a step further and create reproducible project reports:

knitr and Rmarkdown. Both packages allow you to integrate chunks of R code into your text documents (such as a lab notebook or manuscript), so you can describe your data and analysis steps. Then, you can render (also known as "knit") your document, which runs all R code and outputs the result in a variety of formats such as HTML or PDF (using LaTeX). Images, tables, and other output created by your R code will also appear in your finalized rendered document. Each document greatly improves reproducibility by integrating code and explanation (an approach inspired by Donald Knuth's *literate programming*, which was discussed in "Unix Data Tools and the Unix One-Liner Approach: Lessons from Programming Pearls" on page 125).

While we don't have the space to cover these packages in this chapter, both are easy to learn on your own with resources online. If you're just beginning R, I'd recommend starting with Rmarkdown. Rmarkdown is well integrated into RStudio and is very easy to use. R code is simply woven between Markdown text using a simple syntax:

```
The following code draws 100 random normally distributed
value and finds their mean:

```{r}
set.seed(0)
x <- rnorm(100)
mean(x)
```
```

Then, you can save and call Rmarkdown's render() on this file—creating an HTML version containing the documentation, code, and the results of the R block between ```{r} and ```. There are numerous other options; the best introductions are:

- RStudio's Rmarkdown tutorial (*http://bit.ly/rstudio-rmkdwn*)
- Karl Broman's knitr in a knutshell (*http://kbroman.org/knitr_knutshell/*)
- Minimal Examples of Knitr (*http://yihui.name/knitr/demo/minimal/*)

You can run R scripts from R using the function source(). For example, to execute an R script named *my_analysis.R* use:

```
> source("my_analysis.R")
```

As discussed in "Loading Data into R" on page 194, it's important to mind R's working directory. Scripts should *not* use setwd() to set their working directory, as this is not portable to other systems (which won't have the same directory structure). For the same reason, use *relative* paths like *data/achievers.txt* when loading in data, and *not* absolute paths like */Users/jlebowski/data/achievers.txt* (as point also made in "Project Directories and Directory Structures" on page 21). Also, it's a good idea to indicate (either in comments or a *README* file) which directory the user should set as their working directory.

Alternatively, we can execute a script in batch mode from the command line with:

```
$ Rscript --vanilla my_analysis.R
```

This comes in handy when you need to rerun an analysis on different files or with different arguments provided on the command line. I recommend using `--vanilla` because by default, `Rscript` will restore any past saved environments and save its current environment after the execution completes. Usually we don't want R to restore any past state from previous runs, as this can lead to irreproducible results (because how a script runs depends on files only on your machine). Additionally, saved environments can make it a nightmare to debug a script. See `R --help` for more information.

Reproducibility and sessionInfo()

Versions of R and any R packages installed change over time. This can lead to reproducibility headaches, as the results of your analyses may change with the changing version of R and R packages. Solving these issues is an area of ongoing development (see, for example, the packrat package (*http://rstudio.github.io/packrat/*)). At the very least, you should always record the versions of R and any packages you use for an analysis. R actually makes this incredibly easy to do—just call the `sessionInfo()` function:

```
> sessionInfo()
R version 3.1.2 (2014-10-31)
Platform: x86_64-apple-darwin14.0.0 (64-bit)

locale:
[1] en_US.UTF-8/en_US.UTF-8/en_US.UTF-8/C/en_US.UTF-8/en_US.UTF-8

[...]

loaded via a namespace (and not attached):
 [1] assertthat_0.1   colorspace_1.2-4 DBI_0.3.1       digest_0.6.4
 [5] gtable_0.1.2     labeling_0.3     lattice_0.20-29 lazyeval_0.1.10
[17] scales_0.2.4     stringr_0.6.2    tools_3.1.2

[...]
```

Lastly, if you want to retrieve command-line arguments passed to your script, use R's `commandArgs()` with `trailingOnly=TRUE`. For example, this simple R script just prints all arguments:

```
## args.R -- a simple script to show command line args
args <- commandArgs(TRUE)
print(args)
```

We run this with:

```
$ Rscript --vanilla args.R arg1 arg2 arg3
[1] "arg1" "arg2" "arg3"
```

Workflows for Loading and Combining Multiple Files

In bioinformatics projects, sometimes loading data into R can be half the battle. Loading data is nontrivial whenever data resides in large files and when it's scattered across many files. We've seen some tricks to deal with large files in "Loading Data into R" on page 194 and we'll see a different approach to this problem when we cover databases in Chapter 13. In this section, we'll learn some strategies and workflows for loading and combining multiple files. These workflows tie together many of the R tools we've learned so far: lapply(), do.call(), rbind(), and string functions like sub().

Let's step through how to load and combine multiple tab-delimited files in a directory. Suppose you have a directory containing genome-wide hotspot data separated into different files by chromosome. Data split across many files is a common result of pipelines that have been parallelized by chromosome. I've created example data of this nature in this chapter's GitHub repository under the *hotspots/* directory:

```
$ ls -l hotspots
[vinceb]% ls -l hotspots
total 1160
-rw-r--r--  1 vinceb  staff  42041 Feb 11 13:54 hotspots_chr1.bed
-rw-r--r--  1 vinceb  staff  26310 Feb 11 13:54 hotspots_chr10.bed
-rw-r--r--  1 vinceb  staff  24760 Feb 11 13:54 hotspots_chr11.bed
[...]
```

The first step to loading this data into R is to programmatically access these files from R. If your project uses a well-organized directory structure and consistent filenames (see Chapter 2), this is easy. R's function list.files() lists all files in a specific directory. Optionally, list.files() takes a regular-expression pattern used to select matching files. In general, it's wise to use this pattern argument to be as restrictive as possible so as to prevent problems if a file accidentally ends up in your data directory. For example, to load in all chromosomes' *.bed* files:

```
> list.files("hotspots", pattern="hotspots.*\\.bed")
 [1] "hotspots_chr1.bed"  "hotspots_chr10.bed" "hotspots_chr11.bed"
 [4] "hotspots_chr12.bed" "hotspots_chr13.bed" "hotspots_chr14.bed"
 [...]
```

Note that our regular expression needs to use two backslashes to escape the period in the *.bed* extension. list.files() also has an argument that returns full relative paths to each file:

```
> hs_files <- list.files("hotspots", pattern="hotspots.*\\.bed", full.names=TRUE)
> hs_files
 [1] "hotspots/hotspots_chr1.bed"  "hotspots/hotspots_chr10.bed"
```

```
    [3] "hotspots/hotspots_chr11.bed" "hotspots/hotspots_chr12.bed"
    [...]
```

We'll use the `hs_files` vector because it includes the path to each file. `list.files()` has other useful arguments; see `help(list.files)` for more information.

With `list.files()` programmatically listing our files, it's then easy to `lapply()` a function that loads each file in. For example:

```
> bedcols <- c("chr", "start", "end")
> loadFile <- function(x) read.delim(x, header=FALSE, col.names=bedcols)
> hs <- lapply(hs_files, loadFile)
> head(hs[[1]])
   chr  start     end
1 chr1 1138865 1161866
2 chr1 2173749 2179750
3 chr1 2246749 2253750
[...]
```

Often, it's useful to name each list item with the filename (sans full path). For example:

```
> names(hs) <- list.files("hotspots", pattern="hotspots.*\\.bed")
```

Now, we can use `do.call()` and `rbind` to merge this data together:

```
> hsd <- do.call(rbind, hs)
> head(hsd)
                      chr  start     end
hotspots_chr1.bed.1 chr1 1138865 1161866
hotspots_chr1.bed.2 chr1 2173749 2179750
hotspots_chr1.bed.3 chr1 2246749 2253750
[...]
> row.names(hsd) <- NULL
```

`rbind()` has created some row names for us (which aren't too helpful), but you can remove them with `row.names(hsd) <- NULL`.

Often, we need to include a column in our dataframe containing meta-information about files that's stored in each file's filename. For example, if you had to load in multiple samples' data like *sampleA_repl01.txt*, *sampleA_repl02.txt*, …, *sampleC_repl01.txt* you will likely want to extract the sample name and replicate information and attach these as columns in your dataframe. As a simple example of this, let's pretend we *did not* have the column chr in our hotspot files and needed to extract this information from each filename using `sub()`. We'll modify our load File() function accordingly:

```
loadFile <- function(x) {
  # read in a BED file, extract the chromosome name from the file,
  # and add it as a column
  df <- read.delim(x, header=FALSE, col.names=bedcols)
  df$chr_name <- sub("hotspots_([^\\.]+)\\.bed", "\\1", basename(x))
```

```
    df$file <- x
    df
}
```

This version of loadFile() uses sub() to extract part of each file's name. Because we don't care about the path of each file, we use basename() to extract the nondirectory part of each filepath (but depending on your directory structure, you may need to extract information from the full path!). basename() works like:

```
> hs_files[[1]]
[1] "hotspots/hotspots_chr1.bed"
> basename(hs_files[[1]])
[1] "hotspots_chr1.bed"
```

Now, applying this new version of loadFile() to our data:

```
> hs <- lapply(hs_files, loadFile)
> head(hs[[1]])
   chr   start     end chr_name                           file
1 chr1 1138865 1161866     chr1 hotspots/hotspots_chr1.bed
2 chr1 2173749 2179750     chr1 hotspots/hotspots_chr1.bed
3 chr1 2246749 2253750     chr1 hotspots/hotspots_chr1.bed
[...]
```

Just as there's more than one way to pet a cat, there are many ways to bulk load data into R. Alternatively, one might use mapply() or Map() to loop over both file paths (to load in files) and filenames (to extract relevant metadata from a filename).

Lastly, for projects with many large files, it may not be feasible to load all data in at once and combine it into a single large dataframe. In this case, you might need to use lapply() to apply a per-file summarizing function to this data. As a simple example, let's build a workflow that summarizes the number and average length of hotspots by chromosome file. Using lapply(), this is actually quite easy:

```
loadAndSummarizeFile <- function(x) {
  df <- read.table(x, header=FALSE, col.names=bedcols)
  data.frame(chr=unique(df$chr), n=nrow(df), mean_len=mean(df$end - df$start))
}
```

After sourcing this function, let's run it:

```
> chr_hs_summaries <- lapply(hs_files, loadAndSummarizeFile)
> chr_hs_summaries[1:2]
[[1]]
   chr    n mean_len
1 chr1 1753 10702.44

[[2]]
    chr    n mean_len
1 chr10 1080 10181.56
```

If needed, you could use do.call(rbind, chr_hs_summaries) to merge this data into a single dataframe.

Processing files this way is very convenient for large files, and is even more powerful because it's possible to parallelize data processing by simply replacing lapply() with mclapply().

Exporting Data

At some point during an analysis, you'll need to export data from R. We can export dataframes to plain-text files using the R function write.table(). Unfortunately, write.table() has some poorly chosen defaults that we usually need to adjust. For example, if we wanted to write our dataframe mtfs to a tab-delimited file named *hotspot_motifs.txt*, we would use:

```
> write.table(mtfs, file="hotspot_motifs.txt", quote=FALSE,
             sep="\t", row.names=FALSE, col.names=TRUE)
```

write.table()'s first two arguments are the dataframe (or matrix) to write to file and the file path to save it to. By default, write.table() quotes factor and character columns and includes rownames—we disable these with quote=FALSE and row.names=FALSE. We also can set the column separators to tabs with the sep argument. A header can be included or excluded with the col.names argument.

Given how we saw in Chapters 6 and 7 that many Unix tools work well with Gzipped files, it's worth mentioning you can write a compressed version of a file directly from R. In addition to a string file path, write.table's file argument also handles open file connections. So to write our dataframe to a compressed file, we open a gzipped file connect with gzfile(). For example:

```
> hs_gzf <- gzfile("hotspot_motifs.txt.gz")
> write.table(mtfs, file=hs_gzf, quote=FALSE, sep="\t", row.names=FALSE,
             col.names=TRUE)
```

Functions like read.delim() and write.table() are for reading and writing plain-text data. Throughout the book, we've seen how plain text is preferable to specialty formats because it's more portable, easy to read and understand, and works well with Unix data tools. While plain-text formats are the preferable way to share tabular datasets, it's not the best way to save complex R data structures like lists or special R objects. In these cases, it's usually preferable to save R objects *as* R objects. Encoding and saving objects to disk in a way that allows them to be restored as the original object is known as *serialization*. R's functions for saving and loading R objects are save() and load(). For the sake of this example, let's create a simple R data structure using a list containing a vector and a dataframe and save this to file, remove the original object, and then reload it:

```
> tmp <- list(vec=rnorm(4), df=data.frame(a=1:3, b=3:5))
> save(tmp, file="example.Rdata")
> rm(tmp) # remove the original 'tmp' list
> load("example.Rdata") # this fully restores the 'tmp' list from file
> str(tmp)
List of 2
 $ vec: num [1:4] -0.655 0.274 -1.724 -0.49
 $ df :'data.frame':    3 obs. of  2 variables:
  ..$ a: int [1:3] 1 2 3
  ..$ b: int [1:3] 3 4 5
```

The function save() has an analogous function called save.image() that saves *all* objects in your workspace rather than just the objects you specify. Combined with savehistory(), save.image() can be a quick way to store your past work in R in a rush (personally, save.history() has saved my skin a few times when I've needed to record past interactive work right before a server has crashed!).

Further R Directions and Resources

In this rapid R introduction, we've covered many key features of the R language and seen how they can be used to explore, visualize, and understand data. But we've only scratched the surface of R's capabilities. As you continue to use and learn R, I recommend the following texts:

- Hadley Wickham's *Advanced R* (*http://adv-r.had.co.nz/*) (Chapman & Hall, 2014)
- Joseph Adler's *R in a Nutshell* (O'Reilly, 2010)
- Hadley Wickham's *ggplot2: Elegant Graphics for Data Analysis* (Springer, 2010)
- Winston Chang's *R Graphics Cookbook* (O'Reilly, 2012)
- Norman Matloff's *The Art of R Programming: A Tour of Statistical Software Design* (No Starch Press, 2011)

One reason R is popular (and increasingly so) are the numerous R packages available to scientists. New statistical methods are routinely published with R packages implementations released on CRAN. In Chapter 9, we'll use packages created and maintained by the Bioconductor (*http://bioconductor.org/*) project extensively. Bioconductor is an open source R software project focused on developing tools for high-throughput genomics and molecular biology data. Another useful project is rOpenSci (*http://ropensci.org/*), which develops packages that simplify accessing data from and submitting data to public repositories, mining journal articles, and visualizing data. These R software projects have each fostered healthy communities of scientists and statisticians creating and using R packages, which continue to improve and advance R's capabilities in the sciences.

## CHAPTER 9

Working with Range Data

Here is a problem related to yours and solved before. Could you use it? Could you use its result? Could you use its method?

— *How to Solve It* George Pólya (1945)

Luckily for bioinformaticians, every genome from every branch of life on earth consists of chromosome sequences that can be represented on a computer in the same way: as a set of nucleotide sequences (genomic variation and assembly uncertainty aside). Each separate sequence represents a reference DNA molecule, which may correspond to a fully assembled chromosome, or a scaffold or contig in a partially assembled genome. Although nucleotide sequences are linear, they may also represent biologically circular chromosomes (e.g., with plasmids or mitochondria) that have been cut. In addition to containing nucleotide sequences (the As, Ts, Cs, and Gs of life), these reference sequences act as our *coordinate system* for describing the location of everything in a genome. Moreover, because the units of these chromosomal sequences are individual base pairs, there's no finer resolution we could use to specify a location on a genome.

Using this coordinate system, we can describe location or region on a genome as a *range* on a linear chromosome sequence. Why is this important? Many types of genomic data are linked to a specific genomic region, and this region can be represented as a range containing consecutive positions on a chromosome. Annotation data and genomic features like gene models, genetic variants like SNPs and insertions/deletions, transposable elements, binding sites, and statistics like pairwise diversity and GC content can all be represented as ranges on a linear chromosome sequence. Sequencing read alignment data resulting from experiments like whole genome resequencing, RNA-Seq, ChIP-Seq, and bisulfite sequencing can also be represented as ranges.

Once our genomic data is represented as ranges on chromosomes, there are numerous range operations at our disposal to tackle tasks like finding and counting overlaps, calculating coverage, finding nearest ranges, and extracting nucleotide sequences from specific ranges. Specific problems like finding which SNPs overlap coding sequences, or counting the number of read alignments that overlap an exon have simple, general solutions once we represent our data as ranges and reshape our problem into one we can solve with range operations.

As we'll see in this chapter, there are already software libraries (like R's GenomicRanges) and command-line tools (bedtools) that implement range operations to solve our problems. Under the hood, these implementations rely on specialized data structures like interval trees to provide extremely fast range operations, making them not only the easiest way to solve many problems, but also the fastest.

A Crash Course in Genomic Ranges and Coordinate Systems

So what are ranges exactly? Ranges are integer intervals that represent a subsequence of consecutive positions on a sequence like a chromosome. We use *integer* intervals because base pairs are discrete—we can't have a fractional genomic position like 50,403,503.53. Ranges alone only specify a region along a single sequence like a chromosome; to specify a genomic region or position, we need three necessary pieces of information:

Chromosome name
> This is also known as *sequence name* (to allow for sequences that aren't fully assembled, such as scaffolds or contigs). Each genome is made up of a set of chromosome sequences, so we need to specify which one a range is on. Rather unfortunately, there is no standard naming scheme for chromosome names across biology (and this will cause you headaches). Examples of chromosome names (in varying formats) include "chr17," "22," "chrX," "Y," and "MT" (for mitochondrion), or scaffolds like "HE667775" or "scaffold_1648." These chromosome names are always with respect to some particular genome assembly version, and may differ between versions.

Range
> For example, 114,414,997 to 114,693,772 or 3,173,498 to 3,179,449. Ranges are how we specify a single subsequence on a chromosome sequence. Each range is composed of a *start position* and an *end position*. As with chromosome names, there's no standard way to represent a range in bioinformatics. The technical details of ranges are quite important, so we'll discuss them in more detail next.

Strand

Because chromosomal DNA is double-stranded, features can reside on either the forward (positive) or reverse (negative) strand. Many features on a chromosome are strand-specific. For example, because protein coding exons only make biological sense when translated on the appropriate strand, we need to specify which strand these features are on.

These three components make up a *genomic range* (also know as a *genomic interval*). Note that because reference genomes are our coordinate system for ranges, *ranges are completely linked to a specific genome version*. In other words, genomic locations are relative to reference genomes, so when working with and speaking about ranges we need to specify the version of genome they're relative to.

To get an idea of what ranges on a linear sequence look like, Figure 9-1 depicts three ranges along a stretch of chromosome. Ranges x and y overlap each other (with a one base pair overlap), while range z is not overlapping any other range (and spans just a single base pair). Ranges x and z are both on the forward DNA strand (note the directionality of the arrows), and their underlying nucleotide sequences are ACTT and C, respectively; range y is on the reverse strand, and its nucleotide sequence would be AGCCTTCGA.

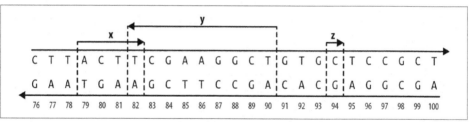

Figure 9-1. Three ranges on an imaginary stretch of chromosome

Reference Genome Versions

Assembling and curating reference genomes is a continuous effort, and reference genomes are perpetually changing and improving. Unfortunately, this also means that our coordinate system will often change between genome versions, so a genomic region like chr15:27,754,876-27,755,076 will not refer to the same genomic location across different genome versions. For example, this 200bp range on human genome version GRCh38 are at chr15:28,000,022-28,000,222 on version GRCh37/hg19, chr15:25,673,617-25,673,817 on versions NCBI36/hg18 and NCBI35/hg17, and chr15: 25,602,381-25,602,581 on version NCBI34/hg16! Thus genomic locations are *always* relative to specific reference genomes versions. For reproducibility's sake (and to make your life easier later on), it's vital to specify which version of reference genome you're working with (e.g., human genome version GRCh38, *Heliconius mel-*

pomene v1.1, or *Zea mays* AGPv3). It's also imperative that you and collaborators use the same genome version so any shared data tied to a genomic regions is comparable.

At some point, you'll need to remap genomic range data from an older genome version's coordinate system to a newer version's coordinate system. This would be a tedious undertaking, but luckily there are established tools for the task:

- CrossMap (*http://crossmap.sourceforge.net/*) is a command-line tool that converts many data formats (BED, GFF/GTF, SAM/BAM, Wiggle, VCF) between coordinate systems of different assembly versions.
- NCBI Genome Remapping Service (*http://bit.ly/remap-tool*) is a web-based tool supporting a variety of genomes and formats.
- LiftOver (*http://bit.ly/liftover*) is also a web-based tool for converting between genomes hosted on the UCSC Genome Browser's site (*http://genome.ucsc.edu/*).

Despite the convenience that comes with representing and working with genomic ranges, there are unfortunately some gritty details we need to be aware of. First, there are two different flavors of range systems used by bioinformatics data formats (see Table 9-1 for a reference) and software programs:

- *0-based* coordinate system, with *half-closed, half-open intervals*.
- *1-based* coordinate system, with *closed intervals*.

With 0-based coordinate systems, the first base of a sequence is position 0 and the last base's position is the *length of the sequence - 1*. In this 0-based coordinate system, we use *half-closed, half-open intervals*. Admittedly, these half-closed, half-open intervals can be a little unintuitive at first—it's easiest to borrow some notation from mathematics when explaining these intervals. For some `start` and `end` positions, half-closed, half-open intervals are written as [`start, end`). Brackets indicate a position is *included* in the interval range (in other words, the interval is *closed* on this end), while parentheses indicate that a position is *excluded* in the interval range (the interval is *open* on this end). So a half-closed, half-open interval like [1, 5) includes the bases at positions 1, 2, 3, and 4 (illustrated in Figure 9-2). You may be wondering why on earth we'd ever use a system that excludes the end position, but we'll come to that after discussing 1-based coordinate systems. In fact, if you're familiar with Python, you've already seen this type of interval system: Python's strings (and lists) are 0-indexed and use half-closed, half-open intervals for indexing portions of a string:

```
>>> "CTTACTTCGAAGGCTG"[1:5]
'TTAC'
```

The second flavor is 1-based. As you might have guessed, with 1-based systems the first base of a sequence is given the position 1. Because positions are counted as we do natural numbers, the last position in a sequence is always equal to its length. With the 1-based systems we encounter in bioinformatics, ranges are represented as *closed intervals*. In the notation we saw earlier, this is simply [start, end], meaning both the start and end positions are included in our range. As Figure 9-2 illustrates, the same bases that cover the 0-based range [1, 5) are covered in the 1-based range [2, 5]. R uses 1-based indexing for its vectors and strings, and extracting a portion of a string with substr() uses closed intervals:

```
> substr("CTTACTTCGAAGGCTG", 2, 5)
[1] "TTAC"
```

If your head is spinning a bit, don't worry too much—this stuff is indeed confusing. For now, the important part is that you are aware of the two flavors and note which applies to the data you're working with.

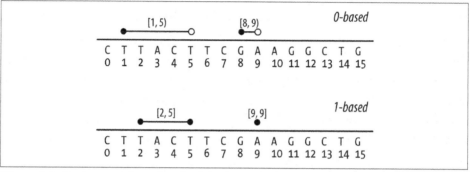

Figure 9-2. Ranges on 0-based and 1-based coordinate systems (lines indicate ranges, open circles indicate open interval endpoints, and closed circles indicate closed endpoints)

Because most of us are accustomed to counting in natural numbers (i.e., 1, 2, 3, etc.), there is a tendency to lean toward the 1-based system initially. Yet both systems have advantages and disadvantages. For example, to calculate how many bases a range spans (sometimes known as the range *width*) in the 0-based system, we use end - start. This is simple and intuitive. With the 1-based system, we'd use the less intuitive end - start + 1. Another nice feature of the 0-based system is that it supports zero-width features, whereas with a 1-based system the smallest supported width is 1 base (though sometimes ranges like [30,29] are used for zero-width features). Zero-width features are useful if we need to represent features *between* bases, such as where a restriction enzyme would cut a DNA sequence. For example, a restriction enzyme that cut at position [12, 12) in Figure 9-2 would leave fragments CTTACTTCGAAGG and CTG.

Table 9-1. Range types of common bioinformatics formats

Format/library	Type
BED	0-based
GTF	1-based
GFF	1-based
SAM	1-based
BAM	0-based
VCF	1-based
BCF	0-based
Wiggle	1-based
GenomicRanges	1-based
BLAST	1-based
GenBank/EMBL Feature Table	1-based

The second gritty detail we need to worry about is strand. There's little to say except: *you need to mind strand* in your work. Because DNA is double stranded, genomic features can lie on either strand. Across nearly all range formats (BLAST results being the exception), a range's coordinates are given *on the forward strand* of the reference sequence. However, a genomic feature can be either on the forward or reverse strand. For genomic features like protein coding regions, strand matters and *must* be specified. For example, a range representing a protein coding region only makes biological sense given the appropriate strand. If the protein coding feature is on the forward strand, the nucleotide sequence underlying this range is the mRNA created during transcription. In contrast, if the protein coding feature is on the reverse strand, the *reverse complement* of the nucleotide sequence underlying this range is the mRNA sequence created during transcription.

We also need to mind strand when comparing features. Suppose you've aligned sequencing reads to a reference genome, and you want to count how many reads overlap a specific gene. Each aligned read creates a range over the region it aligns to, and we want to count how many of these aligned read ranges overlap a gene range. However, information about which strand a sequencing read came from is lost during sequencing (though there are now strand-specific RNA-seq protocols). Aligned reads will map to both strands, and which strand they map to is uninformative. Consequently, when computing overlaps with a gene region that we want to *ignore* strand, an overlap should be counted regardless of whether the aligned read's strand and gene's strand are identical. Only counting overlapping aligned reads that have the same strand as the gene would lead to an underestimate of the reads that likely came from this gene's region.

An Interactive Introduction to Range Data with GenomicRanges

To get a feeling for representing and working with data as ranges on a chromosome, we'll step through creating ranges and using range operations with the Bioconductor packages IRanges and GenomicRanges. Like those in Chapter 8, these examples will be interactive so you grow comfortable exploring and playing around with your data. Through interactive examples, we'll also see subtle gotchas in working with range operations that are important to be aware of.

Installing and Working with Bioconductor Packages

Before we get started with working with range data, let's learn a bit about Bioconductor and install its requisite packages. Bioconductor (*http://www.bioconductor.org*) is an open source software project that creates R bioinformatics packages and serves as a repository for them; it emphasizes tools for high-throughput data. In this section, we'll touch on some of Bioconductor's core packages:

GenomicRanges
> Used to represent and work with genomic ranges

GenomicFeatures
> Used to represent and work with ranges that represent gene models and other features of a genome (genes, exons, UTRs, transcripts, etc.)

Biostrings *and* BSgenome
> Used for manipulating genomic sequence data in R (we'll cover the subset of these packages used for extracting sequences from ranges)

rtracklayer
> Used for reading in common bioinformatics formats like BED, GTF/GFF, and WIG

Bioconductor's package system is a bit different than those on the Comprehensive R Archive Network (CRAN). Bioconductor packages are released on a set schedule, twice a year. Each release is coordinated with a version of R, making Bioconductor's versions tied to specific R versions. The motivation behind this strict coordination is that it allows for packages to be thoroughly tested before being released for public use. Additionally, because there's considerable code re-use within the Bioconductor project, this ensures that all package versions within a Bioconductor release are compatible with one another. For users, the end result is that packages work as expected and have been rigorously tested before you use it (this is good when your scientific results depend on software reliability!). If you need the cutting-edge version of a package for some reason, it's always possible to work with their development branch.

When installing Bioconductor packages, we use the `biocLite()` function. `bio cLite()` installs the correct version of a package for your R version (and its corresponding Bioconductor version). We can install Bioconductor's primary packages by running the following (be sure your R version is up to date first, though):

```
> source("http://bioconductor.org/biocLite.R")
> biocLite()
```

One package installed by the preceding lines is `BiocInstaller`, which contains the function `biocLite()`. We can use `biocLite()` to install the `GenomicRanges` package, which we'll use in this chapter:

```
> biocLite("GenomicRanges")
```

This is enough to get started with the ranges examples in this chapter. If you wish to install other packages later on (in other R sessions), load the `BiocInstaller` package with `library(BiocInstaller)` first. `biocLite()` will notify you when some of your packages are out of date and need to be upgraded (which it can do automatically for you). You can also use `biocUpdatePackages()` to manually update Bioconductor (and CRAN) packages . Because Bioconductor's packages are all tied to a specific version, you can make sure your packages are consistent with `biocValid()`. If you run into an unexpected error with a Bioconductor package, it's a good idea to run `biocUp datePackages()` and `biocValid()` before debugging.

In addition to a careful release cycle that fosters package stability, Bioconductor also has extensive, excellent documentation. The best, most up-to-date documentation for each package will always be at Bioconductor (*http://bioconductor.org*). Each package has a full reference manual covering all functions and classes included in a package, as well as one or more in-depth vignettes. Vignettes step through many examples and common workflows using packages. For example, see the `GenomicRanges` reference manual and vignettes (*http://bit.ly/genomicranges*). I highly recommend that you read the vignettes for all Bioconductor packages you intend to use—they're extremely well written and go into a lot of useful detail.

Storing Generic Ranges with IRanges

Before diving into working with genomic ranges, we're going to get our feet wet with generic ranges (i.e., ranges that represent a contiguous subsequence of elements over any type of sequence). Beginning this way allows us to focus more on thinking abstractly about ranges and how to solve problems using range operations. The real power of using ranges in bioinformatics doesn't come from a specific range library implementation, but in tackling problems using the range abstraction (recall Pólya's quote at the beginning of this chapter). To use range libraries to their fullest potential in real-world bioinformatics, you need to master this abstraction and "range thinking."

The purpose of the first part of this chapter is to teach you range thinking through the use of use Bioconductor's IRanges package. This package implements data structures for generic ranges and sequences, as well as the necessary functions to work with these types of data in R. This section will make heavy use of visualizations to build your intuition about what range operations do. Later in this chapter, we'll learn about the GenomicRanges package, which extends IRanges by handling biological details like chromosome name and strand. This approach is common in software development: implement a more general solution than the one you need, and then extend the general solution to solve a specific problem (see xkcd's "The General Problem" comic (*http://xkcd.com/974/*) for a funny take on this).

Let's get started by creating some ranges using IRanges. First, load the IRanges package. The IRanges package is a dependency of the GenomicRanges package we installed earlier with biocLite(), so it should already be installed:

```
> library(IRanges) # you might see some package startup
                   # messages when you run this
```

The ranges we create with the IRanges package are called IRanges objects. Each IRanges object has the two basic components of any range: a start and end position. We can create ranges with the IRanges() function. For example, a range starting at position 4 and ending at position 13 would be created with:

```
> rng <- IRanges(start=4, end=13)
> rng
IRanges of length 1
    start end width
[1]     4  13    10
```

The most important fact to note: IRanges (and GenomicRanges) is 1-based, and uses closed intervals. The 1-based system was adopted to be consistent with R's 1-based system (recall the first element in an R vector has index 1).

You can also create ranges by specifying their width, and either start or end position:

```
> IRanges(start=4, width=3)
IRanges of length 1
    start end width
[1]     4   6    3
> IRanges(end=5, width=5)
IRanges of length 1
```

```
    start end width
[1]     1   5     5
```

Also, the IRanges() *constructor* (a function that creates a new object) can take vector arguments, creating an IRanges object containing many ranges:

```
> x <- IRanges(start=c(4, 7, 2, 20), end=c(13, 7, 5, 23))
> x
IRanges of length 4
    start end width
[1]     4  13    10
[2]     7   7     1
[3]     2   5     4
[4]    20  23     4
```

Like many R objects, each range can be given a name. This can be accomplished by setting the names argument in IRanges, or using the function names():

```
> names(x) <- letters[1:4]
> x
IRanges of length 4
    start end width names
[1]     4  13    10     a
[2]     7   7     1     b
[3]     2   5     4     c
[4]    20  23     4     d
```

These four ranges are depicted in Figure 9-3. If you wish to try plotting your ranges, the source for the function I've used to create these plots, plotIRanges(), is available in this chapter's directory in the book's GitHub repository.

While on the outside x may look like a dataframe, it's not—it's a special object with class IRanges. In "Factors and classes in R" on page 191, we learned that an object's class determines its behavior and how we interact with it in R. Much of Bioconductor is built from objects and classes. Using the function class(), we can see it's an IRanges object:

```
> class(x)
[1] "IRanges"
attr(,"package")
[1] "IRanges"
```

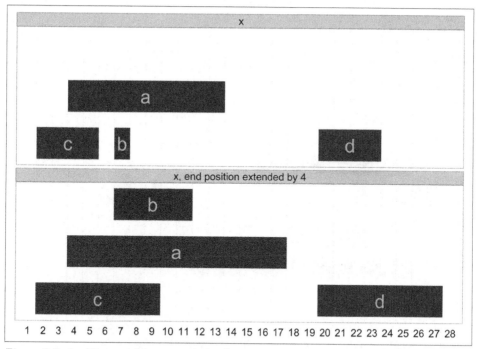

Figure 9-3. An IRanges object containing four ranges

IRanges objects contain all information about the ranges you've created internally. If you're curious what's under the hood, call str(x) to take a peek. Similar to how we used the accessor function levels() to access a factor's levels ("Factors and classes in R" on page 191), we use accessor functions to get parts of an IRanges object. For example, you can access the start positions, end positions, and widths of each range in this object with the methods start(), end(), and width():

```
> start(x)
[1]  4  7  2 20
> end(x)
[1] 13  7  5 23
> width(x)
[1] 10  1  4  4
```

These functions also work with <- to set start, end, and width position. For example, we could increment a range's end position by 4 positions with:

```
> end(x) <- end(x) + 4
> x
IRanges of length 4
    start end width names
[1]    4  17    14     a
[2]    7  11     5     b
```

```
[3]    2   9     8    c
[4]   20  27     8    d
```

Figure 9-4 shows this `IRanges` object before and after extending the end position. Note that the y position in these plots is irrelevant; it's chosen so that ranges can be visualized clearly.

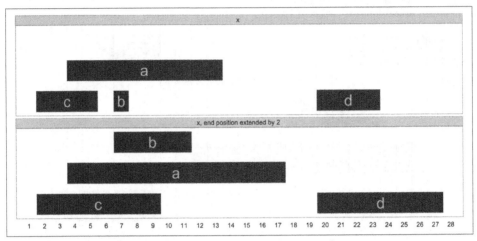

Figure 9-4. Before and after extending the range end position by 4

The `range()` method returns the span of the ranges kept in an `IRanges` object:

```
> range(x)
IRanges of length 1
    start end width
[1]    2  27    26
```

We can subset `IRanges` just as we would any other R objects (vectors, dataframes, matrices), using either numeric, logical, or character (name) index:

```
> x[2:3]
IRanges of length 2
    start end width names
[1]    7  11     5     b
[2]    2   9     8     c
> start(x) < 5
[1]  TRUE FALSE  TRUE FALSE
> x[start(x) < 5]
IRanges of length 2
    start end width names
[1]    4  17    14     a
[2]    2   9     8     c
> x[width(x) > 8]
IRanges of length 1
    start end width names
[1]    4  17    14     a
```

```
> x['a']
IRanges of length 1
    start end width names
[1]     4  17    14     a
```

As with dataframes, indexing using logical vectors created by statements like width(x) > 8 is a powerful way to select the subset of ranges you're interested in.

Ranges can also be easily merged using the function c(), just as we used to combine vectors:

```
> a <- IRanges(start=7, width=4)
> b <- IRanges(start=2, end=5)
> c(a, b)
IRanges of length 2
    start end width
[1]     7  10     4
[2]     2   5     4
```

With the basics of IRanges objects under our belt, we're now ready to look at some basic range operations.

Basic Range Operations: Arithmetic, Transformations, and Set Operations

In the previous section, we saw how IRanges objects conveniently store generic range data. So far, IRanges may look like nothing more than a dataframe that holds range data; in this section, we'll see why these objects are so much more. The purpose of using a special class for storing ranges is that it allows for methods to perform specialized operations on this type of data. The methods included in the IRanges package to work with IRanges objects simplify and solve numerous genomics data analysis tasks. These same methods are implemented in the GenomicRanges package, and work similarly on GRanges objects as they do generic IRanges objects.

First, IRanges objects can be grown or shrunk using arithmetic operations like +, -, and * (the division operator, /, doesn't make sense on ranges, so it's not supported). Growing ranges is useful for adding a buffer region. For example, we might want to include a few kilobases of sequence up and downstream of a coding region rather than just the coding region itself. With IRanges objects, addition (subtraction) will grow (shrink) a range symmetrically by the value added (subtracted) to it:

```
> x <- IRanges(start=c(40, 80), end=c(67, 114))
> x + 4L
IRanges of length 2
    start end width
[1]     36  71    36
[2]     76 118    43
> x - 10L
IRanges of length 2
```

```
     start end width
[1]    50  57     8
[2]    90 104    15
```

The results of these transformations are depicted in Figure 9-5. Multiplication transforms with the width of ranges in a similar fashion. Multiplication by a positive number "zooms in" to a range (making it narrower), while multiplication by a negative number "zooms out" (making it wider). In practice, most transformations needed in genomics are more easily expressed by adding or subtracting constant amounts.

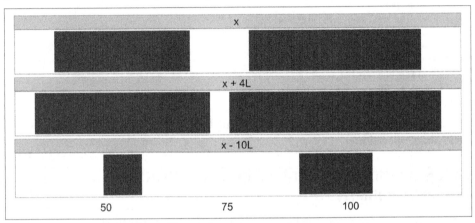

Figure 9-5. Ranges transformed by arithemetic operations

Sometimes, rather than growing ranges by some amount, we want to restrict ranges within a certain bound. The IRanges package method restrict() cuts a set of ranges such that they fall inside of a certain bound (pictured in Figure 9-6):

```
> y <- IRanges(start=c(4, 6, 10, 12), width=13)
> y
IRanges of length 4
     start end width
[1]     4  16    13
[2]     6  18    13
[3]    10  22    13
[4]    12  24    13
> restrict(y, 5, 10)
IRanges of length 3
     start end width
[1]     5  10     6
[2]     6  10     5
[3]    10  10     1
```

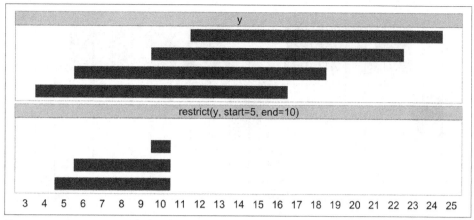

Figure 9-6. Ranges transformed by restrict

Another important transformation is flank(), which returns the regions that flank (are on the side of) each range in an IRanges object. flank() is useful in creating ranges upstream and downstream of protein coding genes that could contain promoter sequences. For example, if our ranges demarcate the transition start site (TSS) and transcription termination site (TTS) of a set of genes, flank() can be used to create a set of ranges upstream of the TSS that contain promoters. To make the example (and visualization) clearer, we'll use ranges much narrower than real genes:

```
> x
IRanges of length 2
    start end width
[1]    40  67    28
[2]    80 114    35
> flank(x, width=7)
IRanges of length 2
    start end width
[1]    33  39     7
[2]    73  79     7
```

By default, flank() creates ranges width positions *upstream* of the ranges passed to it. Flanking ranges downstream can be created by setting start=FALSE:

```
> flank(x, width=7, start=FALSE)
IRanges of length 2
    start end width
[1]    68  74     7
[2]   115 121     7
```

Both upstream and downstream flanking by 7 positions are visualized in Figure 9-7. flank() has many other options; see help(flank) for more detail.

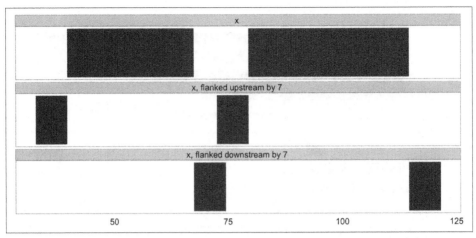

Figure 9-7. Ranges that have been flanked by 7 elements, both upstream and down-stream

Another common operation is reduce(). the reduce() operation takes a set of possibly overlapping ranges and reduces them to a set of nonoverlapping ranges that cover the same positions. Any overlapping ranges are merged into a single range in the result. reduce() is useful when all we care about is what regions of a sequence are covered (and not about the specifics of the ranges themselves). Suppose we had many ranges corresponding to read alignments and we wanted to see which regions these reads cover. Again, for the sake of clarifying the example, we'll use simple, small ranges (here, randomly sampled):

```
> set.seed(0) # set the random number generator seed
> alns <- IRanges(start=sample(seq_len(50), 20), width=5)
> head(alns, 4)
IRanges of length 4
    start end width
[1]    45  49     5
[2]    14  18     5
[3]    18  22     5
[4]    27  31     5
> reduce(alns)
IRanges of length 3
    start end width
[1]     3  22    20
[2]    24  36    13
[3]    40  53    14
```

See Figure 9-8 for a visualization of how reduce() transforms the ranges alns.

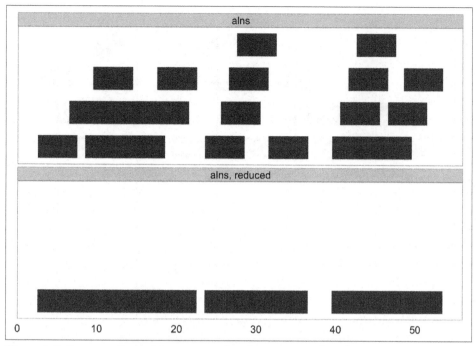

Figure 9-8. Ranges collapsed into nonoverlapping ranges with reduce

A similar operation to reduce() is gaps(), which returns the gaps (uncovered portions) between ranges. gaps() has numerous applications in genomics: creating intron ranges between exons ranges, finding gaps in coverage, defining intragenic regions between genic regions, and more. Here's an example of how gaps() works (see Figure 9-9 for an illustration):

```
> gaps(alns)
IRanges of length 2
    start end width
[1]    23  23     1
[2]    37  39     3
```

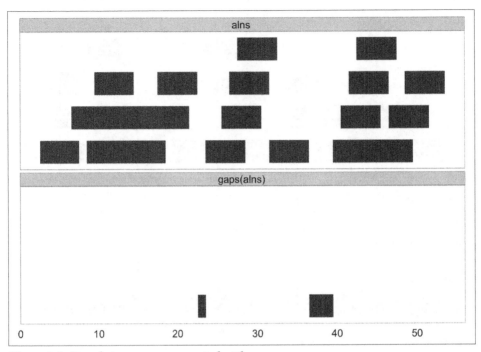

Figure 9-9. Gaps between ranges created with gaps

By default, gaps() only returns the gaps between ranges, and does *not* include those from the beginning of the sequence to the start position of the first range, and the end of the last range to the end of the sequence. IRanges has a good reason for behaving this way: IRanges doesn't know where your sequence starts and ends. If you'd like gaps() to include these gaps, specify the start and end positions in gaps (e.g., gaps(alns, start=1, end=60)).

Another class of useful range operations are analogous to set operations. Each range can be thought of as a set of consecutive integers, so an IRange object like IRange(start=4, end=7) is simply the integers 4, 5, 6, and 7. This opens up the ability to think about range operations as set operations like difference (setdiff()), intersection (intersect()), union (union()), and complement (which is simply the function gaps() we saw earlier)—see Figure 9-10 for an illustration:

```
> a <- IRanges(start=4, end=13)
> b <- IRanges(start=12, end=17)
> intersect(a, b)
IRanges of length 1
    start end width
[1]    12  13     2
> setdiff(a, b)
IRanges of length 1
```

```
          start end width
[1]     4   11    8
> union(b, a)
IRanges of length 1
          start end width
[1]     4   17    14
> union(a, b)
IRanges of length 1
          start end width
[1]     4   17    14
```

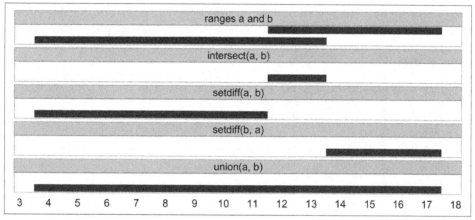

Figure 9-10. Set operations with ranges

Sets operations operate on IRanges with multiple ranges (rows) as if they've been collapsed with reduce() first (because mathematically, overlapping ranges make up the same set). IRanges also has a group of set operation functions that act pairwise, taking two equal-length IRanges objects and working range-wise: psetdiff(), pintersect(), punion(), and pgap(). To save space, I've omitted covering these in detail, but see help(psetdiff) for more information.

The wealth of functionality to manipulate range data stored in IRanges should convince you of the power of representing data as IRanges. These methods provide the basic generalized operations to tackle common genomic data analysis tasks, saving you from having to write custom code to solve specific problems. All of these functions work with genome-specific range data kept in GRanges objects, too.

Finding Overlapping Ranges

Finding overlaps is an essential part of many genomics analysis tasks. Computing overlaps is how we connect experimental data in the form of aligned reads, inferred variants, or peaks of alignment coverage to annotated biological features of the genome like gene regions, methylation, chromatin status, evolutionarily conserved regions, and so on. For tasks like RNA-seq, overlaps are how we quantify our cellular activity like expression and identify different transcript isoforms. Computing overlaps also exemplifies why it's important to use existing libraries: there are advanced data structures and algorithms that can make the computationally intensive task of comparing numerous (potentially billions) ranges to find overlaps efficient. There are also numerous very important technical details in computing overlaps that can have a drastic impact on the end result, so it's vital to understand the different types of overlaps and consider which type is most appropriate for a specific task.

We'll start with the basic task of finding overlaps between two sets of IRanges objects using the findOverlaps() function. findOverlaps() takes query and subject IRanges objects as its first two arguments. We'll use the following ranges (visualized in Figure 9-11):

```
> qry <- IRanges(start=c(1, 26, 19, 11, 21, 7), end=c(16, 30, 19, 15, 24, 8),
                 names=letters[1:6])
> sbj <- IRanges(start=c(1, 19, 10), end=c(5, 29, 16), names=letters[24:26])
> qry
IRanges of length 6
      start end width names
[1]       1  16    16     a
[2]      26  30     5     b
[3]      19  19     1     c
[4]      11  15     5     d
[5]      21  24     4     e
[6]       7   8     2     f
> sbj
IRanges of length 3
      start end width names
[1]       1   5     5     x
[2]      19  29    11     y
[3]      10  16     7     z
```

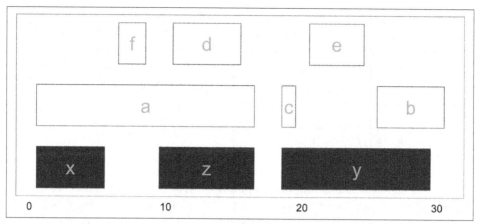

Figure 9-11. Subject ranges (x, y, and z) depicted in gray and query ranges (a through f) depicted in white

Using the IRanges qry and sbj, we can now find overlaps. Calling findOver laps(qry, sbj) returns an object with class Hits, which stores these overlaps:

```
> hts <- findOverlaps(qry, sbj)
> hts
Hits of length 6
queryLength: 6
subjectLength: 3
  queryHits subjectHits
   <integer>   <integer>
  1         1           1
  2         1           3
  3         2           2
  4         3           2
  5         4           3
  6         5           2
```

Thinking abstractly, overlaps represent a *mapping* between query and subject. Depending on how we find overlaps, each query can have many hits in different subjects. A single subject range will always be allowed to have many query hits. Finding qry ranges that overlap sbj ranges leads to a mapping similar to that shown in Figure 9-12 (check that this follows your intuition from Figure 9-13).

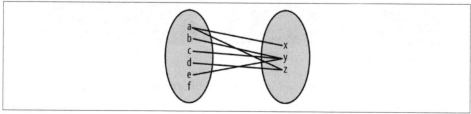

Figure 9-12. Mapping between qry and sbj ranges representing any overlap

The `Hits` object from `findOverlaps()` has two columns on indices: one for the query ranges and one for the subject ranges. Each row contains the index of a query range that overlaps a subject range, and the index of the subject range it overlaps. We can access these indices by using the accessor functions `queryHits()` and `subjectHits()`. For example, if we wanted to find the names of each query and subject range with an overlap, we could do:

```
> names(qry)[queryHits(hts)]
[1] "a" "a" "b" "c" "d" "e"
> names(sbj)[subjectHits(hts)]
[1] "x" "z" "y" "y" "z" "y"
```

Figure 9-13 shows which of the ranges in `qry` overlap the ranges in `sbj`. From this graphic, it's easy to see how `findOverlaps()` is computing overlaps: a range is considered to be overlapping if *any* part of it overlaps a subject range. This type of overlap behavior is set with the `type` argument to `findOverlaps()`, which is `"any"` by default. Depending on our biological task, `type="any"` may not be the best form of overlap. For example, we could limit our overlap results to only include query ranges that fall entirely within subject ranges with `type=within` (Figure 9-14):

```
> hts_within <- findOverlaps(qry, sbj, type="within")
> hts_within
Hits of length 3
queryLength: 6
subjectLength: 3
  queryHits subjectHits
   <integer>   <integer>
  1        3           2
  2        4           3
  3        5           2
```

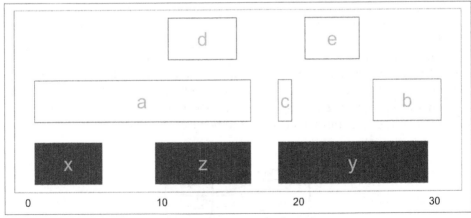

Figure 9-13. Ranges in qry that overlap sbj using findOverlaps

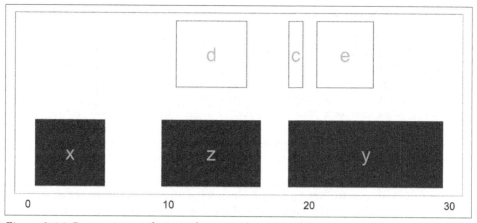

Figure 9-14. Ranges in qry that overlap entirely within sbj

While `type="any"` and `type="within"` are the most common options in day-to-day work, `findOverlaps()` supports other overlap types. See `help(findOverlaps)` to see the others (and much more information about `findOverlaps()` and related functions).

Another `findOverlaps()` parameter that we need to consider when computing overlaps is `select`, which determines how `findOverlaps()` handles cases where a single query range overlaps more than one subject range. For example, the range named a in `qry` overlaps both x and y. By default, `select="all"`, meaning that *all* overlapping ranges are returned. In addition, `select` allows the options `"first"`, `"last"`, and `"arbitrary"`, which return the first, last, and an arbitrary subject hit, respectively. Because the options `"first"`, `"last"`, and `"arbitrary"` all lead `findOverlaps()` to return *only one* overlapping subject range per query (or `NA` if no overlap is found), results are returned in an integer vector where each element corresponds to a query range in `qry`:

```
> findOverlaps(qry, sbj, select="first")
[1] 1  2  2  3  2 NA
> findOverlaps(qry, sbj, select="last")
[1] 3  2  2  3  2 NA
> findOverlaps(qry, sbj, select="arbitrary")
[1] 1  2  2  3  2 NA
```

Mind Your Overlaps (Part I)

What an overlap "is" may seem obvious at first, but the specifics can matter a lot in real-life genomic applications. For example, allowing for a query range to overlap any part of a subject range makes sense when we're looking for SNPs in exons. However, classifying a 1kb genomic window as coding because it overlaps a single base of a gene may make less sense (though depends on the application). It's important to always relate your quantification methods to the underlying biology of what you're trying to understand.

The intricacies of overlap operations are *especially* important when we use overlaps to quantify something, such as expression in an RNA-seq study. For example, if two transcript regions overlap each other, a single alignment could overlap both transcripts and be counted twice—not good. Likewise, if we count how many alignments overlap exons, it's not clear how we should aggregate overlaps to obtain transcript or gene-level quantification. Again, different approaches can lead to sizable differences in statistical results. The take-home lessons are as follows:

- Mind what your code is considering an overlap.

- For quantification tasks, simple overlap counting is best thought of as an approximation (and more sophisticated methods do exist).

See Trapnell, et al., 2013 for a really nice introduction of these issues in RNA-seq quantification.

Counting many overlaps can be a computationally expensive operation, especially when working with many query ranges. This is because the naïve solution is to take a query range, check to see if it overlaps any of the subject ranges, and then repeat across all other query ranges. If you had Q query ranges and S subject ranges, this would entail $Q \times S$ comparisons. However, there's a trick we can exploit: ranges are naturally *ordered* along a sequence. If our query range has an end position of 230,193, there's no need to check if it overlaps subject ranges with start positions larger than 230,193—it won't overlap. By using a clever data structure that exploits this property, we can avoid having to check if each of our Q query ranges overlap our S subject ranges. The clever data structure behind this is the *interval tree*. It takes time to build an interval tree from a set of subject ranges, so interval trees are most appropriate for tasks that involve finding overlaps of many query ranges against a fixed set of subject ranges. In these cases, we can build the subject interval tree once and then we can use it over and over again when searching for overlaps with each of the query ranges.

Implementing interval trees is an arduous task, but luckily we don't have to utilize their massive computational benefits. IRanges has an IntervalTree class that uses interval trees under the hood. Creating an IntervalTree object from an IRanges object is simple:

```
> sbj_it <- IntervalTree(sbj)
> sbj_it
IntervalTree of length 3
    start end width
[1]     1   5     5
[2]    19  29    11
[3]    10  16     7
> class(sbj_it)
[1] "IntervalTree"
attr(,"package")
[1] "IRanges"
```

Using this sbj_it object illustrates we can use findOverlaps() with IntervalTree objects just as we would a regular IRanges object—the interfaces are identical:

```
> findOverlaps(qry, sbj_it)
Hits of length 6
queryLength: 6
subjectLength: 3
  queryHits subjectHits
  <integer>   <integer>
1         1           1
2         1           3
3         2           2
4         3           2
5         4           3
6         5           2
```

Note that in this example, we won't likely realize any computational benefits from using an interval tree, as we have few subject ranges.

After running findOverlaps(), we need to work with Hits objects to extract information from the overlapping ranges. Hits objects support numerous helpful methods in addition to the queryHits() and subjectHits() accessor functions (see help(queryHits) for more information):

```
> as.matrix(hts) ❶
     queryHits subjectHits
[1,]         1           1
[2,]         1           3
[3,]         2           2
[4,]         3           2
[5,]         4           3
[6,]         5           2
> countQueryHits(hts) ❷
[1] 2 1 1 1 1 0
```

```
> setNames(countQueryHits(hts), names(qry))
a b c d e f
2 1 1 1 1 0
> countSubjectHits(hts) ❸
[1] 1 3 2
> setNames(countSubjectHits(hts), names(sbj))
x y z
1 3 2
> ranges(hts, qry, sbj) ❹
IRanges of length 6
    start end width
[1]     1   5     5
[2]    10  16     7
[3]    26  29     4
[4]    19  19     1
[5]    11  15     5
[6]    21  24     4
```

❶ Hits objects can be coerced to matrix using as.matrix().

❷ countQueryHits() returns a vector of how many subject ranges each query IRanges object overlaps. Using the function setNames(), I've given the resulting vector the same names as our original ranges on the next line so the result is clearer. Look at Figure 9-11 and verify that these counts make sense.

❸ The function countSubjectHits() is like countQueryHits(), but returns how many query ranges overlap the subject ranges. As before, I've used setNames() to label these counts with the subject ranges' names so these results are clearly labelled.

❹ Here, we create a set of ranges for overlapping regions by calling the ranges() function using the Hits object as the first argument, and the same query and subject ranges we passed to findOverlaps() as the second and third arguments. These intersecting ranges are depicted in Figure 9-15 in gray, alongside the original subject and query ranges. Note how these overlapping ranges differ from the set created by intersect(qry, sbj): while intersect() would create one range for the regions of ranges a and d that overlap z, using ranges() with a Hits object creates two separate ranges.

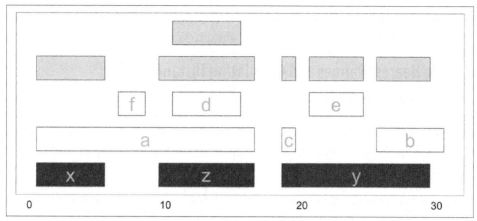

Figure 9-15. Overlapping ranges created from a Hits object using the function ranges

A nice feature of working with ranges in R is that we can leverage R's full array of data analysis capabilities to explore these ranges. For example, after using ranges(hts, qry, sbj) to create a range corresponding to the region shared between each overlapping query and subject range, you could use summary(width(ranges(hts, qry, sbj))) to get a summary of how large the overlaps are, or use ggplot2 to plot a histogram of all overlapping widths. This is one of the largest benefits of working with ranges within R—you can interactively explore and understand your results immediately after generating them.

The functions subsetByOverlaps() and countOverlaps() simplify some of the most common operations performed on ranges once overlaps are found: keeping only the subset of queries that overlap subjects, and counting overlaps. Both functions allow you to specify the same type of overlap to use via the type argument, just as findOverlaps() does. Here are some examples using the objects qry and sbj we created earlier:

```
> countOverlaps(qry, sbj) ❶
a b c d e f
2 1 1 1 1 0
> subsetByOverlaps(qry, sbj) ❷
IRanges of length 5
    start end width names
[1]     1  16    16     a
[2]    26  30     5     b
[3]    19  19     1     c
[4]    11  15     5     d
[5]    21  24     4     e
```

❶ countOverlaps is similar to the solution using countQueryOverlaps() and set Names().

❷ subsetByOverlaps returns the same as qry[unique(queryHits(hts))]. You can verify this yourself (and think through why unique() is necessary).

Finding Nearest Ranges and Calculating Distance

Another common set of operations on ranges focuses on finding ranges that neighbor query ranges. In the IRanges package, there are three functions for this type of operation: nearest(), precede(), and follow(). The nearest() function returns the nearest range, regardless of whether it's upstream or downstream of the query. precede() and follow() return the nearest range that the query is upstream of or downstream of, respectively. Each of these functions take the query and subject ranges as their first and second arguments, and return an index to which subject matches (for each of the query ranges). This will be clearer with examples and visualization:

```
> qry <- IRanges(start=6, end=13, name='query')
> sbj <- IRanges(start=c(2, 4, 18, 19), end=c(4, 5, 21, 24), names=1:4)
> qry
IRanges of length 1
    start end width names
[1]     6  13     8 query
> sbj
IRanges of length 4
    start end width names
[1]     2   4     3     1
[2]     4   5     2     2
[3]    18  21     4     3
[4]    19  24     6     4
> nearest(qry, sbj)
[1] 2
> precede(qry, sbj)
[1] 3
> follow(qry, sbj)
[1] 1
```

To keep precede() and follow() straight, remember that these functions are with respect to the query: precede() finds ranges that the query *precedes* and follow() finds ranges that the query *follows*. Also, illustrated in this example (seen in Figure 9-16), the function nearest() behaves slightly differently than precede() and follow(). Unlike precede() and follow(), nearest() will return the nearest range even if it *overlaps* the query range. These subtleties demonstrate how vital it is to carefully read all function documentation before using libraries.

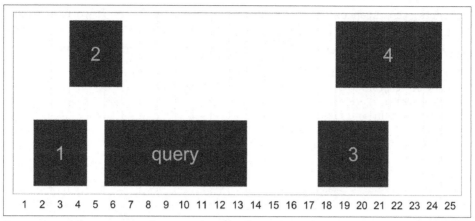

Figure 9-16. The ranges used in nearest, precede, and follow example

Note too that these operations are all vectorized, so you can provide a query IRanges object with multiple ranges:

```
> qry2 <- IRanges(start=c(6, 7), width=3)
> nearest(qry2, sbj)
[1] 2 2
```

This family of functions for finding nearest ranges also includes distan ceToNearest() and distance(), which return the distance to the nearest range and the pairwise distances between ranges. We'll create some random ranges to use in this example:

```
> qry <- IRanges(sample(seq_len(1000), 5), width=10)
> sbj <- IRanges(sample(seq_len(1000), 5), width=10)
> qry
IRanges of length 5
    start end width
[1]   897 906    10
[2]   266 275    10
[3]   372 381    10
[4]   572 581    10
[5]   905 914    10
> sbj
IRanges of length 5
    start end width
[1]   202 211    10
[2]   898 907    10
[3]   943 952    10
[4]   659 668    10
[5]   627 636    10
```

Now, let's use `distanceToNearest()` to find neighboring ranges. It works a lot like `findOverlaps()`—for each query range, it finds the closest subject range, and returns everything in a `Hits` object with an additional column indicating the distance:

```
> distanceToNearest(qry, sbj)
Hits of length 5
queryLength: 5
subjectLength: 5
  queryHits subjectHits  distance
  <integer>   <integer> <integer>
1         1           2         0
2         2           1        54
3         3           1       160
4         4           5        45
5         5           2         0
```

The method `distance()` returns each pairwise distance between query and subject ranges:

```
> distance(qry, sbj)
[1] 685 622 561  77 268
```

Run Length Encoding and Views

The generic ranges implemented by `IRanges` can be ranges over any type of sequence. In the context of genomic data, these ranges' coordinates are based on the underlying nucleic acid sequence of a particular chromosome. Yet, many other types of genomic data form a sequence of numeric values over each position of a chromosome sequence. Some examples include:

- *Coverage*, the depth of overlap of ranges across the length of a sequence. Coverage is used extensively in genomics, from being an important factor in how variants are called to being used to discover coverage peaks that indicate the presence of some feature (as in a ChIP-seq study).

- *Conservation tracks*, which are base-by-base evolutionary conservation scores between species, generated by a program like phastCons (see Siepel et al., 2005 as an example).

- Per-base pair estimates of population genomics summary statistics like nucleotide diversity.

In this section, we'll take a closer look at working with coverage data, creating ranges from numeric sequence data, and a powerful abstraction called *views*. Each of these concepts provides a powerful new way to manipulate sequence and range data. However, this section is a bit more advanced than earlier ones; if you're feeling overwhelmed, you can skim the section on coverage and then skip ahead to "Storing Genomic Ranges with GenomicRanges" on page 299.

Run-length encoding and coverage()

Long sequences can grow quite large in memory. For example, a track containing numeric values over each of the 248,956,422 bases of chromosome 1 of the human genome version GRCh38 would be 1.9Gb in memory. To accommodate working with data this size in R, IRanges can work with sequences compressed using a clever trick: it compresses runs of the same value. For example, imagine a sequence of integers that represent the coverage of a region in a chromosome:

```
4 4 4 3 3 2 1 1 1 1 1 0 0 0 0 0 0 0 0 1 1 1 4 4 4 4 4 4 4
```

Data like coverage often exhibit *runs*: consecutive stretches of the same value. We can compress these runs using a scheme called *run-length encoding*. Run-length encoding compresses this sequence, storing it as: 3 fours, 2 threes, 1 two, 5 ones, 7 zeros, 3 ones, 7 fours. Let's see how this looks in R:

```
> x <- as.integer(c(4, 4, 4, 3, 3, 2, 1, 1, 1, 1, 1, 0, 0, 0,
                     0, 0, 0, 0, 1, 1, 1, 4, 4, 4, 4, 4, 4, 4))
> xrle <- Rle(x)
> xrle
integer-Rle of length 28 with 7 runs
  Lengths: 3 2 1 5 7 3 7
  Values : 4 3 2 1 0 1 4
```

The function Rle() takes a vector and returns a run-length encoded version. Rle() is a function from a low-level Bioconductor package called S4Vectors, which is automatically loaded with IRanges. We can revert back to vector form with as.vector():

```
> as.vector(xrle)
 [1] 4 4 4 3 3 2 1 1 1 1 1 0 0 0 0 0 0 0 0 1 1 1 4 4 4 4 4 4
```

Run-length encoded objects support most of the basic operations that regular R vectors do, including subsetting, arithemetic and comparison operations, summary functions, and math functions:

```
> xrle + 4L
integer-Rle of length 28 with 7 runs
  Lengths: 3 2 1 5 7 3 7
  Values : 8 7 6 5 4 5 8
> xrle/2
numeric-Rle of length 28 with 7 runs
  Lengths:   3   2   1   5   7   3   7
  Values :   2 1.5   1 0.5   0 0.5   2
> xrle > 3
logical-Rle of length 28 with 3 runs
  Lengths:    3   18    7
  Values :  TRUE FALSE  TRUE
> xrle[xrle > 3]
numeric-Rle of length 11 with 3 runs
  Lengths:    3   1   7
  Values :    4 100   4
> sum(xrle)
```

```
[1] 56
> summary(xrle)
   Min. 1st Qu.  Median    Mean 3rd Qu.    Max.
   0.00    0.75    1.00    2.00    4.00    4.00
> round(cos(xrle), 2)
numeric-Rle of length 28 with 7 runs
  Lengths:       3     2     1     5     7     3     7
  Values :   -0.65 -0.99 -0.42  0.54     1  0.54 -0.65
```

We can also access an `Rle` object's lengths and values using the functions `run Lengths()` and `runValues()`:

```
> runLength(xrle)
[1] 3 2 1 5 7 3 7
> runValue(xrle)
[1] 4 3 2 1 0 1 4
```

While we don't save any memory by run-length encoding vectors this short, run-length encoding really pays off with genomic-sized data.

One place where we encounter run-length encoded values is in working with `cover age()`. The `coverage()` function takes a set of ranges and returns their coverage as an `Rle` object (to the end of the rightmost range). Simulating 70 random ranges over a sequence of 100 positions:

```
> set.seed(0)
> rngs <- IRanges(start=sample(seq_len(60), 10), width=7)
> names(rngs)[9] <- "A" # label one range for examples later
> rngs_cov <- coverage(rngs)
> rngs_cov
integer-Rle of length 63 with 18 runs
  Lengths: 11  4  3  3  1  6  4  2  5  2  7  2  3  3  1  3  2  1
  Values :  0  1  2  1  2  1  0  1  2  1  0  1  2  3  4  3  2  1
```

These ranges and coverage can be seen in Figure 9-17 (as before, the y position of the ranges does not mean anything; it's chosen so they can be viewed without overlapping).

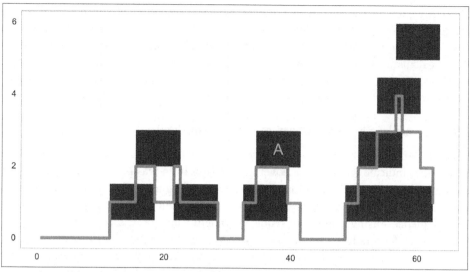

Figure 9-17. Ranges and their coverage plotted

In Chapter 8, we saw how powerful R's subsetting is at allowing us to extract and work with specific subsets of vectors, matrices, and dataframes. We can work with subsets of a run-length encoded sequence using similar semantics:

```
> rngs_cov > 3   # where is coverage greater than 3?
logical-Rle of length 63 with 3 runs
  Lengths:    56     1     6
  Values : FALSE   TRUE FALSE
> rngs_cov[as.vector(rngs_cov) > 3]   # extract the depths that are greater than 3
integer-Rle of length 1 with 1 run
  Lengths: 1
  Values : 4
```

Additionally, we also have the useful option of using `IRanges` objects to extract subsets of a run-length encoded sequence. Suppose we wanted to know what the coverage was in the region overlapping the range labeled "A" in Figure 9-17. We can subset `Rle` objects directly with `IRanges` objects:

```
> rngs_cov[rngs['A']]
integer-Rle of length 7 with 2 runs
  Lengths: 5 2
  Values : 2 1
```

If instead we wanted the mean coverage within this range, we could simply pass the result to `mean()`:

```
> mean(rngs_cov[rngs['A']])
[1] 1.714286
```

Numerous analysis tasks in genomics involve calculating a summary of some sequence (coverage, GC content, nucleotide diversity, etc.) for some set of ranges (repetitive regions, protein coding sequences, low-recombination regions, etc.). These calculations are trivial once our data is expressed as ranges and sequences, and we use the methods in IRanges. Later in this chapter, we'll see how GenomicRanges provides nearly identical methods tailored to these tasks on genomic data.

Going from run-length encoded sequences to ranges with slice()

Earlier, we used rngs_cov > 3 to create a run-length encoded vector of TRUE/FALSE values that indicate whether the coverage for a position was greater than 3. Suppose we wanted to now create an IRanges object containing all regions where coverage is greater than 3. What we want is an operation that's the inverse of using ranges to sub-set a sequence—using a subset of sequence to define new ranges. In genomics, we use these types of operations that define new ranges quite frequently—for example, taking coverage and defining ranges corresponding to extremely high-coverage peaks, or a map of per-base pair recombination estimates and defining a recombinational hot-spot region.

It's very easy to create ranges from run-length encoded vectors. The function slice() takes a run-length encoded numeric vector (e.g., of coverage) as its argument and sli-ces it, creating a set of ranges where the run-length encoded vector has some minimal value. For example, we could take our coverage Rle object rngs_cov and slice it to create ranges corresponding to regions with more than 2x coverage:

```
> min_cov2 <- slice(rngs_cov, lower=2)
> min_cov2
Views on a 63-length Rle subject

views:
    start end width
[1]    16  18     3 [2 2 2]
[2]    22  22     1 [2]
[3]    35  39     5 [2 2 2 2 2]
[4]    51  62    12 [2 2 2 3 3 3 4 3 3 3 2 2]
```

This object that's returned is called a *view*. Views combine a run-length encoded vec-tors and ranges, such that each range is a "view" of part of the sequence. In this case, each view is a view on to the part of the sequence that has more than 2x coverage. The numbers to the right of the ranges are the underlying elements of the run-length encoded vector in this range. If you're simply interested in ranges, it's easy to extract out the underlying ranges:

```
> ranges(min_cov2)
IRanges of length 4
    start end width
[1]    16  18     3
[2]    22  22     1
```

```
[3]    35  39     5
[4]    51  62    12
```

The slice() method is quite handy when we need to define coverage peaks—regions where coverage of experimental data like aligned reads is high such that could indicate something biologically interesting. For example, after looking at a histogram of genome-wide coverage, we could define a coverage threshold that encapsulates outliers, use slice() to find the regions with high coverage, and then see where these regions fall and if they overlap other interesting biological features.

Advanced IRanges: Views

Before we go any further, the end of this section goes into some deeper, slightly more complex material. If you're struggling to keep up at this point, it may be worth skipping to "Storing Genomic Ranges with GenomicRanges" on page 299 and coming back later to this section.

OK, intrepid reader, let's dig a bit deeper into these Views objects we saw earlier. While they may seem a bit strange at first, views are incredibly handy. By combining a sequence vector and ranges, views simplify operations that involve aggregating a sequence vector by certain ranges. In this way, they're similar to calculating per-group summaries as we did in Chapter 8, but groups are ranges.

For example, we could summarize the views we created earlier using slice() using functions like viewMeans(), viewMaxs(), and even viewApply(), which applies an arbitrary function to views:

```
> viewMeans(min_cov2)
[1] 2.000000 2.000000 2.000000 2.666667
> viewMaxs(min_cov2)
[1] 2 2 2 4
> viewApply(min_cov2, median)
[1] 2 2 2 3
```

Each element of these returned vectors is a summary of a range's underlying run-length encoded vector (in this case, our coverage vector min_cov2 summarized by the ranges we carved out using slice()). Also, there are a few other built-in view summarization methods; see help(viewMeans) for a full list.

Using Views, we can also create summaries of sequences by window/bin. In the views lingo, we create a set of ranges for each window along a sequence and then summarize the views onto the underlying sequence these windows create. For example, if we wanted to calculate the average coverage for windows 5-positions wide:

```
> length(rngs_cov) ❶
[1] 63
> bwidth <- 5L ❷
> end <- bwidth * floor(length(rngs_cov) / bwidth) ❸
> windows <- IRanges(start=seq(1, end, bwidth), width=bwidth) ❹
```

```
> head(windows)
IRanges of length 6
    start end width
[1]     1   5     5
[2]     6  10     5
[3]    11  15     5
[4]    16  20     5
[5]    21  25     5
[6]    26  30     5
> cov_by_wnd <- Views(rngs_cov, windows) ❺
> head(cov_by_wnd)
Views on a 63-length Rle subject

views:
    start end width
[1]     1   5     5 [0 0 0 0 0]
[2]     6  10     5 [0 0 0 0 0]
[3]    11  15     5 [0 1 1 1 1]
[4]    16  20     5 [2 2 2 1 1]
[5]    21  25     5 [1 2 1 1 1]
[6]    26  30     5 [1 1 1 0 0]
> viewMeans(cov_by_wnd) ❻
 [1] 0.0 0.0 0.8 1.6 1.2 0.6 0.8 1.8 0.2 0.4 2.4 3.2
```

There's a bit of subtle arithmetic going on here, so let's step through piece by piece.

❶   First, note that our coverage vector is 63 elements long. We want to create consecutive windows along this sequence, with each window containing 5 elements. If we do so, we'll have 3 elements of the coverage vector hanging off the end (63 divided by 5 is 12, with a remainder of 3). These overhanging ends are a common occurrence when summarizing data by windows, and it's common to just ignore these last elements. While cutting these elements off seems like a strange approach, a summary calculated over a smaller range will have a higher variance that can lead to strange results. Dropping this remainder is usually the simplest and best option.

❷   We'll set bwidth to be our bin width.

❸   Now, we compute the end position of our window. To do so, we divide our coverage vector length by the bin width, and chop off the remainder using the floor() function. Then, we multiply by the bin width to give the end position.

❹   Next, we create our windows using IRanges. We use seq() to generate the start positions: a start position from 1 to our end (60, as we just programmatically calculated), moving by 5 each time. If we wanted a different window step width, we could change the third (by) argument of seq() here. With our start position specified, we simply set width=bwidth to give each window range a width of 5.

❺ With our run-length encoded coverage vector and our windows as IRanges objects, we create our Views object. Views effectively groups each element of the coverage vector rngs_cov inside a window.

❻ Finally, we compute summaries on these Views. Here we use viewMeans() to get the mean coverage per window. We could use any other summarization view method (e.g., viewMaxs(), viewSums(), etc.) or use viewApply() to apply any function to each view.

Summarizing a sequence of numeric values by window over a sequence such as a chromosome is a common task in genomics. The techniques used to implement the generic solution to this problem with ranges, run-length encoded vectors, and views are directly extensible to tackling this problem with real genomics data.

Because GenomicRanges extends IRanges, everything we've learned in the previous sections can be directly applied to the genomic version of an IRanges object, GRanges. None of the function names nor behaviors differ much, besides two added complications: dealing with multiple chromosomes and strand. As we'll see in the next sections, GenomicRanges manages these complications and greatly simplifies our lives when working with genomic data.

Storing Genomic Ranges with GenomicRanges

The GenomicRanges package introduces a new class called GRanges for storing genomic ranges. The GRanges builds off of IRanges. IRanges objects are used to store ranges of genomic regions on a single sequence, and GRanges objects contain the two other pieces of information necessary to specify a genomic location: sequence name (e.g., which chromosome) and strand. GRanges objects also have *metadata columns*, which are the data linked to each genomic range. We can create GRanges objects much like we did with IRanges objects:

```
> library(GenomicRanges)
> gr <- GRanges(seqname=c("chr1", "chr1", "chr2", "chr3"),
            ranges=IRanges(start=5:8, width=10),
            strand=c("+", "-", "-", "+"))
> gr
GRanges with 4 ranges and 0 metadata columns:
      seqnames    ranges strand
         <Rle> <IRanges>  <Rle>
  [1]     chr1  [5, 14]      +
  [2]     chr1  [6, 15]      -
  [3]     chr2  [7, 16]      -
  [4]     chr3  [8, 17]      +
  ---
  seqlengths:
```

```
chr1 chr2 chr3
 NA   NA   NA
```

Using the GRanges() constructor, we can also add arbitrary metadata columns by specifying additional named arguments:

```
> gr <- GRanges(seqname=c("chr1", "chr1", "chr2", "chr3"),
            ranges=IRanges(start=5:8, width=10),
            strand=c("+", "-", "-", "+"), gc=round(runif(4), 3))
> gr
GRanges with 4 ranges and 1 metadata column:
      seqnames    ranges strand |        gc
         <Rle> <IRanges>  <Rle> | <numeric>
  [1]     chr1   [5, 14]      + |     0.897
  [2]     chr1   [6, 15]      - |     0.266
  [3]     chr2   [7, 16]      - |     0.372
  [4]     chr3   [8, 17]      + |     0.573
  ---
  seqlengths:
   chr1 chr2 chr3
     NA   NA   NA
```

This illustrates the structure of GRanges objects: genomic location specified by sequence name, range, and strand (on the left of the dividing bar), and metadata columns (on the right). Each row of metadata corresponds to a range on the same row.

All metadata attached to a GRanges object are stored in a DataFrame, which behaves identically to R's base data.frame, but supports a wider variety of column types. For example, DataFrames allow for run-length encoded vectors to save memory (whereas R's base data.frame does not). Whereas in the preceding example metadata columns are used to store numeric data, in practice we can store any type of data: identifiers and names (e.g., for genes, transcripts, SNPs, or exons), annotation data (e.g., conservation scores, GC content, repeat content, etc.), or experimental data (e.g., if ranges correspond to alignments, data like mapping quality and the number of gaps). As we'll see throughout the rest of this chapter, the union of genomic location with any type of data is what makes GRanges so powerful.

Also, notice seqlengths in the gr object we've just created. Because GRanges (and genomic range data in general) is always with respect to a particular genome version, we usually know beforehand what the length of each sequence/chromosome is. Knowing the length of chromosomes is necessary when computing coverage and gaps (because we need to know where the end of the sequence is, not just the last range). We can specify the sequence lengths in the GRanges constructor, or set it after the object has been created using the seqlengths() function:

```
> seqlens <- c(chr1=152, chr2=432, chr3=903)
> gr <- GRanges(seqname=c("chr1", "chr1", "chr2", "chr3"),
            ranges=IRanges(start=5:8, width=10),
            strand=c("+", "-", "-", "+"),
```

```
                gc=round(runif(4), 3),
                seqlengths=seqlens)
> seqlengths(gr) <- seqlens # another way to do the same as above
> gr
GRanges with 4 ranges and 1 metadata column:
        seqnames    ranges strand |        gc
          <Rle> <IRanges>  <Rle> | <numeric>
  [1]      chr1  [5, 14]      + |     0.897
  [2]      chr1  [6, 15]      - |     0.266
  [3]      chr2  [7, 16]      - |     0.372
  [4]      chr3  [8, 17]      + |     0.573
  ---
  seqlengths:
   chr1 chr2 chr3
    152  432  903
```

We access data in GRanges objects much like we access data from IRanges objects: with accessor functions. Accessors for start position, end position, and width are the same as with IRanges object:

```
> start(gr)
[1] 5 6 7 8
> end(gr)
[1] 14 15 16 17
> width(gr)
[1] 10 10 10 10
```

For the GRanges-specific data like sequence name and strand, there are new accessor functions—seqnames and strand:

```
> seqnames(gr)
factor-Rle of length 4 with 3 runs
  Lengths:    2    1    1
  Values : chr1 chr2 chr3
Levels(3): chr1 chr2 chr3
> strand(gr)
factor-Rle of length 4 with 3 runs
  Lengths: 1 2 1
  Values : + - +
Levels(3): + - *
```

The returned objects are all run-length encoded. If we wish to extract all IRanges ranges from a GRanges object, we can use the ranges accessor function:

```
> ranges(gr)
IRanges of length 4
    start end width
  [1]     5  14    10
  [2]     6  15    10
  [3]     7  16    10
  [4]     8  17    10
```

Like most objects in R, GRanges has a length that can be accessed with length(), and supports names:

```
> length(gr)
[1] 4
> names(gr) <- letters[1:length(gr)]
> gr
GRanges with 4 ranges and 1 metadata column:
    seqnames    ranges strand |        gc
       <Rle> <IRanges>  <Rle> | <numeric>
  a    chr1  [5, 14]       + |     0.897
  b    chr1  [6, 15]       - |     0.266
  c    chr2  [7, 16]       - |     0.372
  d    chr3  [8, 17]       + |     0.573
  ---
  seqlengths:
   chr1 chr2 chr3
    100  100  100
```

The best part of all is that GRanges objects support the same style of subsetting you're already familiar with (i.e., from working with other R objects like vectors and data-frames). For example, if you wanted all ranges with a start position greater than 7:

```
> start(gr) > 7
[1] FALSE FALSE FALSE  TRUE
> gr[start(gr) > 7]
GRanges with 1 range and 1 metadata column:
    seqnames    ranges strand |        gc
       <Rle> <IRanges>  <Rle> | <numeric>
  d    chr3  [8, 17]       + |     0.573
  ---
  seqlengths:
   chr1 chr2 chr3
    100  100  100
```

Once again, there's no magic going on; GRanges simply interprets a logical vector of TRUE/FALSE values given by start(gr) > 7 as which rows to include/exclude. Using the seqname() accessor, we can count how many ranges there are per chromosome and then subset to include only ranges for a particular chromosome:

```
> table(seqnames(gr))

chr1 chr2 chr3
   2    1    1
> gr[seqnames(gr) == "chr1"]
GRanges with 2 ranges and 1 metadata column:
    seqnames    ranges strand |        gc
       <Rle> <IRanges>  <Rle> | <numeric>
  a    chr1  [5, 14]       + |     0.897
  b    chr1  [6, 15]       - |     0.266
  ---
  seqlengths:
```

```
      chr1 chr2 chr3
       100  100  100
```

The `mcols()` accessor is used access metadata columns:

```
> mcols(gr)
DataFrame with 4 rows and 1 column
          gc
   <numeric>
1      0.897
2      0.266
3      0.372
4      0.573
```

Because this returns a `DataFrame` and `DataFrame` objects closely mimic `data.frame`, `$`
works to access specific columns. The usual syntactic shortcut for accessing a column
works too:

```
> mcols(gr)$gc
[1] 0.897 0.266 0.372 0.573
> gr$gc
[1] 0.897 0.266 0.372 0.573
```

The real power is when we combine subsetting with the data kept in our metadata
columns. Combining these makes complex queries trivial. For example, we could
easily compute the average GC content of all ranges on `chr1`:

```
> mcols(gr[seqnames(gr) == "chr1"])$gc
[1] 0.897 0.266
> mean(mcols(gr[seqnames(gr) == "chr1"])$gc)
[1] 0.5815
```

If we wanted to find the average GC content for all chromosomes, we would use the
same split-apply-combine strategy we learned about in Chapter 9. We'll see this later
on.

Grouping Data with GRangesList

In Chapter 8, we saw how R's lists can be used to group data together, such as after
using `split()` to split a dataframe by a factor column. Grouping data this way is useful for both organizing data and processing it in chunks. `GRanges` objects also have
their own version of a list, called `GRangesList`, which are similar to R's lists. `GRanges`
`Lists` can be created manually:

```
> gr1 <- GRanges(c("chr1", "chr2"), IRanges(start=c(32, 95), width=c(24, 123)))
> gr2 <- GRanges(c("chr8", "chr2"), IRanges(start=c(27, 12), width=c(42, 34)))
> grl <- GRangesList(gr1, gr2)
> grl
GRangesList of length 2:
[[1]]
GRanges with 2 ranges and 0 metadata columns:
```

```
         seqnames    ranges strand
            <Rle> <IRanges>  <Rle>
     [1]     chr1 [32,  55]      *
     [2]     chr2 [95, 217]      *

  [[2]]
  GRanges with 2 ranges and 0 metadata columns:
         seqnames    ranges strand
     [1]     chr8 [27,  68]      *
     [2]     chr2 [12,  45]      *

  ---
  seqlengths:
   chr1 chr2 chr8
     NA   NA   NA
```

GRangesList objects behave almost identically to R's lists:

```
> unlist(grl) ❶
GRanges with 4 ranges and 0 metadata columns:
       seqnames    ranges strand
          <Rle> <IRanges>  <Rle>
   [1]     chr1 [32,  55]      *
   [2]     chr2 [95, 217]      *
   [3]     chr8 [27,  68]      *
   [4]     chr2 [12,  45]      *
   ---
   seqlengths:
    chr1 chr2 chr8
      NA   NA   NA

> doubled_grl <- c(grl, grl) ❷
> length(doubled_grl)
[1] 4
```

❶ unlist() combines all GRangesList elements into a single GRanges object (much
   like unlisting an R list of vectors to create one long vector).

❷ We can combine many GRangesList objects with c().

Accessing certain elements works exactly as it did with R's lists. Single brackets return
GRangesList objects, and double brackets return what's in a list element—in this case,
a GRanges object:

```
> doubled_grl[2]
GRangesList of length 1:
[[1]]
GRanges with 2 ranges and 0 metadata columns:
       seqnames    ranges strand
          <Rle> <IRanges>  <Rle>
   [1]     chr8 [27,  68]      *
   [2]     chr2 [12,  45]      *
```

```
 ---
seqlengths:
 chr1 chr2 chr8
   NA   NA   NA
> doubled_grl[[2]]
GRanges with 2 ranges and 0 metadata columns:
       seqnames     ranges strand
          <Rle>  <IRanges>  <Rle>
  [1]      chr8   [27, 68]      *
  [2]      chr2   [12, 45]      *
 ---
seqlengths:
 chr1 chr2 chr8
   NA   NA   NA
```

Like lists, we can also give and access list element names with the function names().
GRangesList objects also have some special features. For example, accessor functions
for GRanges data (e.g., seqnames(), start(), end(), width(), ranges(), strand(),
etc.) also work on GRangesList objects:

```
> seqnames(grl)
RleList of length 2
[[1]]
factor-Rle of length 2 with 2 runs
  Lengths:    1    1
  Values : chr1 chr2
Levels(3): chr1 chr2 chr8

[[2]]
factor-Rle of length 2 with 2 runs
  Lengths:    1    1
  Values : chr8 chr2
Levels(3): chr1 chr2 chr8

> start(grl)
IntegerList of length 2
[[1]] 32 95
[[2]] 27 12
```

Note the class of object Bioconductor uses for each of these: RleList and Integer
List. While these are classes we haven't seen before, don't fret—both are analogous to
GRangesList: a list for a specific type of data. Under the hood, both are specialized,
low-level data structures from the S4Vectors package. RleList are lists for run-
length encoded vectors, and IntegerList objects are lists for integers (with added
features). Both RleList and IRangesList are a bit advanced for us now, but suffice to
say they behave a lot like R's lists and they're useful for intermediate and advanced
GenomicRanges users. I've included some resources about these in the *README* file
in this chapter's directory on GitHub.

In practice, we're usually working with too much data to create GRanges objects manually with GRangesList(). More often, GRangesLists come about as the result of using the function split() on GRanges objects. For example, I'll create some random GRanges data, and demonstrate splitting by sequence name:

```
> chrs <- c("chr3", "chr1", "chr2", "chr2", "chr3", "chr1")
> gr <- GRanges(chrs, IRanges(sample(1:100, 6, replace=TRUE),
                width=sample(3:30, 6, replace=TRUE)))
> head(gr)
GRanges with 6 ranges and 0 metadata columns:
      seqnames     ranges strand
         <Rle>  <IRanges>  <Rle>
  [1]     chr3  [90,  93]      *
  [2]     chr1  [27,  34]      *
  [3]     chr2  [38,  44]      *
  [4]     chr2  [58,  79]      *
  [5]     chr3  [91, 103]      *
  [6]     chr1  [21,  44]      *
  ---
  seqlengths:
   chr3 chr1 chr2
     NA   NA   NA

> gr_split <- split(gr, seqnames(gr))
> gr_split[[1]]
GRanges with 4 ranges and 0 metadata columns:
      seqnames     ranges strand
         <Rle>  <IRanges>  <Rle>
  [1]     chr3  [90,  93]      *
  [2]     chr3  [91, 103]      *
  [3]     chr3  [90, 105]      *
  [4]     chr3  [95, 117]      *
  ---
  seqlengths:
   chr3 chr1 chr2
     NA   NA   NA
> names(gr_split)
[1] "chr3" "chr1" "chr2"
```

Bioconductor also provides an unsplit() method to rejoin split data on the same factor that was used to split it. For example, because we created gr_split by splitting on seqnames(gr), we could unsplit gr_split with unsplit(gr_split, seqnames(gr)).

So why split GRanges objects into GRangesList objects? The primary reason is that GRangesList objects, like R's base lists, are a natural way to group data. For example, if we had a GRanges object containing all exons, we may want to work with exons grouped by what gene or transcript they belong to. With all exons grouped in a GRangesList object, exons for a particular gene or transcript can be returned by accessing a particular list element.

Grouped data is also the basis of the split-apply-combine pattern (covered in "Working with the Split-Apply-Combine Pattern" on page 239). With R's base lists, we could use lapply() and sapply() to iterate through all elements and apply a function. Both of these functions work with GRangesLists objects, too:

```
> lapply(gr_split, function(x) order(width(x))) ❶
$chr3
[1] 1 2 3 4

$chr1
[1] 1 2

$chr2
[1] 1 4 2 3
> sapply(gr_split, function(x) min(start(x))) ❷
chr3 chr1 chr2
  90   21   38
> sapply(gr_split, length) ❸
chr3 chr1 chr2
   4    2    4
> elementLengths(gr_split) ❹
chr3 chr1 chr2
   4    2    4
```

❶ Return the order of widths (smallest range to largest) of each GRanges element in a GRangesList.

❷ Return the start position of the earliest (leftmost) range.

❸ The number of ranges in every GRangesList object can be returned with this R idiom.

❹ However, a faster approach to calculating element lengths is with the specialized function elementLengths().

lapply() and sapply() (as well as mapply()) give you the most freedom to write and use your own functions to apply to data. However, for many overlap operation functions (e.g., reduce(), flank(), coverage(), and findOverlaps()), we don't need to explicitly apply them—they can work directly with GRangesList objects. For example, reduce() called on a GRangesList object automatically works at the list-element level:

```
> reduce(gr_split)
GRangesList of length 3:
$chr3
GRanges with 1 range and 0 metadata columns:
      seqnames     ranges strand
         <Rle>  <IRanges>  <Rle>
  [1]     chr3 [90, 117]      *
```

```
$chr1
GRanges with 1 range and 0 metadata columns:
      seqnames   ranges strand
  [1]     chr1 [21, 44]      *

$chr2
GRanges with 2 ranges and 0 metadata columns:
      seqnames   ranges strand
  [1]     chr2 [38, 44]      *
  [2]     chr2 [58, 96]      *

---
seqlengths:
 chr3 chr1 chr2
   NA   NA   NA
```

reduce() illustrates an important (and extremely helpful) property of GRangesList objects: many methods applied to GRangesList objects work at the grouped-data level automatically. Had this list contained exons grouped by transcript, only overlapping exons *within a list element (transcript)* would be collapsed with reduce(). findOverlaps() behaves similarly; overlaps are caclulated at the list-element level. We'll see a more detailed example of findOverlaps() with GRangesList objects once we start working with real annotation data in the next section.

Working with Annotation Data: GenomicFeatures and rtracklayer

We've been working a lot with toy data thus far to learn basic range concepts and operations we can perform on ranges. Because the GenomicRanges package shines when working interactively with moderately large amounts of data, let's switch gears and learn about two Bioconductor packages for importing and working with external data. Both packages have different purposes and connect with GenomicRanges. The first, GenomicFeatures, is designed for working with transcript-based genomic annotations. The second, rtracklayer, is designed for importing and exporting annotation data into a variety of different formats. As with other software covered in this book, both of these packages have lots of functionality that just can't be covered in a single section; I highly recommend that you consult both packages' vignettes.

GenomicFeatures is a Bioconductor package for creating and working with transcript-based annotation. GenomicFeatures provides methods for creating and working with TranscriptDb objects. These TranscriptDb objects wrap annotation data in a way that allows genomic features, like genes, transcripts, exons, and coding sequences (CDS), to be extracted in a consistent way, regardless of the organism and origin of the annotation data. In this section, we'll use a premade TranscriptDb object, contained in one of Bioconductor's transcript annotation packages. Later on, we'll see some functions `GenomicFeatures` has for creating `TranscriptDb` objects (as well as transcript annotation packages) from external annotation data.

R Packages for Data

While it may sound strange to use an R package that contains data rather than R code, it's actually a clever and appropriate use of an R package. Bioconductor uses packages for many types of data, including transcript and organism annotation data, experimental data, compressed reference genomes, and microarray and SNP platform details. Packages are a terrific way to unite data from multiple sources into a single easily loaded and explicitly versioned shared resource. Using data packages can eliminate the hassle of coordinating which files and what versions (and from what websites) collaborators need to download and use for an analysis. Overall, working with data from packages facilitates reproducibility; if you're working with annotation data in R, use the appropriate package if it exists.

Let's start by installing `GenomicFeatures` and the transcript annotation package for mouse, *Mus musculus*. We've already installed Bioconductor's package installer, so we can install `GenomicFeatures` with:

```
> library(BiocInstaller)
> biocLite("GenomicFeatures")
```

Now, we need to install the appropriate *Mus musculus* transcript annotation package. We can check which annotation packages are available on the Bioconductor annotation package page (*http://bit.ly/anno-data*). There are a few different packages for *Mus musculus*. At the time I'm writing this, `TxDb.Mmusculus.UCSC.mm10.ensGene` is the most recent. Let's install this version:

```
> biocLite("TxDb.Mmusculus.UCSC.mm10.ensGene")
```

While this is installing, notice the package's naming scheme. All transcript annotation packages use the same consistent naming scheme—that is, `TxDb.<organism>.<annotation-source>.<annotation-version>`. This annotation is for mouse genome version mm10 (Genome Reference Consortium version GRCm38), and the annotation comes from UCSC's Ensembl track. Once these packages have installed, we're ready to load them and start working with their data:

```
> library(TxDb.Mmusculus.UCSC.mm10.ensGene)
> txdb <- TxDb.Mmusculus.UCSC.mm10.ensGene
> txdb
TranscriptDb object:
| Db type: TranscriptDb
| Supporting package: GenomicFeatures
| Data source: UCSC
```

```
| Genome: mm10
| Organism: Mus musculus
| UCSC Table: ensGene
| Resource URL: http://genome.ucsc.edu/
| Type of Gene ID: Ensembl gene ID
| Full dataset: yes
| miRBase build ID: NA
| transcript_nrow: 94647
| exon_nrow: 348801
| cds_nrow: 226282
| Db created by: GenomicFeatures package from Bioconductor
| Creation time: 2014-03-17 16:22:04 -0700 (Mon, 17 Mar 2014)
| GenomicFeatures version at creation time: 1.15.11
| RSQLite version at creation time: 0.11.4
| DBSCHEMAVERSION: 1.0
> class(txdb)
[1] "TranscriptDb"
attr(,"package")
[1] "GenomicFeatures"
```

Loading TxDb.Mmusculus.UCSC.mm10.ensGene gives us access to a transcriptDb object with the same name as the package. The package name is quite long and would be a burden to type, so it's conventional to alias it to txdb. When we look at the tran scriptDb object txdb, we get *a lot* of metadata about this annotation object's version (how it was created, when it was created, etc.). Under the hood, this object simply represents a SQLite database contained inside this R package (we'll learn more about these in Chapter 13). We don't need to know any SQLite to interact with and extract data from this object; the GenomicFeatures package provides all methods we'll need. This may sound a bit jargony now, but will be clear after we look at a few examples.

First, suppose we wanted to access all gene regions in *Mus musculus* (in this version of Ensembl annotation). There's a simple accessor function for this, unsurprisingly named genes():

```
> mm_genes <- genes(txdb)
> head(mm_genes)
> head(mm_genes)
GRanges with 6 ranges and 1 metadata column:
                      seqnames                 ranges strand |           gene_id
                         <Rle>              <IRanges>  <Rle> |   <CharacterList>
  ENSMUSG00000000001      chr3 [108107280, 108146146]      - | ENSMUSG00000000001
  ENSMUSG00000000003      chrX [ 77837901,  77853623]      - | ENSMUSG00000000003
  ENSMUSG00000000028     chr16 [ 18780447,  18811987]      - | ENSMUSG00000000028
  ENSMUSG00000000031      chr7 [142575529, 142578143]      - | ENSMUSG00000000031
  ENSMUSG00000000037      chrX [161117193, 161258213]      + | ENSMUSG00000000037
  ENSMUSG00000000049     chr11 [108343354, 108414396]      + | ENSMUSG00000000049
  ---
  seqlengths:
                  chr1             chr2 ...   chrUn_JH584304
             195471971        182113224 ...           114452
```

```
> length(mm_genes)
[1] 39017
```

GenomicFeatures returns the data in a GRanges object, so all the tricks we've learned for working with GRanges can be used to work with this data.

Note that the GRanges object containing all mouse genes has 39,014 ranges. This includes coding genes, short noncoding genes, long noncoding genes, and pseudogenes—everything that comprises the entire ensGene track on UCSC (and all genes with an Ensembl gene identifier). It's always a good idea to make sure you know what you're getting with gene annotation; you should also validate that the totals make sense against an external source. For example, I've validated that the total number of genes is consistent with Ensembl's gene annotation summary page for mouse genome version mm10 (*http://bit.ly/ensembl-mouse*).

GenomicFeatures has other functions for retrieving all transcripts, exons, coding sequences (CDS), and promoters—the functions transcripts(), exons(), cds(), and promoters(). Consult the documentation for this family of functions for extracting information from transcriptDb objects at help(transcripts).

It's often more natural to work with a GRangesList object of these types of features grouped by some other type of feature than working with a massive GRanges list object of everything. For example, we might want to retrieve all exons grouped by transcript or gene:

```
> mm_exons_by_tx <- exonsBy(txdb, by="tx")
> mm_exons_by_gn <- exonsBy(txdb, by="gene")
> length(mm_exons_by_tx)
[1] 94647
> length(mm_exons_by_gn)
[1] 39017
```

These functions that extract grouped features all take the transcriptDb object as their first argument and which type of feature to group by (e.g., *gene*, *tx*, *exon*, or *cds*) as their second argument. There are variety of these types of functions—for example, transcriptsBy(), exonsBy(), cdsBy(), intronsBy(), fiveUTRsByTranscript(), and threeUTRsByTranscript() (see help(transcriptsBy) for more information).

GenomicFeatures also provides functions for extracting subsets of features that overlap a specific chromosome or range. We can limit our queries to use a subset of chromosomes by setting which sequences our transcriptDb should query using the following approach:

```
> seqlevels(txdb, force=TRUE) <- "chr1"
> chr1_exons <- exonsBy(txdb, "tx")
> all(unlist(seqnames(chr1_exons)) == "chr1")
[1] TRUE
> txdb <- restoreSeqlevels(txdb) # restore txdb so it queries all sequences
```

To extract feature data that only overlaps a specific region, use the following family of functions: `transcriptsByOverlaps()`, `exonsByOverlaps()`, and `cdsByOverlaps()` (see `help(transcriptByOverlaps()`) for more information). For example, say a QTL study has identified a quantitative trait loci in the region roughly on chromosome 8, from 123,260,562 to 123,557,264. Our coordinates are rough, so we'll add 10kbp. Recall that with `IRanges`, we grow ranges by a fixed number of bases by adding that number of bases to the object (from "Basic Range Operations: Arithmetic, Transformations, and Set Operations" on page 275); the same method is used to resize `GRanges` objects. So we can get all genes within this expanded region with:

```
> qtl_region <- GRanges("chr8", IRanges(123260562, 123557264))
> qtl_region_expanded <- qtl_region + 10e3
> transcriptsByOverlaps(txdb, qtl_region_expanded)
GRanges with 73 ranges and 2 metadata columns:
       seqnames               ranges strand |    tx_id             tx_name
          <Rle>            <IRanges>  <Rle> | <integer>         <character>
   [1]     chr8 [119910841, 124345722]     + |    47374 ENSMUST00000127664
   [2]     chr8 [123254195, 123269745]     + |    47530 ENSMUST00000001092
   [3]     chr8 [123254271, 123257636]     + |    47531 ENSMUST00000150356
   [4]     chr8 [123254284, 123269743]     + |    47532 ENSMUST00000156896
   [5]     chr8 [123254686, 123265070]     + |    47533 ENSMUST00000154450
   ...      ...                  ...   ... ...       ...                 ...
  [69]     chr8 [123559201, 123559319]     - |    49320 ENSMUST00000178208
  [70]     chr8 [123560888, 123561006]     - |    49321 ENSMUST00000179143
  [71]     chr8 [123562595, 123562713]     - |    49322 ENSMUST00000178297
  [72]     chr8 [123564286, 123564404]     - |    49323 ENSMUST00000179019
  [73]     chr8 [123565969, 123566087]     - |    49324 ENSMUST00000179081
  ---
  seqlengths:
               chr1               chr2 ...      chrUn_JH584304
          195471971          182113224 ...              114452
```

`transcriptByOverlaps()` returns all transcripts overlapping this range. All functions in this family also take a `maxgap` argument, which can be used to specify how large a gap between ranges is considered an overlap (0 by default). Setting the `maxgap` argument to 10kbp has the same effect as widening our ranges and then extracting elements as we did in the preceding example.

Creating TranscriptDb Objects

If there isn't a transcript annotation package containing a `transcriptDb` object for your organism, annotation track, or genome version of choice, `GenomicFeatures` provides a multitude of methods to create one. If the annotation track you'd like to use is on the UCSC Genome Browser or a *BioMart* (e.g., a data management system used by databases Ensembl and WormBase), `GenomicFeatures` contains the functions `make TranscriptDbFromUCSC()` and `makeTranscriptDbFromBiomart()` for creating `transcriptDb` from these databases. For some nonmodel systems, annotation only exists as a Gene Transfer Format (GTF) or Gene Feature Format (GFF) file. In this case, a `transcriptDb` object can be created with the `makeTranscriptDbFromGFF()` method. Once you've created a `transcriptDb` object using one of these methods, you can save the underlying SQLite database with `saveDb()` and load it again with `loadDb()`. As a demonstration of how easy this is, the following line downloads all required annotation data from Ensembl to create a `transcriptDb` object for Platypus:

```
> species <- "oanatinus_gene_ensembl"
> platypus_txdb <- makeTranscriptDbFromBiomart("ensembl", species)
```

Although creating `transcriptDb` directly and saving these objects as SQLite databases certainly works, `GenomicFeatures` makes it easy to create a transcript annotation package directly from tracks from the UCSC Genome Browser, Biomart, or from a `transcriptDb` object. See the functions `makeTxDbPackageFromUCSC()`, `makeTxDbPackageFromBiomart()`, and `makeTxDbPackage()`, and the `GenomicFeatures` vignette (*http://bit.ly/genom-feat*) or documentation for more detail on how to use these functions.

The `transcriptDb` objects and the methods provided by the `GenomicFeatures` package provide a consistent representation of transcript-based annotation and consistent functions for interacting with these objects. However, like everything in life, there are trade-offs. Convenience and consistency can come at the cost of flexibility. The `rtracklayer` package includes flexible functions for importing and exporting data that stores ranges from a variety of formats like GTF/GFF, BED, BED Graph, and Wiggle. These functions automatically convert entries to `GRanges` objects and handle technicalities like missing values and assigning columns in the file to metadata columns—features that general solutions like `read.delim()` don't have. Let's look at how the `rtracklayer` function `import()` loads the *Mus_musculus.GRCm38.75_chr1.gtf.gz* file (available on this book's GitHub page):

```
> mm_gtf <- import('Mus_musculus.GRCm38.75_chr1.gtf.gz')
> colnames(mcols(mm_gtf)) # metadata columns read in
 [1] "source"            "type"               "score"
 [4] "phase"             "gene_id"            "gene_name"
 [7] "gene_source"       "gene_biotype"       "transcript_id"
[10] "transcript_name"   "transcript_source"  "tag"
[13] "exon_number"       "exon_id"            "ccds_id"
[16] "protein_id"
```

The function `import()` detects the file type (with hints from the file extension— another reason to use the proper extension) and imports all data as a `GRanges` object. There are also specific functions (e.g., `import.bed()`, `import.gff()`, `import.wig()`, etc.) that can you can use if you want to specify the format.

The `rtracklayer` package also provides export methods, for taking range data and saving it to a variety of common range formats. For example, suppose we wanted to write five random pseudogenes to a GTF file. We could use:

```
> set.seed(0)
> pseudogene_i <- which(mm_gtf$gene_biotype == "pseudogene" &
  mm_gtf$type == "gene")
> pseudogene_sample <- sample(pseudogene_i, 5)
> export(mm_gtf[pseudogene_sample], con="five_random_pseudogene.gtf",
  format="GTF")
```

If we didn't care about the specifics of these ranges (e.g., the information stored in the metadata columns), the BED file format may be more appropriate. BED files require at a minimum three columns: chromosomes (or sequence name), start position, and end position (sometimes called the BED3 format). The easiest way to save only this information would be:

```
> bed_data <- mm_gtf[pseudogene_sample]
> mcols(bed_data) <- NULL # clear out metadata columns
> export(bed_data, con="five_random_pseudogene.bed", format="BED")
```

Finally, it's worth noting that we're just scratching the surface of `rtracklayer`'s capabilities. In addition to its `import`/`export` functions, `rtracklayer` also interfaces with genome browsers like UCSC's Genome Browser (*http://genome.ucsc.edu*). Using `rtracklayer`, one can create tracks for UCSC's browser directly from `GRanges` objects and send these to a UCSC Genome Browser web session directly from R. If you find yourself using the UCSC Genome Browser frequently, it's worth reading the `rtracklayer` vignette (*http://bit.ly/rtracklayer*) and learning how to interact with it through R.

Retrieving Promoter Regions: Flank and Promoters

Now, let's start seeing how the range operations in `GenomicRanges` can solve real bioinformatics problems. For example, suppose we wanted to grab the promoter regions of all protein-coding genes from the GRCh38 *Mus musculus* Ensembl GTF annotation track for chromosome 1 we loaded in using `rtracklayer` in the previous section. We'll use this object rather than extracting genes from a `transcriptDb` object because it contains additional information about the type of transcript (such as the `gene_bio type` and `type` columns we used in the previous section to find pseudogenes). So, first we could find the subset of genes we're interested in—in this case, let's say all protein coding genes:

```
> table(mm_gtf$gene_biotype) ❶
```

```
           antisense                    lincRNA                    miRNA
                 480                        551                      354
           misc_RNA polymorphic_pseudogene     processed_transcript
                  93                         61                      400
```

```
   protein_coding              pseudogene                rRNA
           77603                     978                   69
   sense_intronic        sense_overlapping              snoRNA
              21                       4                  297
           snRNA
             315
> chr1_pcg <- mm_gtf[mm_gtf$type == "gene" & ❷
                     mm_gtf$gene_biotype == "protein_coding"]
> summary(width(chr1_pcg))
   Min. 1st Qu.  Median    Mean 3rd Qu.    Max.
     78    9429   25750   60640   62420 1076000
> length(chr1_pcg)
[1] 1240
> chr1_pcg_3kb_up <- flank(chr1_pcg, width=3000) ❸
>
```

❶ First, let's do some basic EDA. We know we want to select protein coding genes, which we can do by subsetting the gene_bioype column. Calling table() on it returns the number of features of each biotype. A full list of biotypes can be found on the GENCODE website (*http://bit.ly/gencode-bio*). We see that there are around 78,000 features with the protein coding biotype.

❷ Next, we can subset all features that have type "gene" (rather than exon, CDS, transcript, etc.) and biotype "protein_coding." As a sanity check, we make sure that the length distribution and number of features makes sense (remember, this is just chromosome 1 data).

❸ Then, we can use flank to grab 3kbp upstream of each feature. Read help(flank) to refresh your memory of this method. You first question should be, "how does flank() handle strand?" Looking at the documentation, we see that by default flank() takes strand into consideration (option ignore.strand=FALSE), so we just specify the width of our flanking region. Note that our promoter regions are those defined from the start of our gene region to 3kbp upstream.

Extracting promoter regions is such a common operation that GenomicRanges packages have a convenience function to make it even simpler: promoters(). promoters() default arguments extract 3kbp upstream of each range, and 200bp downstream. For example, we could mimic our flank() call and show that the results are identical with:

```
> chr1_pcg_3kb_up2 <- promoters(chr1_pcg, upstream=3000, downstream=0)
> identical(chr1_pcg_3kb_up, chr1_pcg_3kb_up2)
[1] TRUE
```

Retrieving Promoter Sequence: Connection GenomicRanges with Sequence Data

Once we've created promoter ranges using flank() (or promoters()), we can use these to grab the promoter nucleotide sequences from a genome. There are two different ways we could do this:

- Entirely through Bioconductor's packages (as we'll see in this section)
- By exporting the GenomicRanges objects to a range file format like BED, and using a command-line tool like BEDTools

Either method works, but the Bioconductor approach requires that genome sequences be stored in a special R package (similar to the annotation packages we encountered earlier). If your organism doesn't have a premade genome package, it may be faster to write the promoter ranges to a file and use BEDTools. Because our example promoter regions are in mouse (a model organism), Bioconductor has a premade genome package we can download and use:

```
> library(BiocInstaller)
> biocLite("BSgenome")
# Note: this file is about 712MB, so be ensure you have enough
# disk space before installing!
> biocLite("BSgenome.Mmusculus.UCSC.mm10")
```

This is a *BSgenome* package, where *BS* stands for *Biostrings*, a Bioconductor package that contains classes for storing sequence data and methods for working with it. BSgenome packages contain the full reference genome for a particular organism, compressed and wrapped in a user-friendly package with common accessor methods. We'll just scratch the surface of these packages in this section, so it's definitely worth reading their vignettes on Bioconductor's website.

Let's first load the BSgenome.Mmusculus.UCSC.mm10 package and poke around:

```
> library(BSgenome.Mmusculus.UCSC.mm10)
> mm_gm <- BSgenome.Mmusculus.UCSC.mm10
> organism(mm_gm)
[1] "Mus musculus"
> providerVersion(mm_gm)
[1] "mm10"
> provider(mm_gm)
[1] "UCSC"
```

organisim(), providerVersion(), and provider() are all accessor functions to extract information from BSgenome packages.

We can use the accessor function seqinfo() to look at sequence information. BSgenome packages contain sequences for each chromosome, stored in a list-like structure we can access using indexing:

```
> seqinfo(mm_gm)
Seqinfo of length 66
seqnames        seqlengths isCircular genome
chr1             195471971      FALSE   mm10
chr2             182113224      FALSE   mm10
[...]
> mm_gm$chrM
  16299-letter "DNAString" instance
seq: GTTAATGTAGCTTAATAACAAAGCAAAGCACTGAAA...TCTAATCATACTCTATTACGCAATAAACATTAACAA
> mm_gm[[22]]
  16299-letter "DNAString" instance
seq: GTTAATGTAGCTTAATAACAAAGCAAAGCACTGAAA...TCTAATCATACTCTATTACGCAATAAACATTAACAA
```

While we won't go into this topic in detail, it's worth mentioning that BSgenome objects can be searched using the string-matching and alignment functions in the Bio strings packages. These are meant for a few, quick queries (certainly not large-scale alignment or mapping!). For example:

```
> library(Biostrings)
> matchPattern("GGCGCGCC", mm_gm$chr1)
  Views on a 195471971-letter DNAString subject
subject: NNNNNNNNNNNNNNNNNNNNNNNNNNNNNNNNNNNN...NNNNNNNNNNNNNNNNNNNNNNNNNNNNNNNNNNNN
views:
        start       end width
  [1]  4557138  4557145     8 [GGCGCGCC]
  [2]  4567326  4567333     8 [GGCGCGCC]
  [3]  6960128  6960135     8 [GGCGCGCC]
  [4]  7397441  7397448     8 [GGCGCGCC]
  [5]  7398352  7398359     8 [GGCGCGCC]
```

Using these promoter regions and the *Mus musculus* promoter regions we created in the previous section, we're ready to extract promoter sequences. Unfortunately, we need to adjust our promoter GRanges object first because the genome annotation file *Mus_musculus.GRCm38.75_chr1.gtf.gz* uses chromosome names like "1", "2", etc. while the *BSgenome.Mmusculus.UCSC.mm10* package uses names like "chr1", "chr2", etc. Having to remap one chromosome naming scheme to another is quite a common operation as there is no standard chromosome naming scheme. Bioconductor provides some nice functions to make this as simple and safe as possible. We'll see the problem, how to fix is manually, and how to fix it using convenience functions:

```
> all(seqlevels(chr1_pcg_3kb_up) %in% seqlevels(mm_gm)) ❶
[1] FALSE
> gr <- GRanges(c("chr1", "chr2"), IRanges(start=c(3, 4), width=10)) ❷
> seqlevels(gr)
[1] "chr1" "chr2"
> seqlevels(gr) <- c("1", "2") ❸
> seqlevels(gr)
[1] "1" "2"

> seqlevelsStyle(chr1_pcg_3kb_up) ❹
[1] "NCBI"
```

```
> seqlevelsStyle(mm_gm)
[1] "UCSC"
> seqlevelsStyle(chr1_pcg_3kb_up) <- "UCSC"  ❺
> all(seqlevels(chr1_pcg_3kb_up) %in% seqlevels(mm_gm))
[1] TRUE
```

❶ First, always check that all sequences we want to grab are in the BSgenome
    object. Because the *BSgenome.Mmusculus.UCSC.mm10* file uses the UCSC chro-
    mosome name style and our annotation file uses Ensembl/NCBI style, this is not
    the case.

❷ Let's create a test GRanges object so we can show how we can manually change
    chromosome names. Bioconductor packages treat sequence names much like the
    factors we saw in "Factors and classes in R" on page 191. We can access and set
    the names of these levels using the seqlevels() function.

❸ Here, we change the sequence levels using seqlevels(). We provide the
    sequence level names in the same order as the original sequence levels.

❹ Because having to switch between the style "chr1" (UCSC style) and "1"
    (Ensembl/NCBI style) is common, Bioconductor provides a convenience func-
    tion seqlevelsStyle(). Here, we see that the style of chr1_pcg_3kb_up is indeed
    "NCBI".

❺ Now, we set the style to "UCSC". After (as demonstrated on the next line), the
    sequence levels are consistent between the two objects.

With our chromosome names consistent between our GRanges promoter regions and
the mouse BSgenome package, it's easy to grab the sequences for particular regions
kept in a GRanges object:

```
> chr1_3kb_seqs <- getSeq(mm_gm, chr1_pcg_3kb_up)
> chr1_3kb_seqs
  A DNAStringSet instance of length 1240
        width seq
   [1]   3000 ATTCTGAGATGTGGTTACTAGATCAATGGGAT...CGGCTAGCCGGGCCCAGCGCCCAGCCCCGCGG
   [2]   3000 GAAGTGGTATATCTGCCTAGTCTAGGTGTGCA...GCTGTACTTAATCTGTGAGCACACATGCTAGT
   [3]   3000 CTTAAAAACCTAGATATTCTATTTTTTTTTTT...CTTTGATAACGTCGTGAGCTCGGCTTCCAACA
   [4]   3000 GAATTGGCACAGTTTCACATGATTGGTCCATT...GTACGGCCGCTGCAGCGCGACAGGGGCCGGGC
   [5]   3000 AAATATAAAGTTAACATACAAAAACTAGTCGC...TCGGGGCGCGAGCTCGGGGCCGAACGCGAGGA
   ...    ... ...
[1236]   3000 CAACATGGGTAGTAGTGGGGGAGCTTTAGTTC...GAGGGGCTGGCCTCACCAAGACGCAACAGGGA
[1237]   3000 AGGTGTGTTATATAATAATTGGTTTGACACTG...CTTAAAACTTGCTCTCTGGCTTCCTGGCGCCC
[1238]   3000 TTGGCCAGGTGATTGATCTTGTCCAACTGGAA...GTAAGGCCGGGCTATATGCAAACCGAGTTCCC
[1239]   3000 GGCATTCCCCTATACTGGGGCATAGAACCTTC...ATTTAAGGGTCTGCTCCCCACTGCTTACAGCC
[1240]   3000 GTAAATTTTCAGGTATATTTCTTTCTACTCTT...CTTTGATATTTCTGTGGTCCTTATTTCTAGGT
```

The method getSeq() takes a BSgenome object and a GRanges object, and returns the sequences for each range. We could then write these sequences to a FASTA file using:

```
> writeXStringSet(chr1_3kb_seqs, file="mm10_chr1_3kb_promoters.fasta",
    format="fasta")
```

It's worth mentioning that Bioconductor has many other packages for working with promoter sequences, extracting motifs, and creating sequence logo plots. This functionality is well documented in a detailed workflow on the Bioconductor website (*http://bit.ly/gene-binding*).

Getting Intergenic and Intronic Regions: Gaps, Reduce, and Setdiffs in Practice

As another example, let's look at gaps() applied to GRanges objects. The gaps() method illustrates a rather important point about applying range operations to genomic data: *we need to consider how range operations will work with strand and work across difference chromosomes/sequences.* With IRanges, gaps were simple: they're just the areas of a sequence with a range on them. With genomic ranges, gaps are calculated on every combination of strand and sequence. Here's an example of this on a toy GRanges object so it's clear what's happening:

```
> gr2 <- GRanges(c("chr1", "chr2"), IRanges(start=c(4, 12), width=6),
                strand=c("+", "-"), seqlengths=c(chr1=21, chr2=41))
> gr2 # so we can see what these ranges look like
GRanges with 2 ranges and 0 metadata columns:
      seqnames     ranges strand
         <Rle>  <IRanges>  <Rle>
  [1]     chr1   [ 4,  9]      +
  [2]     chr2   [12, 17]      -
  ---
  seqlengths:
   chr1 chr2
     21   41
> gaps(gr2)
GRanges with 8 ranges and 0 metadata columns:
      seqnames     ranges strand
         <Rle>  <IRanges>  <Rle>
  [1]     chr1   [ 1,  3]      +
  [2]     chr1   [10, 21]      +
  [3]     chr1   [ 1, 21]      -
  [4]     chr1   [ 1, 21]      *
  [5]     chr2   [ 1, 41]      +
  [6]     chr2   [ 1, 11]      -
  [7]     chr2   [18, 41]      -
  [8]     chr2   [ 1, 41]      *
  ---
  seqlengths:
   chr1 chr2
     21   41
```

That's a lot of gaps—a lot more than we'd expect from two ranges! What's going on? This function is dealing with the complexities of strands. When applied to GRanges, gaps() creates ranges for all sequences (chr1 and chr2 here) and all strands (+, -, and ambiguous strand *). For sequence-strand combinations without any ranges, the gap spans *the entire chromosome*. GRanges does this because this is the safest way to return gaps: anything less would lead to a loss of information. If we just care about gaps irrespective of strand, we could take the original gr2 ranges, ditch strand information, and then create gaps with these (ignoring the gaps across entire chromosome-strand combinations):

```
> gr3 <- gr2
> strand(gr3) <- "*"
> gaps(gr3)[strand(gaps(gr3)) == "*"]
GRanges with 4 ranges and 0 metadata columns:
      seqnames    ranges strand
         <Rle> <IRanges>  <Rle>
  [1]     chr1  [ 1,  3]      *
  [2]     chr1  [10, 21]      *
  [3]     chr2  [ 1, 11]      *
  [4]     chr2  [18, 41]      *
  ---
  seqlengths:
   chr1 chr2
     21   41
```

Replacing strand with the ambiguous strand * is a common trick when we don't care about keeping strand information. With gaps, we usually don't care about the specifics of strand—we usually say a region is covered by a range, or it's a gap.

Another approach to creating gaps using range operations is to use the set operations covered in "Basic Range Operations: Arithmetic, Transformations, and Set Operations" on page 275. A good illustration of the advantage of using set operations on GRanges objects is creating intergenic regions from all transcripts. This can be thought of as taking a set of ranges that represent entire chromosomes, and taking the set difference of these and all transcripts:

```
> chrom_grngs <- as(seqinfo(txdb), "GRanges")  ❶
> head(chrom_grngs, 2)
GRanges with 2 ranges and 0 metadata columns:
       seqnames           ranges strand
          <Rle>        <IRanges>  <Rle>
  chr1     chr1 [1, 195471971]       *
  chr2     chr2 [1, 182113224]       *
  ---
  seqlengths:
                    chr1                 chr2 ...      chrUn_JH584304
               195471971            182113224 ...              114452
> collapsed_tx <- reduce(transcripts(txdb))  ❷
```

```
> strand(collapsed_tx) <- "*"  ❸
> intergenic <- setdiff(chrom_grngs, collapsed_tx)  ❹
```

❶ First, we use the `as()` method to coerce the `TranscriptDb` object's chromosome information (such as names, lengths, etc.) into a `GRanges` object representing the ranges for entire chromosomes. In general, Bioconductor packages include many helpful coercion methods through the `as()` method. As the next line demonstrates, the result of this is indeed a `GRanges` object representing entire chromosomes.

❷ Next, we take all transcripts from our `txdb` `TranscriptDb` object and reduce them, so overlapping transcripts are collapsed into a single range. We collapse these overlapping ranges because we don't care about the gaps between each transcript, but rather any region *covered* by a transcript (which is what `reduce()` returns).

❸ As before, we ditch strand because we want all regions uncovered by a transcript on either strand.

❹ Finally, we take the set difference between ranges representing an entire chromosome, and those that represent transcripts on those ranges. This leaves all regions without a transcript, which are the intergenic regions. Technically, this is a more inclusive set of intergenic ranges, as we're saying a "gene" is any transcript, not just protein coding transcripts. This is yet another example where a common bioinformatics task can be easily accomplished with range operations and range-thinking. Try drawing the reduced transcript ranges and chromosome ranges out and seeing what `setdiff()` does to understand what's going on.

Now, let's look at how to create `GRanges` objects representing the introns of transcripts. We're going to do this two ways: first, using a simple convenience function appropriately named `intronsByTranscripts()`, then using range set operations. The former method is simple and fast (it gives you a fish), while the latter method teaches you really important range manipulations that will allow you to solve many other range problems (it teaches you to fish). First, let's consider the simple solution that uses the `TranscriptDb` object `txdb` we loaded earlier:

```
> mm_introns <- intronsByTranscript(txdb)
> head(mm_introns[['18880']], 2) # get first two introns for transcript 18880
GRanges with 2 ranges and 0 metadata columns:
      seqnames                 ranges strand
         <Rle>              <IRanges>  <Rle>
  [1]     chr3 [113556174, 113558092]      -
  [2]     chr3 [113558219, 113558321]      -
  ---
  seqlengths:
```

| chr1 | chr2 ... | chrUn_JH584304 |
| 195471971 | 182113224 ... | 114452 |

We'll now look at the manual approach that uses range set operations. This is a bit advanced, so if you're feeling lost, you can skip ahead and come back to this later (but it does illustrate some awesome range manipulation tricks). We'll make this example simpler by only creating the introns for a single gene, amylase 1 (which from the Ensembl website has gene identifier ENSMUSG00000074264). The set operation approach we'll take considers introns as the set difference between transcripts range and the exons' ranges for these transcripts. Each set difference is computed pairwise between the exons of a particular transcript and the transcript's range. Let's first get the transcripts for the amylase 1 gene:

```
> amy1 <- transcriptsBy(txdb, 'gene')$ENSMUSG00000074264
> amy1
GRanges with 5 ranges and 2 metadata columns:
        seqnames                 ranges strand |     tx_id             tx_name
           <Rle>              <IRanges>  <Rle> | <integer>         <character>
   [1]      chr3 [113555710, 113577830]     - |     18879 ENSMUST00000067980
   [2]      chr3 [113555953, 113574762]     - |     18880 ENSMUST00000106540
   [3]      chr3 [113556149, 113562018]     - |     18881 ENSMUST00000172885
   [4]      chr3 [113562690, 113574272]     - |     18882 ENSMUST00000142505
   [5]      chr3 [113564987, 113606699]     - |     18883 ENSMUST00000174147
   ---
   seqlengths:
                 chr1            chr2 ...   chrUn_JH584304
            195471971       182113224 ...           114452
```

There are five transcripts in the amylase 1 gene that we need to create introns for. Each of these transcripts contains a different set of exons. We'll extract all exons from the TranscriptDb object first, and then subset out the ones we need for these transcripts later:

```
> mm_exons <- exonsBy(txdb, "tx")
> mm_exons[[18881]] # an example exon GRanges object to see what it looks like
GRanges object with 5 ranges and 3 metadata columns:
        seqnames                 ranges strand |   exon_id   exon_name exon_rank
           <Rle>              <IRanges>  <Rle> | <integer> <character> <integer>
   [1]      chr3 [113561824, 113562018]     - |     68132        <NA>         1
   [2]      chr3 [113561632, 113561731]     - |     68130        <NA>         2
   [3]      chr3 [113558322, 113558440]     - |     68129        <NA>         3
   [4]      chr3 [113558093, 113558218]     - |     68128        <NA>         4
   [5]      chr3 [113556149, 113556173]     - |     68127        <NA>         5
   -------
   seqinfo: 66 sequences (1 circular) from mm10 genome
```

mm_exons contains *all* mouse exons, so we first need to extract only the exons belonging to our amylase 1 gene transcripts. A nice feature of GRangesList objects created by the split() method is that each list element is given the name of the vector used to split the ranges. Similarly, each list element created by exonsBy(txdb, "tx") is also named, using the transcript names (because we used by="tx"). Thus, we can easily match up our transcripts and exons by transcript identifiers, which are the element names of both the GRangesList objects amy1_tx and mm_exons. Matching these two GRangesList objects is then just a matter of using match():

```
> amy1_exons <- mm_exons[match(names(amy1_tx), names(mm_exons))]
```

Here's what amy1_exons now looks like (only one transcript's exons shown):

```
> amy1_exons
GRangesList object of length 5:
$18879
GRanges object with 11 ranges and 3 metadata columns:
        seqnames               ranges strand |    exon_id   exon_name [...]
           <Rle>            <IRanges>  <Rle> |  <integer> <character> [...]
   [1]      chr3 [113577701, 113577830]     - |      68142        <NA> [...]
   [2]      chr3 [113569845, 113570057]     - |      68139        <NA> [...]
   [3]      chr3 [113569382, 113569528]     - |      68138        <NA> [...]
   [4]      chr3 [113564869, 113565066]     - |      68136        <NA> [...]
   [5]      chr3 [113563445, 113563675]     - |      68135        <NA> [...]
   [...]
```

Now, we want to process each transcript and its exons *pairwise*. Our exons are already grouped by transcript, but our transcripts are in a single GRanges object (because these are transcripts grouped by gene, and we're looking at one gene). So we split up transcript ranges into a GRangeList object, this time grouping by transcript identifier:

```
> amy1_tx <- split(amy1, amy1$tx_id)
> amy1_tx
GRangesList object of length 5:
$18879
GRanges object with 1 range and 2 metadata columns:
        seqnames               ranges strand |     tx_id             tx_name
           <Rle>            <IRanges>  <Rle> | <integer>         <character>
   [1]      chr3 [113555710, 113577830]     - |     18879 ENSMUST00000067980

$18880
GRanges object with 1 range and 2 metadata columns:
        seqnames               ranges strand | tx_id             tx_name
   [1]      chr3 [113555953, 113574762]     - | 18880 ENSMUST00000106540
   [...]
```

With both our exons and transcripts grouped by transcript, we can finally take the pairwise set difference (with psetdiff()) which creates the set of introns for each transcript. Remember, it's imperative when using pairwise set functions to make sure your two objects are correctly matched up!

```
> all(names(amy1_tx) == names(amy1_exons))  # check everything's matched up
[1] TRUE
> amy1_introns <- psetdiff(amy1_tx, amy1_exons)
> head(amy1_introns[['18880']], 2) # the first two introns of amylase
```

```
                              # 1 transcript 18880
GRanges with 2 ranges and 0 metadata columns:
       seqnames                      ranges strand
          <Rle>                   <IRanges>  <Rle>
   [1]      chr3 [113556174, 113558092]        -
   [2]      chr3 [113558219, 113558321]        -
   ---
   seqlengths:
                      chr1                  chr2 ...       chrUn_JH584304
                 195471971             182113224 ...               114452
```

Are the introns we created manually for this gene identical to those created by the function `intronsByTranscripts()`? It's always a good idea to validate your methods through comparison to another method or visual inspection using a genome browser. Here, we compare to the introns found with `intronsByTranscript()`:

```
> identical(mm_introns[names(amy1_tx)], amy1_introns)
[1] TRUE
```

Indeed, the manual method using `psetdiff()` works.

Finding and Working with Overlapping Ranges

Finding and counting overlaps are probably the most important operations in working with `GRanges` objects and ranges in general. Bioconductor's `GenomicRanges` package has functions for finding overlaps, subsetting by overlaps, and counting overlaps that are nearly identical to methods found in the `IRanges` package. There are a couple exceptions:

- Overlap methods that work with `GRanges` objects.
- Only consider overlaps of ranges on the same chromosome or sequence.
- You have the choice to consider how strands are handled.

Mind Your Overlaps (Part II)

Overlaps get quite complex very quickly (as discussed in a warning earlier). Nowhere is this more apparent than with RNA-seq, where many technical issues can make the simple act of estimating transcript abundance (by counting how many aligned sequencing reads overlap a transcript region) incredibly complicated. First, counting overlaps in RNA-seq yields quantitative results that are used in downstream statistical analyses (e.g., the abundance of estimates for a particular transcript). This means that bias and noise that enter the range overlap quantification process could lead to inaccuracies in differential expression tests. Second, sequencing reads may align ambiguously—to multiple spots in the genome equally well. Do we count these multiple mapping reads (sometimes known as "multireads")? Or discard them entirely? Some

---

modern RNA-seq quantification methods like RSEM (Li et al., 2011) attempt to rescue these multireads. Third, some reads may align uniquely, but overlap an exon that's
shared by two or more transcripts' isoforms. Should we ignore this read, count it once
for each transcript (leading to double counting), or assign it to a transcript? All of
these rather technical decisions in finding and counting overlaps can unfortunately
lead to different biological results. Accurate RNA-seq quantification is still an actively
researched area, and methods are still being developed and improving. In general, the
methods in this section are only appropriate for simple overlap operations, and may
not be appropriate for quantification tasks like estimating transcript abundance for
RNA-seq.

Some popular RNA-seq quantification tools are:

- RSEM, or RNA-Seq by Expectation Maximization (*http://dewey
  lab.biostat.wisc.edu/rsem/*) (Li et al., 2011).
- TopHat (*http://bit.ly/TopHat-tool*) and The TopHat and Cufflinks (*http://
  cufflinks.cbcb.umd.edu/*) suite of tools (Trapnell et al., 2009; Trapnell et al., 2012).
- HTSeq (*http://bit.ly/HTSeq*) (Anders et al., 2014).
- The GenomicAlignments Bioconductor package (*http://bit.ly/genom-align*) (Lawrence et al., 2013). This new package from Bioconductor has special structures
  for storing genomic alignments and working with these alignments. This package
  also contains a summarizeOverlaps() method, which supports specific overlap
  options (and works directly on BAM files).

To demonstrate how findOverlaps() can be used with GRanges objects, we'll load in
a BED file of dbSNP (build 137) variants (in addition to SNPs, dbSNP also includes
other types of variants like insertions/deletions, short tandem repeats, multi-
nucleotide polymorphisms) for mouse chromosome 1. This BED file is available in
the book's GitHub repository, in the directory for this chapter. Using rtracklayer,
we'll load these in:

```
> library(rtracklayer)
> dbsnp137 <- import("mm10_snp137_chr1_trunc.bed.gz")
```

Suppose we want to find out how many of these variants fall into exonic regions, and
how many do not. Using the mouse TranscriptDb object we loaded earlier (txdb) we
can extract and collapse all overlapping exons with reduce(). We'll also subset so that
we're only looking at chromosome 1 exons (because our variants are only from chromosome 1):

```
> collapsed_exons <- reduce(exons(txdb), ignore.strand=TRUE)
> chr1_collapsed_exons <- collapsed_exons[seqnames(collapsed_exons) == "chr1"]
```

Let's explore our dbsnp137 object before looking for overlaps (remember the Golden
Rule: don't trust your data). Let's look at the length distribution of our variants:

```
> summary(width(dbsnp137))
   Min. 1st Qu.  Median    Mean 3rd Qu.    Max.
  0.000   1.000   1.000   1.138   1.000 732.000
> dbsnp137$name[which.max(width(dbsnp137))]
[1] "rs232497063"
```

The variant that's 732 bases long is a bit large, so we find out which entry it is and grab its RS identifier. Checking on the dbSNP website (*http://bit.ly/ref-snp*), we see that this is indeed a real variant—a rather large insertion/deletion on chromosome 1. From our summary(), we see that the vast majority of variants are 1 nucleotide long, which are either SNPs or 1 basepair insertons/deletions. Note that the minimum width is zero—there are also variants with zero widths. Using dbsnp137[width(dbsnp137) == 0], we can take look at a few of these. In most cases, these correspond to insertions into the reference genome (these can be easily verified with the dbSNP or UCSC Genome Browser websites). Zero-width ranges will not overlap *any* feature, as they don't have any region to overlap another range (create some range with zero widths and validate this). These technical details are rather annoying, but illustrate why it's important to inspect and understand our data and not just blindly apply functions. To count these zero-width features too, we'll resize using the resize() function:

```
> dbsnp137_resized <- dbsnp137
> zw_i <- width(dbsnp137_resized) == 0
> dbsnp137_resized[zw_i] <- resize(dbsnp137_resized[zw_i], width=1)
```

With this set of ranges, it's easy now to find out how many variants overlap our chromosome 1 exon regions. We'll use findOverlaps() to create a Hits object. We'll tell findOverlaps() to ignore strand:

```
> hits <- findOverlaps(dbsnp137_resized, chr1_collapsed_exons,
  ignore.strand=TRUE)
> hits
Hits of length 58346
queryLength: 2700000
subjectLength: 15048
      queryHits subjectHits
      <integer>   <integer>
 1        1250           1
 2        1251           1
 3        1252           1
 4        1253           1
 5        1254           1
[...]
> length(unique(queryHits(hits)))
[1] 58343
> length(unique(queryHits(hits)))/length(dbsnp137_resized)
[1] 0.02160852
```

The number of dbSNP variants on chromosome 1 that overlap our exons is given by the number of unique query hits in the Hits object: 118,594. This represents about

---

2% of the variants on chromosome 1, a figure that makes sense given that exonic regions make up small proportion of chromosome 1 (you should be able to work out what proportion of this is using GenomicRanges methods fairly easily).

Suppose we now wanted to look at the variants that do overlap these exons. We could do this by using the indices in the Hits object, but a simpler method is to use the method subsetByOverlaps():

```
> subsetByOverlaps(dbsnp137_resized, chr1_collapsed_exons, ignore.strand=TRUE)
GRanges with 118594 ranges and 2 metadata columns:
        seqnames              ranges strand |        name     score
           <Rle>           <IRanges>  <Rle> | <character> <numeric>
   [1]     chr1 [3054535, 3054535]        + |  rs30525614         0
   [2]     chr1 [3054556, 3054556]        + | rs233033126         0
   [3]     chr1 [3054666, 3054666]        + |  rs51218981         0
   [4]     chr1 [3054674, 3054674]        + |  rs49979360         0
   [5]     chr1 [3054707, 3054707]        + | rs108510965         0
   [...]
```

Note that the length of this GRanges object matches up with the number of overlapping variants that overlap exons earlier.

GenomicRanges also includes a method for counting overlaps, countOverlaps(). So suppose we wanted to count the number of variants that overlap each exonic region. Because we want the counts to be based on the exonic regions (which were the subject ranges in these operations), we reverse the order of the arguments:

```
> var_counts <- countOverlaps(chr1_collapsed_exons, dbsnp137_resized,
    ignore.strand=TRUE)
> head(var_counts)
[1]  6  0 35 48  2  5
```

To make these easier to follow, let's append them to our chromosome 1 exonic regions GRanges object as a metadata column:

```
> chr1_collapsed_exons$num_vars <- var_counts
> chr1_collapsed_exons
GRanges with 15048 ranges and 1 metadata column:
        seqnames              ranges strand |  num_vars
           <Rle>           <IRanges>  <Rle> | <integer>
   [1]     chr1 [3054233, 3054733]        * |         6
   [2]     chr1 [3102016, 3102125]        * |         0
   [3]     chr1 [3205901, 3207317]        * |        35
   [4]     chr1 [3213439, 3216968]        * |        48
   [5]     chr1 [3421702, 3421901]        * |         2
   [...]
```

At this point, it wouldn't be a bad idea to visit some of these regions (like "chr1:3054233-3054733") on the mouse UCSC Genome Browser and check that the number of variants on the dbSNP track matches up with the number we see here (6).

Calculating Coverage of GRanges Objects

The same coverage methods we saw with IRanges also work with GRanges and GRangesList objects. Let's generate some random fake 150bp reads on chromosome 19 (because it's the smallest chromosome) of our mouse genome and then empirically measure their coverage. We'll target 5x coverage. Using the famous Lander-Waterman coverage equation for coverage (C = LN/G, where C is coverage, L is the read length, N is the sequence length, and N is the number of reads), we see that for 150bp reads, a chromosome length of 61,431,566bp, and a target of 5x coverage, we need: 5*61,431,566/150 = 2,047,719 reads. Let's generate these using R's sample() function:

```
> set.seed(0)
> chr19_len <- seqlengths(txdb)['chr19']
> chr19_len
   chr19
61431566
> start_pos <- sample(1:(chr19_len-150), 2047719, replace=TRUE)
> reads <- GRanges("chr19", IRanges(start=start_pos, width=150))
```

Now, let's use the coverage() method from GenomicRanges to calculate the coverage of these random reads:

```
> cov_reads <- coverage(reads)
```

Coverage is calculated per every chromosome in the ranges object, and returned as a run-length encoded list (much like sequence names are returned by seqnames() from a GRangesList). We can calculate mean coverage per chromosome easily:

```
> mean(cov_reads)
   chr19
5.000001
```

It's also easy to calculate how much of this chromosome has no reads covering it (this will happen with shotgun sequencing, due to the random nature of read placement). We can do this two ways. First, we could use == and table():

```
> table(cov_reads == 0)
         FALSE      TRUE
chr19 61025072    406487
```

Or, we could use some run-length encoding tricks (these are faster, and scale to larger data better):

```
> sum(runLength(cov_reads)[runValue(cov_reads) == 0])
  chr19
406487
> 406487/chr19_len
     chr19
0.006616908
```

So, about 0.6% of our chromosome 19 remains uncovered. Interestingly, this is very close to the proportion of uncovered bases expected under Lander-Waterman, which is Poisson distributed with $\lambda = c$ (where c is coverage). Under the Poisson distribution, the expected proportion of uncovered genome is e^{-c} where c is coverage, which in this case is $e^{-c} \sim 0.0067$ (where e is Napier/Euler's constant).

Working with Ranges Data on the Command Line with BEDTools

While Bioconductor's IRanges, GenomicRanges, and GenomicFeatures packages are all powerful libraries for working with genomic interval data in R, for some tasks working with range data on the command line can be more convenient. While GenomicRanges offers the huge advantage of allowing interactive work, there's a cost to loading all data into memory at once so we can work with it. For example, suppose you wanted to calculate the coverage across a genome for each of 300 BED files. A single command that reads a BED file and calculates coverage would would be easier to run than loading the data into R, writing (simple) code to calculate coverage, and saving the results to file. While a custom solution allows more fine-tuned control, this means more work. A specialized command-line tool may offer less flexibility, but it's easier to get up and running and parallelize in Unix (we'll see how we could parallelize this in Chapter 12).

The BEDTools suite is a set of command-line tools for working directly with data kept in range file formats like BED, GTF, and GFF. Much like Git, BEDTools is a single command that uses a series of subcommands that do specific tasks. Each of these subcommands has similarities with one or more range operations we've seen in IRanges and GenomicRanges. The skills you've learned for manipulating IRanges and GenomicRanges objects are directly applicable to the range operations you'll perform with BEDTools. Because you're likely very familiar with range operations and "range-thinking" by now, we'll move through this material quickly, mostly as a demonstration of how powerful BEDTools is and when it will be useful in your work.

First, let's download the toy data we'll be playing with. All of these files are available in this chapter's directory in the GitHub repository. We'll continue using the data from *Mus musculus*, genome version mm10. Specifically, we'll use the following files:

- *ranges-qry.bed* is a simple BED file containing six ranges. These are the query ranges used in the GenomicRanges findOverlaps examples (except, because these are in BED format, they are 0-indexed).

- *ranges-sbj.bed* is the counterpart of ranges-sbj; these are the subject ranges used in the GenomicRanges findOverlaps examples. Both *ranges-sbj.bed* and *ranges-*

qry.bed are depicted in Figure 9-11 (though, this visualization uses 1-based coordinates).

- *Mus_musculus.GRCm38.75_chr1.gtf.gz* are features on chromosome 1 of *Mus_musculus.GRCm38.75.gtf.gz*. The latter file is Ensembl's annotation file for mm10 (which is the same as GRCh38) and was downloaded from Ensembl's FTP site (*http://bit.ly/mus-dl*).
- *Mus_musculus.GRCm38_genome.txt* is a tab-delimited file of all chromosome names and lengths from the mm10/GRCm38 genome version.

As a Unix command-line tool that works with plain-text range formats like BED, GTF, and GFF, BEDTools is desgined to be used in collaboration with other Unix tools like cut, sort, grep, and awk. For example, while BEDTools has its own sortBed command, it's more efficient to use Unix sort. Also, BEDTools operations that can be applied to many independent files can also be parallelized using Unix tools like xargs or GNU parallel (covered in Chapter 12).

Computing Overlaps with BEDTools Intersect

Overlaps are the most commonly used range-based operations in bioinformatics, so they're a natural place to start with BEDTools. The BEDTools intersect subcommand computes the overlaps between two sets of ranges. Run in the default mode, bedtools intersect works exactly like the GenomicRanges intersect() function. Let's use the BED files *ranges-qry.bed* and *ranges-sbj.bed*, which contain the same ranges used in the findOverlaps() example. Rather than thinking about ranges as *query* and *subject*, BEDTools labels ranges "A" and "B." We specify the input BED files "A" and "B" through the options -a and -b. We'll use BED files, but a nice feature of BEDTools is that its commands also work with GTF/GFF files.

Below, we calculate the intersection using bedtools intersect.

```
$ cat ranges-qry.bed
chr1    0       15      a
chr1    25      29      b
chr1    18      18      c
chr1    10      14      d
chr1    20      23      e
chr1    6       7       f
$ cat ranges-sbj.bed
chr1    0       4       x
chr1    18      28      y
chr1    9       15      z
$ bedtools intersect -a ranges-qry.bed -b ranges-sbj.bed
chr1    0       4       a
chr1    9       15      a
chr1    25      28      b
chr1    18      18      c
```

```
chr1    10      14      d
chr1    20      23      e
```

bedtools intersect returns the intersecting ranges with names from the file passed to the -a argument. Like GenomicRanges intersect(), only the overlap region is returned. For example, only the sections of range a in *ranges-qry.bed* that overlap the x and z ranges in *ranges-sbj.bed* are returned.

Alignment Files with BEDTools

While we're focusing on annotation/track formats like BED and GFF/GTF, BEDTools also has full support to work with BAM files, a binary alignment format we'll talk about in much detail in Chapter 10. For now, note that for subcommands like intersect and genomecov, you can also work with BAM files directly. BEDTools also has a subcommand to convert BAM files to BED files (bed tools bamtobed), but usually it's best to work with BAM files directly: the conversion from BAM to BED loses information stored about each alignment.

Often, we don't want just the intersect of overlapping query and subject ranges, but the *entire* query range. This is akin to using findOverlaps() to create a Hits object, and taking all query ranges with an overlap (covered in "Finding Overlapping Ranges" on page 281). With BEDTools, we can do this using the option -wa, which returns the ranges in "A" that overlap "B":

```
$ bedtools intersect -a ranges-qry.bed -b ranges-sbj.bed -wa
chr1    0       15      a
chr1    0       15      a
chr1    25      29      b
chr1    18      18      c
chr1    10      14      d
chr1    20      23      e
```

Now, note that because the range "a" overlaps *two* subjects (x and z), the full "a" range is reported *twice*, one for each subject. If you don't want these duplicates, you could specify the flag -u (for unique).

Similarly, there's a -wb option too:

```
$ bedtools intersect -a ranges-qry.bed -b ranges-sbj.bed -wb
chr1    0       4       a       chr1    0       4       x
chr1    9       15      a       chr1    9       15      z
chr1    25      28      b       chr1    18      28      y
chr1    18      18      c       chr1    18      28      y
chr1    10      14      d       chr1    9       15      z
chr1    20      23      e       chr1    18      28      y
```

Here, the overlapping region is first reported (this is the same as `bedtools inter sect` without any additional arguments), followed by the full ranges of "B" that "A" overlaps. Flags `-wa` and `-wb` can be combined to return the full "A" and "B" ranges that are overlapping:

```
$ bedtools intersect -a ranges-qry.bed -b ranges-sbj.bed -wa -wb
chr1    0       15      a       chr1    0       4       x
chr1    0       15      a       chr1    9       15      z
chr1    25      29      b       chr1    18      28      y
chr1    18      18      c       chr1    18      28      y
chr1    10      14      d       chr1    9       15      z
chr1    20      23      e       chr1    18      28      y
```

Running `bedtools intersect` on large files can use a lot of your system's memory. For very large files, it can be much more efficient to first sort both "A" and "B" files by chromosome and position, and then compute intersects on these sorted input files. `bedtools intersect` has a special algorithm that works faster (and uses less memory) on sorted input files; we enable this with `-sorted`. For example, using the fake files *query-sorted.bed* and *subject-sorted.bed*:

```
$ bedtools intersect -a query-sorted.bed -b subject-sorted.bed --sorted
```

These BED files would have to be sorted first, as we did in "Sorting Plain-Text Data with Sort" on page 147 with `sort -k1,1 -k2,2n`. Note that the performance benefits pay off most when intersecting large files; with small files, the benefits may be too small to notice or might be swamped out by fixed costs.

BEDTools's `intersect` has lots of functionality beyond just overlaps. For example, you can also return the number of overlapping bases with the `-wo` argument:

```
$ bedtools intersect -a ranges-qry.bed -b ranges-sbj.bed -wo
chr1    0       15      a       chr1    0       4       x       4
chr1    0       15      a       chr1    9       15      z       6
chr1    25      29      b       chr1    18      28      y       3
chr1    18      18      c       chr1    18      28      y       0
chr1    10      14      d       chr1    9       15      z       4
chr1    20      23      e       chr1    18      28      y       3
```

Like `grep -v` (which returns nonmatching lines), `bedtools intersect` also has a `-v` option, which returns all nonoverlapping ranges.

```
$ bedtools intersect -a ranges-qry.bed -b ranges-sbj.bed -v
chr1    6       7       f
```

Getting Help in BEDTools

Much like Git, BEDTools has extensive documentation built into the BEDTools command (and luckily, it's much less technical and easier to understand than Git's documentation). We can see all BEDTools subcommands by simply running bedtools without any arguments. Running a subcommand such as bedtools intersect without any arguments prints out the arguments, options, and useful notes for this command.

This test data isn't stranded, but with real data we need to be concerned with strand. bedtools intersect does not consider strand by default, so if you want to only look for intersects of features on the same strand, you must specify -s.

This is just a quick demonstration of bedtools intersect; see the full documentation (*http://bit.ly/bt-intersect*) for much more information.

BEDTools Slop and Flank

Like GenomicRanges, BEDTools has support for increasing the size of features and extracting flanking sequences. Earlier we saw how we could grow GRanges objects using arithemetic operators like + or the resize() function. With BEDTools, we grow ranges with bedtools slop. Unlike bedtools intersect, slop takes a single input range file through the argument -i. We also need to provide bedtools slop with a tab-delimited genome file that specifies the length of each chromosome—this is what BEDTools uses to ensure that ranges don't grow past the end of the of the chromosome. For our fake test data, we'll create this using echo, but in general we could use bioawk (discussed in "Bioawk: An Awk for Biological Formats" on page 163):

```
$ bioawk -c fastx '{print $name"\t"length($seq)}' your_genome.fastq > genome.txt
```

So for our test data, we'd use:

```
$ echo -e "chr1\t100" > genome.txt
$ bedtools slop -i ranges-qry.bed -g genome.txt -b 4 ❶
chr1    0    19    a
chr1    21   33    b
chr1    14   22    c
chr1    6    18    d
chr1    16   27    e
chr1    2    11    f
$ bedtools slop -i ranges-qry.bed -g genome.txt -l 3 -r 5 ❷
chr1    0    20    a
chr1    22   34    b
chr1    15   23    c
chr1    7    19    d
chr1    17   28    e
chr1    3    12    f
```

❶ Here, the argument `-b 4` tells `bedtools slop` to grow both sides of each range by 4 base pairs. If this runs past the beginning or end of the chromosome (as range a in *ranges-qry.bed* does), this will return 0 or the total chromosome length, respectively.

❷ Optionally, you can specify how many bases to add to the left (`-l`) and right (`-r`) sides of each range. Either can be set to zero to resize only one side.

BEDTools also has a tool for extracting flanking ranges, which is handy for getting promoter sequences. As with `slop`, `flank` needs a genome file containing the lengths of each chromosome sequence. Let's use `bedtools flank` to extract some promoter regions for genes. We'll use the *Mus_musculus.GRCm38.75_chr1.gtf.gz* GTF file. However, this file contains all types of features: exons, transcripts, CDS, noncoding regions, and so on (look at it with `zless` for confirmation). If we only want to use promoters for each protein coding gene, we can use the Unix tricks of Chapter 7 to extract only these ranges:

```
$ bioawk -cgff '{if ($feature == "gene") print $0}' \
           Mus_musculus.GRCm38.75_chr1.gtf.gz | \
    grep 'gene_biotype "protein_coding";' > mm_GRCm38.75_protein_coding_genes.gtf
```

Mind Your Chromosome Names

It's very likely that when working with data (annotation or other data) from other sources, you'll run into the different chromosome naming scheme problem. For example, the UCSC Genome Browser gives *Mus musculus* chromosome names like "chr1," "chrX," etc. while Ensembl uses named like "1," "X," etc. BEDTools is good about warning you when it can't find a sequence length in a genome file, but it won't warn for subcommands that don't require a genome file (like `intersect`). Always check that your chromosome names across different data sources are compatible.

Now, using these protein coding gene feature and the mm10/GRCm38 genome file in this chapter's GitHub repository, we can extract 3kbp left of each range:

```
$ bedtools flank -i mm_GRCm38.75_protein_coding_genes.gtf \
                 -g Mus_musculus.GRCm38_genome.txt \
                 -l 3000 -r 0 > mm_GRCm38_3kb_promoters.gtf
$ cut -f1,4,5,7 mm_GRCm38.75_protein_coding_genes.gtf | head -n 3
1       3205901 3671498 -
1       4343507 4360314 -
1       4490928 4496413 -
$ cut -f1,4,5,7 mm_GRCm38_3kb_promoters.gtf | head -n 3
1       3671499 3674498 -
1       4360315 4363314 -
1       4496414 4499413 -
```

Check that these upstream ranges make sense, minding the very important fact that these are all on the negative strand (a common stumbling block).

We can use the `bedtools getfasta` subcommand to extract sequences for a given set of ranges. For example, we could use `bedtools getfasta` to extract the promoter sequences for the ranges we've just created. Including the entire mouse genome in this book's GitHub repository would be a bit too much, so rather than using the full genome, we'll use the chromosome 1 file, *Mus_musculus.GRCm38.75.dna_rm.toplevel_chr1.fa*. You'll need to unzip this first from *Mus_musculus.GRCm38.75.dna_rm.toplevel_chr1.fa.gz*, then run bedtools getfasta:

```
$ gunzip Mus_musculus.GRCm38.75.dna_rm.toplevel_chr1.fa.gz
$ bedtools getfasta -fi Mus_musculus.GRCm38.75.dna_rm.toplevel_chr1.fa \ ❶
    -bed mm_GRCm38_3kb_promoters.gtf -fo mm_GRCm38_3kb_promoters.fasta    ❷
```

❶ The input FASTA file with `-fi`.

❷ Input BED file is specified with `-bed` and output FASTA file of extracted sequences is specified with `-fo`.

Coverage with BEDTools

BEDTools's `genomecov` subcommand is a versatile tool for summarizing the coverage of features along chromosome sequences. By default, it summarizes the coverage per chromosome sequence (and across the entire genome) as a histogram. A simple example will make this clearer:

```
$ cat ranges-cov.bed
chr1    4       9
chr1    1       6
chr1    8       19
chr1    25      30
chr2    0       20
$ cat cov.txt
chr1    30
chr2    20
$ bedtools genomecov -i ranges-cov.bed -g cov.txt
chr1    0       7       30      0.233333 ❶
chr1    1       20      30      0.666667
chr1    2       3       30      0.1
chr2    1       20      20      1 ❷
genome  0       7       50      0.14 ❸
genome  1       40      50      0.8
genome  2       3       50      0.06
```

❶ By default, `genomecov` summarizes the number of bases covered at a certain depth, per chromosome. The columns are depth, how many bases covered at this depth, total number of bases per chromosome, and proportion of bases covered at this depth. Here, we see that 23% of chr1's bases are not covered (coverage of zero); 7/30 is around 0.23.

❷ chr2 is entirely covered by the only range on this sequence, so its proportion covered at depth 1 is 1.

❸ genomecov also includes genome-wide statistics on coverage.

Sorting Chromosomes

BEDTools genomecov requires that all ranges from a particular chromosome are grouped together—in other words, consecutive in the BED file. Because the first column of a BED file is the chromosome, we can easily group features with Unix sort, using: `sort -k1,1 ranges.bed > ranges.sorted.bed`. See "Sorting Plain-Text Data with Sort" on page 147 for more information on Unix `sort`.

With `bedtools genomecov`, it's also possible to get per-base pair coverage. This is much like GenomicRanges coverage():

```
$ bedtools genomecov -i ranges-cov.bed -g cov.txt -d  | head -n5
chr1    1       0
chr1    2       1
chr1    3       1
chr1    4       1
chr1    5       2
```

These three columns correspond to the chromosome, the position on that chromosome, and the coverage at that position. As you might imagine, this is not the most compact format for per-base coverage statistics. A much better solution is to use the BedGraph format. BedGraph is similar to the run-length encoding we encountered earlier, except runs of the same depth are stored as the ranges. We can tell genomecov to output data in BedGraph format using:

```
$ bedtools genomecov -i ranges-cov.bed -g cov.txt -bg
chr1    1       4       1
chr1    4       6       2
chr1    6       8       1
chr1    8       9       2
chr1    9       19      1
chr1    25      30      1
chr2    0       20      1
```

Other BEDTools Subcommands and pybedtools

There are many other useful subcommands in the BEDTools suite that we don't have the space to discuss. BEDTools subcommands are extensively well documented though, both within the command itself (discussed earlier) and online at the BED-Tools documentation site (*http://bedtools.readthedocs.org*). The following are a few useful subcommands worth reading about:

bedtools annotate

> This is an incredibly useful command that takes a set of files (e.g., BED files for CpG islands, conservation tracks, methylation regions, etc.) and annotates how much coverage each of these files has over another input file (e.g., exonic regions).

bedtools merge

> merge is a lot like the GenomicRanges reduce() method; it merges overlapping ranges into a single range.

bedtools closest

> This subcommand is a lot like the GenomicRanges nearest() method.

bedtools complement

> This is the BEDTools version of gaps() (similar to setdiff() too).

bedtools multicov

> multicov counts the number of alignments in multiple BAM files that overlap a specified BED file. This is an incredibly useful tool for counting overlaps across numerous BED files, which could represent different samples or treatments.

bedtools multiinter

> This subcommand is similar to bedtools intersect, but works with multiple file inputs. multiinter will return whether each input BED file overlaps a given feature. This is useful for comparing whether experimental data (e.g., ChIP-Seq peaks or DNase I hypersensitivity sites) for many samples intersect features differently across samples. Similarly, the subcommand bedtools jaccard can calculate similarity metrics for pairs of datasets to summarize similarities across samples through a simple statistic (Jaccard Similarity).

bedtools unionbedg

> This subcommand merges multiple BedGraph files into a single file.

This is not an exhaustive subcommand list; see the full listing at the BEDTools documentation website (*http://bedtools.readthedocs.org/*).

Finally, it's worth mentioning the Pybedtools library (*http://pythonhosted.org/pybedtools/*) (Dale et al., 2011). Pybedtools is an excellent Python wrapper for BEDTools that allows you to employ BEDTools's range operations from within Python.

## CHAPTER 10

Working with Sequence Data

One of the core issues of Bioinformatics is dealing with a profusion of (often poorly defined or ambiguous) file formats. Some *ad hoc* simple human readable formats have over time attained the status of de facto standards.

—Peter Cock et al. (2010)

Good programmers know what to write. Great ones know what to rewrite (and reuse).

— *The Cathedral and the Bazaar* Eric S. Raymond

Nucleotide (and protein) sequences are stored in two plain-text formats widespread in bioinformatics: FASTA and FASTQ—pronounced fast-ah (or fast-A) and fast-Q, respectively. We'll discuss each format and their limitations in this section, and then see some tools for working with data in these formats. This is a short chapter, but one with an important lesson: beware of common pitfalls when working with *ad hoc* bioinformatics formats. Simple mistakes over minor details like file formats can consume a disproportionate amount of time and energy to discover and fix, so mind these details early on.

The FASTA Format

The FASTA format originates from the FASTA alignment suite, created by William R. Pearson and David J. Lipman. The FASTA format is used to store any sort of sequence data not requiring per-base pair quality scores. This includes reference genome files, protein sequences, coding DNA sequences (CDS), transcript sequences, and so on. FASTA can also be used to store multiple alignment data, but we won't discuss this specialized variant of the format here. We've encountered the FASTA format in earlier chapters, but in this section, we'll cover the format in more detail, look at common pitfalls, and introduce some tools for working with this format.

FASTA files are composed of sequence entries, each containing two parts: a description and the sequence data. The description line begins with a greater than symbol (>) and contains the sequence identifier and other (optional) information. The sequence data begins on the next line after the description, and continues until there's another description line (a line beginning with >) or the file ends. The *egfr_flank.fasta* file in this chapter's GitHub directory is an example FASTA file:

```
$ head -10 egfr_flank.fasta
>ENSMUSG00000020122|ENSMUST00000138518
CCCTCCTATCATGCTGTCAGTGTATCTCTAAATAGCACTCTCAACCCCCGTGAACTTGGT
TATTAAAAACATGCCCAAAGTCTGGGAGCCAGGGCTGCAGGGAAATACCACAGCCTCAGT
TCATCAAAACAGTTCATTGCCCAAAATGTTCTCAGCTGCAGCTTTCATGAGGTAACTCCA
GGGCCCACCTGTTCTCTGGT
>ENSMUSG00000020122|ENSMUST00000125984
GAGTCAGGTTGAAGCTGCCCTGAACACTACAGAGAAGAGAGGCCTTGGTGTCCTGTTGTC
TCCAGAACCCCAATATGTCTTGTGAAGGGCACACAACCCCTCAAAGGGGTGTCACTTCTT
CTGATCACTTTTGTTACTGTTTACTAACTGATCCTATGAATCACTGTGTCTTCTCAGAGG
CCGTGAACCACGTCTGCAAT
```

The FASTA format's simplicity and flexibility comes with an unfortunate downside: the FASTA format is a loosely defined *ad hoc* format (which unfortunately are quite common in bioinformatics). Consequently, you might encounter variations of the FASTA format that can lead to subtle errors unless your programs are robust to these variations. This is why it's usually preferable to use existing FASTA/FASTQ parsing libraries instead of implementing your own; existing libraries have already been vetted by the open source community (more on this later).

Most troubling about the FASTA format is that there's no universal specification for the format of an identifier in the description. For example, should the following FASTA descriptions refer to the same entry?

```
>ENSMUSG00000020122|ENSMUST00000138518
> ENSMUSG00000020122|ENSMUST00000125984
>ENSMUSG00000020122|ENSMUST00000125984|epidermal growth factor receptor
>ENSMUSG00000020122|ENSMUST00000125984|Egfr
>ENSMUSG00000020122|ENSMUST00000125984|11|ENSFM00410000138465
```

Without a standard scheme for identifiers, we can't use simple exact matching to check if an identifier matches a FASTA entry header line. Instead, we would need to rely on fuzzy matching between FASTA descriptions and our identifier. This could get quite messy quickly: how permissive should our pattern be? Do we run the risk of matching the wrong sequence with too permissive of a regular expression? Fundamentally, fuzzy matching is a fragile strategy.

Fortunately, there's a better solution to this problem (and it's quite simple, too): rather than relying on post-hoc fuzzy matching to correct inconsistent naming, start off with a strict naming convention and be consistent. Then, run any data from outside sources through a few sanity checks to ensure it follows your format. These checks don't need to be complex (check for duplicate names, inspect some entries by hand, check for errant spaces between the > and the identifier, check the overlap in names between different files, etc.).

If you need to tidy up outside data, always keep the original file and write a script that writes a corrected version to a new file. This way, the script can be easily rerun on any new version of the original dataset you receive (but you'll still need to check every-thing—don't blindly trust data!).

A common naming convention is to split the description line into two parts at the first space: the identifier and the comment. A sequence in this format would look like:

```
>gene_00284728 length=231;type=dna
GAGAACTGATTCTGTTACCGCAGGGCATTCGGATGTGCTAAGGTAGTAATCCATTATAAGTAACATGCGCGGAATATCCG
GAGGTCATAGTCGTAATGCATAATTATTCCCTCCCTCAGAAGGACTCCCTTGCGAGACGCCAATACCAAAGACTTTCGTA
GCTGGAACGATTGGACGGCCCAACCGGGGGGGAGTCGGCTATACGTCTGATTGCTACGCCTGGACTTCTCTT
```

Here `gene_00284728` is the identifier, and `length=231;type=dna` is the comment. Additionally, the ID should be unique. While certainly not a standard, the convention of treating everything before the first space as identifier and everything after as non-essential is common in bioinformatics programs (e.g., BEDtools, Samtools, and BWA all do this). With this convention in place, finding a particular sequence by identifier is easy—we'll see how to do this efficiently with indexed FASTA files at the end of this chapter.

The FASTQ Format

The FASTQ format extends FASTA by including a numeric quality score to each base in the sequence. The FASTQ format is widely used to store high-throughput sequenc-ing data, which is reported with a per-base quality score indicating the confidence of each base call. Unfortunately like FASTA, FASTQ has variants and pitfalls that can make the seemingly simple format frustrating to work with.

The FASTQ format looks like:

```
@DJB775P1:248:D0MDGACXX:7:1202:12362:49613  ❶
TGCTTACTCTGCGTTGATACCACTGCTTAGATCGGAAGAGCACACGTCTGAA  ❷
+  ❸
JJJJJIIJJJJJJJHIHHHGHFFFFFFCEEEEEDBD?DDDDDDBDDDABDDCA  ❹
@DJB775P1:248:D0MDGACXX:7:1202:12782:49716
CTCTGCGTTGATACCACTGCTTACTCTGCGTTGATACCACTGCTTAGATCGG
+
IIIIIIIIIIIIIIIIHHHHHHFFFFFFEECCCCBCECCCCCCCCCCCCCCCC
```

❶ The description line, beginning with @. This contains the record identifier and other information.

❷ Sequence data, which can be on one or many lines.

❸ The line beginning with +, following the sequence line(s) indicates the end of the sequence. In older FASTQ files, it was common to repeat the description line here, but this is redundant and leads to unnecessarily large FASTQ files.

❹ The quality data, which can also be on one or many lines, but *must* be the same length as the sequence. Each numeric base quality is encoded with ASCII characters using a scheme we'll discuss later ("Base Qualities" on page 344).

As with FASTA, it's a common convention in FASTQ files to split description lines by the first space into two parts: the record identifier and comment.

FASTQ is deceivingly tricky to parse correctly. A common pitfall is to treat every line that begins with @ as a description line. However, @ is also a valid quality character. FASTQ sequence and quality lines can wrap around to the next line, so it's possible that a line beginning with @ could be a quality line and *not* a header line. Consequently, writing a parser that always treats lines beginning with @ as header lines can lead to fragile and incorrect parsing. However, we can use the fact that the number of quality score characters must be equal to the number of sequence characters to reliably parse this format—which is how the readfq parser introduced later on works.

The Ins and Outs of Counting FASTA/FASTQ Entries

As plain-text formats, it's easy to work with FASTQ and FASTA with Unix tools. A common command-line bioinformatics idiom is:

```
$ grep -c "^>" egfr_flank.fasta
5
```

As shown in "Pipes in Action: Creating Simple Programs with Grep and Pipes" on page 47 you must quote the > character to prevent the shell from interpreting it as a redirection operator (and overwriting your FASTA file!). This is a safe way to count FASTA files because, while the format is loosely defined, every sequence has a one-line description, and only these lines start with >.

We might be tempted to use a similar approach with FASTQ files, using @ instead of >:

```
$ grep -c "^@" untreated1_chr4.fq
208779
```

Which tells us *untreated1_chr4.fq* has 208,779 entries. But by perusing *untreated1_chr4.fq*, you'll notice that each FASTQ entry takes up four lines, but the total number of lines is:

```
$ wc -l untreated1_chr4.fq
  817420 untreated1_chr4.fq
```

and 817,420/4 = 204,355 which is quite different from what `grep -c` gave us! What happened? Remember, @ is a valid quality character, and quality lines can begin with

---

this character. You can use `grep "^@" untreated1_chr4.fq | less` to see examples of this.

If you're absolutely positive your FASTQ file uses four lines per sequence entry, you can estimate the number of sequences by estimating the number of lines with `wc -l` and dividing by four. If you're unsure if some of your FASTQ entries wrap across many lines, a more robust way to count sequences is with `bioawk`:

```
$ bioawk -cfastx 'END{print NR}' untreated1_chr4.fq
204355
```

Nucleotide Codes

With the basic FASTA/FASTQ formats covered, let's look at the standards for encoding nucleotides and base quality scores in these formats. Clearly, encoding nucleotides is simple: A, T, C, G represent the nucleotides adenine, thymine, cytosine, and guanine. Lowercase bases are often used to indicate *soft masked* repeats or low complexity sequences (by programs like RepeatMasker (*http://www.repeatmasker.org*) and Tandem Repeats Finder (*http://tandem.bu.edu/trf/trf.html*)). Repeats and low-complexity sequences may also be *hard masked*, where nucleotides are replaced with N (or sometimes an X).

Degenerate (or *ambiguous*) nucleotide codes are used to represent two or more bases. For example, N is used to represent any base. The International Union of Pure and Applied Chemistry (IUPAC) has a standardized set of nucleotides, both unambiguous and ambiguous (see Table 10-1).

Table 10-1. IUPAC nucleotide codes

IUPAC code	Base(s)	Mnemonic
A	Adenine	Adenine
T	Thymine	Thymine
C	Cytosine	Cytosine
G	Guanine	Guanine
N	A, T, C, G	aNy base
Y	C, T	pYrimidine
R	A, G	puRine
S	G, C	Strong bond

IUPAC code	Base(s)	Mnemonic
W	A, T	Weak bond
K	G, T	Keto
M	A, C	aMino
B	C, G, T	All bases except A, B follows A
D	A, G, T	All bases except C, D follows C
H	A, C, T	All bases except G, H follows G
V	A, C, G	All bases except T or U (for Uracil), V follows U

Some bioinformatics programs may handle ambiguous nucleotide differently. For example, the BWA read alignment tool converts ambiguous nucleotide characters in the reference genome to random bases (Li and Durbin, 2009), but with a random seed set so regenerating the alignment index twice will not lead to two different versions.

Base Qualities

Each sequence base of a FASTQ entry has a corresponding numeric quality score in the quality line(s). Each base quality scores is encoded as a single ASCII character. Quality lines look like a string of random characters, like the fourth line here:

```
@AZ1:233:B390NACCC:2:1203:7689:2153
GTTGTTCTTGATGAGCCATGAGGAAGGCATGCCAAATTAAAATACTGGTGCGAATTTAAT
+
CCFFFFHHHHHJJJJJEIFJIJIJJJIJIJJJJCDGHIIIGIGIJIJIIIIJIJJIJIIH
```

(This FASTQ entry is in this chapter's *README* file if you want to follow along.)

Remember, ASCII characters are just represented as integers between 0 and 127 under the hood (see man ascii for more details). Because not all ASCII characters are printable to screen (e.g., character echoing "\07" makes a "ding" noise), qualities are restricted to the printable ASCII characters, ranging from 33 to 126 (the space character, 32, is omitted).

All programming languages have functions to convert from a character to its decimal ASCII representation, and from ASCII decimal to character. In Python, these are the functions ord() and chr(), respectively. Let's use ord() in Python's interactive interpreter to convert the quality characters to a list of ASCII decimal representations:

```
>>> qual = "JJJJJJJJJJJJGJJJJJIIJJJJJIGJJJJJIJJJJJJJIJIJJJJHHHHHFFFDFCCC"
>>> [ord(b) for b in qual]
[74, 74, 74, 74, 74, 74, 74, 74, 74, 74, 74, 74, 71, 74, 74, 74, 74, 74, 73,
 73, 74, 74, 74, 74, 74, 73, 71, 74, 74, 74, 74, 74, 73, 74, 74, 74, 74, 74,
 74, 74, 73, 74, 73, 74, 74, 74, 74, 72, 72, 72, 72, 72, 70, 70, 70, 68, 70,
 67, 67, 67]
```

Unfortunately, converting these ASCII values to meaningful quality scores can be tricky because there are three different quality schemes: Sanger, Solexa, and Illumina (see Table 10-2). The Open Bioinformatics Foundation (OBF), which is responsible for projects like Biopython, BioPerl, and BioRuby, gives these the names fastq-sanger, fastq-solexa, and fastq-illumina. Fortunately, the bioinformatics field has finally seemed to settle on the Sanger encoding (which is the format that the quality line shown here is in), so we'll step through the conversion process using this scheme.

Table 10-2. FASTQ quality schemes (adapted from Cock et al., 2010 with permission)

Name	ASCII character range	Offset	Quality score type	Quality score range
Sanger, Illumina (versions 1.8 onward)	33–126	33	PHRED	0–93
Solexa, early Illumina (before 1.3)	59–126	64	Solexa	5–62
Illumina (versions 1.3–1.7)	64–126	64	PHRED	0–62

First, we need to subtract an *offset* to convert this Sanger quality score to a PHRED quality score. PHRED was an early base caller written by Phil Green, used for fluorescence trace data written by Phil Green. Looking at Table 10-2, notice that the Sanger format's offset is 33, so we subtract 33 from each quality score:

```
>>> phred = [ord(b)-33 for b in qual]
>>> phred
[41, 41, 41, 41, 41, 41, 41, 41, 41, 41, 41, 41, 38, 41, 41, 41, 41, 41, 40,
 40, 41, 41, 41, 41, 41, 40, 38, 41, 41, 41, 41, 41, 40, 41, 41, 41, 41, 41,
 41, 41, 40, 41, 40, 41, 41, 41, 41, 39, 39, 39, 39, 39, 37, 37, 37, 35, 37,
 34, 34, 34]
```

Now, with our Sanger quality scores converted to PHRED quality scores, we can apply the following formula to convert quality scores to the estimated probability the base is correct:

$$P = 10^{-Q/10}$$

To go from probabilities to qualities, we use the inverse of this function:

$$Q = -10 \log_{10} P$$

In our case, we want the former equation. Applying this to our PHRED quality scores:

```
>>> [10**(-q/10) for q in phred]
[1e-05, 1e-05, 1e-05, 1e-05, 1e-05, 1e-05, 1e-05, 1e-05, 1e-05, 1e-05, 1e-05,
 1e-05, 0.0001, 1e-05, 1e-05, 1e-05, 1e-05, 1e-05, 0.0001, 0.0001, 1e-05,
 1e-05, 1e-05, 1e-05, 1e-05, 0.0001, 0.0001, 1e-05, 1e-05, 1e-05, 1e-05,
 1e-05, 0.0001, 1e-05, 1e-05, 1e-05, 1e-05, 1e-05, 1e-05, 1e-05, 0.0001,
 1e-05, 0.0001, 1e-05, 1e-05, 1e-05, 1e-05, 0.0001, 0.0001, 0.0001, 0.0001,
 0.0001, 0.0001, 0.0001, 0.0001, 0.0001, 0.0001, 0.0001, 0.0001, 0.0001]
```

Converting between Illumina (version 1.3 to 1.7) quality data is an identical process, except we use the offset 64 (see Table 10-2). The Solexa conversion is a bit trickier because this scheme doesn't use the PHRED function that maps quality scores to probabilities. Instead, it uses $Q = (10^{P/10} + 1)^{-1}$. See Cock et al., 2010 for more details about this format.

Example: Inspecting and Trimming Low-Quality Bases

Notice how the bases' accuracies decline in the previous example; this a characteristic error distribution for Illumina sequencing. Essentially, the probability of a base being incorrect increases the further (toward the 3' end) we are in a read produced by this sequencing technology. This can have a profound impact on downstream analyses! When working with sequencing data, you should always

- Be aware of your sequencing technology's error distributions and limitations (e.g., whether it's affected by GC content)
- Consider how this might impact your analyses

All of this is experiment specific, and takes considerable planning.

Our Python list of base accuracies is useful as a learning tool to see how to convert qualities to probabilities, but it won't help us much to understand the quality profiles of millions of sequences. In this sense, a picture is worth a thousand words—and there's software to help us see the quality distributions across bases in reads. The most popular is the Java program FastQC (*http://bit.ly/FastQC*), which is easy to run and outputs useful graphics and quality metrics. If you prefer to work in R, you can use a Bioconductor package called qrqc (*http://bit.ly/quick-qc*) (written by yours truly). We'll use qrqc in examples so we can tinker with how we visualize this data ourselves.

Let's first install all the necessary programs for this example. First, install qrqc in R with:

```
> library(BiocInstaller)
> biocLite('qrqc')
```

Next, let's install two programs that will allow us to trim low-quality bases: sickle (*http://github.com/najoshi/sickle*) and seqtk (*http://github.com/lh3/seqtk*). seqtk is a general-purpose sequence toolkit written by Heng Li that contains a subcommand for

---

trimming low-quality bases off the end of sequences (in addition to many other useful functions). Both `sickle` and `seqtk` are easily installable on Mac OS X with Homebrew (e.g., with `brew install seqtk` and `brew install sickle`).

After getting these programs installed, let's trim the *untreated1_chr4.fq* FASTQ file in this chapter's directory in the GitHub repository. This FASTQ file was generated from the *untreated1_chr4.bam* BAM file in the `pasillaBamSubset` Bioconductor package (see the *README* file in this chapter's directory for more information). To keep things simple, we'll use each program's default settings. Starting with `sickle`:

```
$ sickle se -f untreated1_chr4.fq -t sanger -o untreated1_chr4_sickle.fq

FastQ records kept: 202866
FastQ records discarded: 1489
```

`sickle` takes an input file through `-f`, a quality type through `-t`, and trimmed output file with `-o`.

Now, let's run `seqtk trimfq`, which takes a single argument and outputs trimmed sequences through standard out:

```
$ seqtk trimfq untreated1_chr4.fq > untreated1_chr4_trimfq.fq
```

Let's compare these results in R. We'll use `qrqc` to collect the distributions of quality by position in these files, and then visualize these using `ggplot2`. We could load these in one at a time, but a nice workflow is to automate this with `lapply()`:

```
# trim_qual.R -- explore base qualities before and after trimming
library(qrqc)

# FASTQ files
fqfiles <- c(none="untreated1_chr4.fq",
             sickle="untreated1_chr4_sickle.fq",
             trimfq="untreated1_chr4_trimfq.fq")

# Load each file in, using qrqc's readSeqFile
# We only need qualities, so we turn off some of
# readSeqFile's other features.
seq_info <- lapply(fqfiles, function(file) {
                readSeqFile(file, hash=FALSE, kmer=FALSE)
                })

# Extract the qualities as dataframe, and append
# a column of which trimmer (or none) was used. This
# is used in later plots.
quals <- mapply(function(sfq, name) {
                qs <- getQual(sfq)
                qs$trimmer <- name
                qs
                }, seq_info, names(fqfiles), SIMPLIFY=FALSE)
```

```
# Combine separate dataframes in a list into single dataframe
d <- do.call(rbind, quals)

# Visualize qualities
p1 <- ggplot(d) + geom_line(aes(x=position, y=mean, linetype=trimmer))
p1 <- p1 + ylab("mean quality (sanger)") + theme_bw()
print(p1)

# Use qrqc's qualPlot with list produces panel plots
# Only shows 10% to 90% quantiles and lowess curve
p2 <- qualPlot(seq_info, quartile.color=NULL, mean.color=NULL) + theme_bw()
p2 <- p2 + scale_y_continuous("quality (sanger)")
print(p2)
```

This script produces two plots: Figures 10-1 and 10-2. We see the effect both trimming programs have on our data's quality distributions in Figure 10-2: by trimming low-quality bases, we narrow the quality ranges in base positions further in the read. In Figure 10-1, we see this increases mean quality across across the read, but we still see a decline in base quality over the length of the reads.

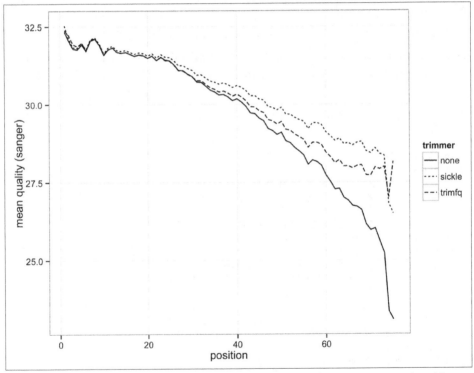

Figure 10-1. Mean base quality by position in the read with no trimming, with sickle and with seqtk trimfq

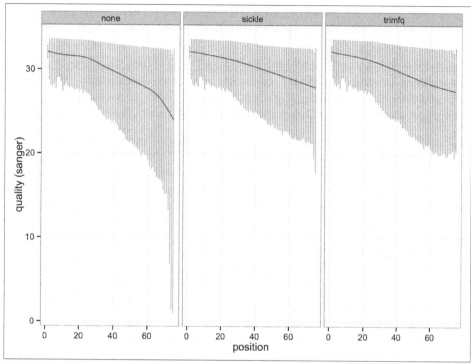

Figure 10-2. 10% to 90% quality range for each base position, with lowess curve for reads with no trimming, with sickle and with seqtk trimfq

In one line, we can trim low-quality bases from the ends of these sequences—running the trimming commands is not difficult. The more important step is to *visualize* what these trimming programs did to our data by comparing the files before and after trimming. Checking how programs change our data rather than trusting they did the right thing is an essential part of robust bioinformatics and abides by the Golden Rule (don't trust your tools). In this example, checking a small subset of data took fewer than 20 lines of code (ignoring blank lines and comments that improve readability) and only a few extra minutes—but it gives us valuable insight in what these programs do to our data and how they differ. If we like, we could also run both quality trimming programs with numerous different settings and compare how these affect our results. Much of careful bioinformatics is this process: run a program, compare output to original data, run a program, compare output, and so on.

A FASTA/FASTQ Parsing Example: Counting Nucleotides

It's not too difficult to write your own FASTA/FASTQ parser and it's a useful educational programming exercise. But when it comes to using a parser for processing real data, it's best to use a reliable existing library. Remember the quote at the beginning of this chapter: great programmers know when to reuse code. There are numerous open source, freely available, reliable parsers that have already been vetted by the open source community and tuned to be as fast as possible. We'll use Heng Li's readfq implementation because it parses both FASTA and FASTQ files, it's simple to use, and is standalone (meaning it doesn't require installing any dependencies). Biopython (*http://biopython.org/wiki/Main_Page*) and BioPerl (*http://www.bioperl.org/wiki/Main_Page*) are two popular libraries with good alternative FASTA/FASTQ parsers.

We'll use the Python implementation of readfq, *readfq.py*. You can obtain this Python file by downloading it from GitHub (*http://github.com/lh3/readfq*) (there's also a copy included in this book's repository). Although we could include *readfq.py*'s FASTA/FASTQ parsing routine using from readfq import readfq, for single file scripts it's simpler to just copy and paste the routine into your script. In this book, we'll use from readfq import readfq to avoid taking up space in examples.

readfq's readfq() function is simple to use. readfq() takes a file object (e.g., a filename that's been opened with open('filename.fa') or sys.stdin) and will generate FASTQ/FASTA entries until it reaches the end of the file or stream. Each FASTQ/FASTA entry is returned by readfq() as a tuple containing that entry's description, sequence, and quality. If readfq is reading a FASTA file, the quality line will be None. That's all there is to reading FASTQ/FASTA lines with readfq().

 If you dig into readfq()'s code, you'll notice a yield statement. This is a hint that readfq() is a *generator function*. If you're not familiar with Python's generators, you might want to read up on these at some point (though you don't need a detailed knowledge of these to use readfq()). I've included some resources in this chapter's *README* file on GitHub.

Let's write a simple program that counts the number of each IUPAC nucleotide in a file:

```
#!/usr/bin/env python
# nuccount.py -- tally nucleotides in a file
import sys
from collections import Counter ❶
from readfq import readfq

IUPAC_BASES = "ACGTRYSWKMBDHVN-." ❷

# intialize counter
counts = Counter() ❸

for name, seq, qual in readfq(sys.stdin): ❹
    # for each sequence entry, add all its bases to the counter
    counts.update(seq.upper()) ❺
```

```
# print the results
for base in IUPAC_BASES:
    print base + "\t" + str(counts[base])  ❻
```

❶ We use Counter in the collections module to count nucleotides. Counter works much like a Python dict (technically it's a subclass of dict), with added features that make counting items easier.

❷ This global variable defines all IUPAC nucleotides, which we'll use later to print bases in a consistent order (because like Python's dictionaries, Counter objects don't maintain order). It's a good practice to put constant variables like IUPAC_BASES in uppercase, and at the top of a script so they're clear to readers.

❸ Create a new Counter object.

❹ This line uses the readfq function in the readfq module to read FASTQ/FASTQ entries from the file handle argument (in this case, sys.stdin, standard in) into the name, seq, and qual variables (through a Python feature known as *tuple unpacking*).

❺ The Counter.update() method takes any iterable object (in this case the string of sequence bases), and adds them to the counter. We could have also used a for loop over each character in seq, incrementing counts with counts[seq.upper()] += 1. Note that we convert all characters to uppercase with the upper() method, so that lowercase soft-masked bases are also counted.

❻ Finally, we iterate over all IUPAC bases and print their counts.

This version takes input through standard in, so after saving this file and adding execute permissions with chmod +x nuccount.py, we could run it with:

```
$ cat contam.fastq | ./nuccount.py
A       103
C       110
G       94
T       109
R       0
Y       0
S       0
W       0
K       0
M       0
B       0
D       0
H       0
V       0
```

```
N    0
-    0
.    0
```

Note that we don't necessarily have to make the Python script executable; another option is to simply launch it with `python nuccount.py`. Either way leads to the same result, so this is ultimately a stylistic choice. But if you do want to make it an executable script, remember to do the following:

- Include the *shebang line* `#!/usr/bin/env python`
- Make the script executable with `chmod +x <scriptname.py>`

There are many improvements we could add to this script: add support for per-sequence base composition statistics, take file arguments rather than input through standard in, count soft-masked (lowercase) characters, count CpG sites, or warn when a non-IUPAC nucleotide is found in a file. Counting nucleotides is simple—the most complex part of the script is `readfq()`. This is the beauty of reusing code: well-written functions and libraries prevent us from having to rewrite complex parsers. Instead, we use software the broader community has collaboratively developed, tested, debugged, and made more efficient (see `readfq`'s GitHub commit history as an example). Reusing software isn't cheating—it's how the experts program.

Indexed FASTA Files

Very often we need efficient random access to subsequences of a FASTA file (given regions). At first glance, writing a script to do this doesn't seem difficult. We could, for example, write a script that iterates through FASTA entries, extracting sequences that overlaps the range specified. However, this is not an efficient method when extracting a few *random* subsequences. To see why, consider accessing the sequence from position chromosome 8 (123,407,082 to 123,410,742) from the mouse genome. This approach would needlessly parse and load chromosomes 1 through 7 into memory, even though we don't need to extract subsequences from these chromosomes. Reading entire chromosomes from disk and copying them into memory can be quite inefficient—we would have to load all 125 megabytes of chromosome 8 to extract 3.6kb! Extracting numerous random subsequences from a FASTA file can be quite computationally costly.

A common computational strategy that allows for easy and fast random access is *indexing* the file. Indexed files are ubiquitous in bioinformatics; in future chapters, we'll index genomes so they can be used as an alignment reference and we'll index files containing aligned reads for fast random access. In this section, we'll look at how indexed FASTA files allow us to quickly and easily extract subsequences. Usually, we often index an entire genome but to simplify the examples in the rest of this chapter, we will work with only chromosome 8 of the mouse genome. Download the

Mus_musculus.GRCm38.75.dna.chromosome.8.fa.gz file from this chapter's GitHub directory and unzip it with `gunzip`.

Indexing Compressed FASTA Files

Although `samtools faidx` *does* work with compressed FASTA files, it only works with BGZF-compressed files (a more advanced topic, which we'll cover in "Fast Access to Indexed Tab-Delimited Files with BGZF and Tabix" on page 425).

We'll index this file using Samtools, a popular suite of tools for manipulating the SAM and BAM alignment formats (which we'll cover in much detail in Chapter 11). You can install `samtools` via a ports or packages manager (e.g., Mac OS X's Hombrew or Ubuntu's `apt-get`). Much like Git, Samtools uses subcommands to do different things. First, we need to index our FASTA file using the `faidx` subcommand:

```
$ samtools faidx Mus_musculus.GRCm38.75.dna.chromosome.8.fa
```

This creates an index file named *Mus_musculus.GRCm38.75.dna.chromosome.8.fa.fai*. We don't have to worry about the specifics of this index file when extracting subsequences—`samtools faidx` takes care of these details. To access the subsequence for a particular region, we use `samtools faidx <in.fa> <region>`, where `<in.fa>` is the FASTA file (you've just indexed) and `<region>` is in the format `chromosome:start-end`. For example:

```
$ samtools faidx Mus_musculus.GRCm38.75.dna.chromosome.8.fa 8:123407082-123410744
>8:123407082-123410744
GAGAAAAGCTCCCTTCTTCTCCAGAGTCCCGTCTACCCTGGCTTGGCGAGGGAAAGGAAC
CAGACATATATCAGAGGCAAGTAACCAAGAAGTCTGGAGGTGTTGAGTTTAGGCATGTCT
[...]
```

Be sure to mind differences in chromosome syntax (e.g., UCSC's `chr8` format versus Ensembl's `8`). If no sequence is returned from `samtools faidx`, this could be why.

`samtools faidx` allows for multiple regions at once, so we could do:

```
$ samtools faidx Mus_musculus.GRCm38.75.dna.chromosome.8.fa \
      8:123407082-123410744 8:123518835-123536649
>8:123407082-123410744
GAGAAAAGCTCCCTTCTTCTCCAGAGTCCCGTCTACCCTGGCTTGGCGAGGGAAAGGAAC
CAGACATATATCAGAGGCAAGTAACCAAGAAGTCTGGAGGTGTTGAGTTTAGGCATGTCT
[...]
>8:123518835-123536649
TCTCGCGAGGATTTGAGAACCAGCACGGGATCTAGTCGGAGTTGCCAGGAGACCGCGCAG
CCTCCTCTGACCAGCGCCCATCCCGGATTAGTGGAAGTGCTGGACTGCTGGCACCATGGT
[...]
```

What Makes Accessing Indexed Files Faster?

In Chapter 3 (see "The Almighty Unix Pipe: Speed and Beauty in One" on page 45), we discussed how reading from and writing to the disk is exceptionally slow compared to data kept in memory. We can avoid needlessly reading the entire file off of the disk by using an index that points to where certain blocks are in the file. In the case of our FASTA file, the index essentially stores the location of where each sequence begins in the file (as well as other necessary information).

When we look up a range like chromosome 8 (123,407,082-123,410,744), `samtools faidx` uses the information in the index to quickly calculate exactly where in the file those bases are. Then, using an operation called a file *seek*, the program jumps to this exact position (called the *offset*) in the file and starts reading the sequence. Having precomputed file offsets combined with the ability to jump to these exact positions is what makes accessing sections of an indexed file fast.

## CHAPTER 11

Working with Alignment Data

In Chapter 9, we learned about range formats such as BED and GTF, which are often used to store genomic range data associated with genomic feature annotation data such as gene models. Other kinds of range-based formats are designed for storing large amounts of *alignment* data—for example, the results of aligning millions (or billions) of high-throughput sequencing reads to a genome. In this chapter, we'll look at the most common high-throughput data alignment format: the Sequence Alignment/Mapping (SAM) format for mapping data (and its binary analog, BAM). The SAM and BAM formats are the standard formats for storing sequencing reads mapped to a reference.

We study SAM and BAM for two reasons. First, a huge part of bioinformatics work is manipulating alignment files. Nearly every high-throughput sequencing experiment involves an alignment step that produces alignment data in the SAM/BAM formats. Because each sequencing read has an alignment entry, alignment data files are massive and require space-efficient complex binary file formats. Furthermore, modern aligners output an incredible amount of useful information about each alignment. It's vital to have the skills necessary to extract this information and explore data kept in these complex formats.

Second, the skills developed through learning to work with SAM/BAM files are extensible and more widely applicable than to these specific formats. It would be unwise to bet that these formats won't change (or even be replaced at some point)— the field of bioinformatics is notorious for inventing new data formats (the same goes with computing in general, see xkcd's "Standards" comic (*http://xkcd.com/927/*)). Some groups are already switching to storing alignments in CRAM format, a closely related alignment data format we'll also discuss. So while learning how to work with specific bioinformatics formats may seem like a lost cause, skills such as following a format specification, manipulating binary files, extracting information from bitflags, and working with application programming interfaces (APIs) are essential skills when working with any format.

Getting to Know Alignment Formats: SAM and BAM

Before learning to work with SAM/BAM, we need to understand the structure of these formats. We'll do this by using *celegans.sam*, a small example SAM file included in this chapter's directory in the GitHub repository.

The *celegans.sam* file was created by aligning reads simulated directly from the *C. elegans* genome (version WBcel235) using the wgsim read simulator. These reads differ slightly from the reference genome through simulated mutations and base errors. Simulating reads, realigning back to the reference, and calling SNPs is a very useful exercise in understanding the limitations of aligners and SNP callers; I encourage you to try this on your own. See the documentation in this chapter's directory on GitHub for more information on how these reads were simulated and why this is a useful exercise.

We'll step through the basic ideas of the SAM/BAM format, but note that as with any well-specified bioinformatics format, the ultimate reference is the original format specification and documentation, which is available on GitHub (*http://samtools.github.io/hts-specs/*). The original SAM Format paper (*http://bit.ly/sam-format*) (Li et al., 2009) is also a good introduction.

The SAM Header

Files in the SAM format consist of a header section and an alignment section. Because SAM files are plain text (unlike their binary counterpart, BAM), we can take a peek at a few lines of the header with head:

```
$ head -n 10 celegans.sam
@SQ     SN:I     LN:15072434  ❶
@SQ     SN:II    LN:15279421
@SQ     SN:III   LN:13783801
@SQ     SN:IV    LN:17493829
@SQ     SN:MtDNA       LN:13794
@SQ     SN:V     LN:20924180
@SQ     SN:X     LN:17718942
@RG     ID:VB00023_L001 SM:celegans-01  ❷
@PG     ID:bwa  PN:bwa  VN:0.7.10-r789 [...]  ❸
I_2011868_2012306_0:0:0_0:0:0_2489 83  I  2012257  40  50M  [...]  ❹
```

Header lines contain vital metadata about the reference sequences, read and sample information, and (optionally) processing steps and comments. Each header line begins with an @, followed by a two-letter code that distinguishes the different type of metadata records in the header. Following this two-letter code are tab-delimited key-value pairs in the format KEY:VALUE (the SAM format specification names these tags and values). The *celegans.sam* file contains the most common header records types you'll encounter in SAM/BAM files. Let's step through some of the header components in more detail:

❶ @SQ header entries store information about the reference sequences (e.g., the chromosomes if you've aligned to a reference genome). The required key-values are SN, which stores the sequence name (e.g., the *C. elegans* chromosome I), and LN, which stores the sequence length (e.g., 15,072,434 bases). All separate sequences in your reference have a corresponding entry in the header.

❷ @RG header entries contain important read group and sample metadata. The read group identifier ID is required and *must be unique*. This ID value contains information about the origin of a set of reads. Some software relies on read groups to indicate a technical groups of reads, to account for *batch effects* (undesirable technical artifacts in data). Consequently, it's beneficial to create read groups related to the specific sequencing run (e.g., ID could be related to the name of the sequencing run and lane).

Although ID is the only required key in @RG headers, in practice your SAM/BAM files should also keep track of sample information using the SM key. Sample information is the metadata about your experiment's samples (e.g., individual, treatment group, replicate, etc.). Finally, it's worth noting that the SAM format specification also allows a PL key for indicating sequencing platform such as ILLUMINA, PACBIO, and so on. (See the specification for a full list of valid values.) Read group, sample, and platform information should be added to your SAM/BAM during alignment (and aligners have options for this).

❸ @PG header entries contain metadata about the programs used to create and process a set of SAM/BAM files. Each program must have a unique ID value, and metadata such as program version number (via the VN key) and the exact command line (via the CL key) can be saved in these header entries. Many programs will add these lines automatically.

❹ This is the first line of the *alignment* section (because this line does not begin with @). We'll cover the alignment section in more detail in the following section.

This is just an introduction to the basics of the SAM format's header section; see the SAM format specification (*http://samtools.github.io/hts-specs/*) for more detail.

Read Groups

One of the best features of the SAM/BAM format is that it supports including an extensive amount of metadata about the samples, the alignment reference, processing steps, etc. to be included within the file. (Note that in contrast, the FASTQ format doesn't provide a standard way to include this metadata; in practice, we use file-names to connect metadata kept in a separate spreadsheet or tab-delimited file.) Many downstream applications make use of the metadata contained in the SAM header (and many programs require it). Given that this metadata is important (and often required), you should add read group and sample metadata when you align reads to a reference.

Luckily, most aligners allow you to specify this important metadata through your alignment command. For example, BWA allows (using made-up files in this example):

```
$ bwa mem -R'@RG\tID:readgroup_id\tSM:sample_id' ref.fa
  in.fq
```

Bowtie2 similarly allows read group and sample information to be set with the --rg-id and --rg options.

Although head works to take a quick peek at the top of a SAM file, keep the following points in mind:

- head won't always provide the entire header.
- It won't work with binary BAM files.

The standard way of interacting with SAM/BAM files is through the SAMtools command-line program (samtools), which we'll use extensively throughout the rest of this chapter. Like Git, samtools has many subcommands. samtools view is the general tool for viewing and converting SAM/BAM files. A universal way to look at an entire SAM/BAM header is with samtools view option -H:

```
$ samtools view -H celegans.sam
@SQ     SN:I    LN:15072434
@SQ     SN:II   LN:15279421
@SQ     SN:III  LN:13783801
[...]
```

This also works with BAM files, without any need to convert beforehand (samtools automatically detects whether the file is SAM or BAM):

```
$ samtools view -H celegans.bam
@SQ     SN:I    LN:15072434
@SQ     SN:II   LN:15279421
@SQ     SN:III  LN:13783801
[...]
```

Of course, all our usual Unix tricks can be combined with samtools commands through piping results to other commands. For example, we could see all read groups with:

```
$ samtools view -H celegans.bam | grep "^@RG"
@RG     ID:VB00023_L001 SM:celegans-01
```

samtools view without any arguments returns the entire alignment section without the header:

```
$ samtools view celegans.sam | head -n 1
I_2011868_2012306_0:0:0_0:0:0_2489      83      I       2012257 40      50M
```

The SAM Alignment Section

The alignment section contains read alignments (and usually includes reads that did not align, but this depends on the aligner and file). Each alignment entry is composed of 11 required fields (and optional fields after this).

 We'll step through the basic structure of an alignment entry, but it would be unnecessarily redundant to include all information in the original SAM format specification in this section. I highly recommend reading the alignment section of the SAM format specification for more detail.

Let's step through an alignment entry's fields. Because these alignment lines are quite lengthy and would overflow the width of this page, I use tr to convert tabs to newlines for a single alignment entry in *celegans.sam*:

```
$ samtools view celegans.sam | tr '\t' '\n' | head -n 11
I_2011868_2012306_0:0:0_0:0:0_2489 ❶
83 ❷
I ❸
2012257 ❹
40 ❺
50M ❻
= ❼
2011868
-439 ❽
CAAAAAATTTTGAAAAAAAAAAATTGAATAAAAATTCACGGATTTCTGGCT ❾
2222222222222222222222222222222222222222222222222222 ❿
```

❶ QNAME, the *query name* (e.g., a sequence read's name).

❷ FLAG, the *bitwise flag*, which contains information about the alignment. Bitwise flags are discussed in "Bitwise Flags" on page 360 in much more detail.

❸ RNAME, the *reference name* (e.g., which sequence the query aligned to, such as a specific chromosome name like "chr1"). The reference name must be in the SAM/BAM header as an SQ entry. If the read is unaligned, this entry may be *.

❹ POS, the *position* on the reference sequence (using 1-based indexing) of the first mapping base (leftmost) in the query sequence. This may be zero if the read does not align.

❺ MAPQ is the *mapping quality*, which is a measure of how likely the read is to actually originate from the position it maps to. Mapping quality is estimated by the aligner (and beware that different aligners have different estimation procedures!). Many tools downstream of aligners filter out reads that map with low mapping quality (because a low mapping quality score indicates the alignment program is not confident about the alignment's position). Mapping qualities are an incredibly important topic that we'll discuss in more depth later. Mapping quality is discussed in more depth in "Mapping Qualities" on page 365.

❻ CIGAR is the *CIGAR* string, which is a specialized format for describing the alignment (e.g., matching bases, insertions/deletions, clipping, etc.). We discuss CIGAR strings in much more detail in "CIGAR Strings" on page 363.

❼ RNEXT and PNEXT (on the next line) are the reference name and position (the R and P in RNEXT and PNEXT) of a paired-end read's partner. The value * indicates RNEXT is not available, and = indicates that RNEXT is the same as RNAME. PNEXT will be 0 when not available.

❽ TLEN is the template length for paired-end reads.

❾ SEQ stores the original read sequence. This sequence will always be in the orientation it aligned in (and this may be the reverse complement of the original read sequence). Thus, if your read aligned to the reverse strand (which is information kept in the bitwise flag field), this sequence will be the reverse complement.

❿ QUAL stores the original read base quality (in the same format as Sanger FASTQ files).

Bitwise Flags

Many important pieces of information about an alignment are encoded using *bitwise flags* (also known as a *bit field*). There's a lot of important information encoded in SAM/BAM bitwise flags, so it's essential you understand how these work. Furthermore, bitwise flags are a very space-efficient and common way to encode attributes, so they're worth understanding because you're very likely to encounter them in other formats.

Bitwise flags are much like a series of toggle switches, each of which can be either on or off. Each switch represents whether a particular attribute of an alignment is true or false, such as whether a read is unmapped, is paired-end, or whether it aligned in the reverse orientation. Table 11-1 shows these bitwise flags and the attributes they encode, but the most up-to-date source will be the SAM format specification or the `samtools flag` command:

```
$ samtools flags

About: Convert between textual and numeric flag representation
Usage: samtools flags INT|STR[,...]

Flags:
        0x1     PAIRED          .. paired-end (or multiple-segment) sequencing
                                   technology
        0x2     PROPER_PAIR     .. each segment properly aligned according to
                                   the aligner
        0x4     UNMAP           .. segment unmapped
[...]
```

Under the hood, each of these toggle switches' values are bits (0 or 1) of a *binary number* (the base-2 system of computing that uses 0s and 1s). Each bit in a bitfield represents a particular attribute about an alignment, with 1 indicating that the attribute is true and 0 indicating it's false.

Table 11-1. SAM bitwise flags

Flag (in hexidecimal)	Meaning
0x1	Paired-end sequence (or multiple-segment, as in strobe sequencing)
0x2	Aligned in proper pair (according to the aligner)
0x4	Unmapped
0x8	Mate pair unmapped (or next segment, if mulitple-segment)
0x10	Sequence is reverse complemented
0x20	Sequence of mate pair is reversed
0x40	The first read in the pair (or segment, if multiple-segment)
0x80	The second read in the pair (or segment, if multiple-segment)
0x100	Secondary alignment

Flag (in hexidecimal)	Meaning
0x200	QC failure
0x400	PCR or optical duplicate
0x800	Supplementary alignment

As an example, suppose you encounter the bitflag 147 (0x93 in hexidecimal) and you want to know what this says about this alignment. In binary this number is represented as 0x1001 0011 (the space is used to make this more readable). We see that the first, second, fifth, and eighth bits are 1 (in our switch analogy, these are the switches that are turned on). These specific bits correspond to the hexidecimal values 0x1, 0x2, 0x10, and 0x80. Looking at Table 11-1, we see these hexidecimal values correspond to the attributes paired-end, aligned in proper pair, the sequence is reverse complemented, and that this is the second read in the pair—which describes how our read aligned.

Converting between Binary, Hexadecimal, and Decimal

I won't cover the details of converting between binary, hexadecimal, and decimal number systems because the command samtools flags can translate bitflags for us. But if you continue to dig deeper into computing, it's a handy and necessary skill. I've included some supplementary resources in the GitHub repository's *README* file for this chapter. Many calculators, such as the OS X calculator in "programmer" mode will also convert these values for you.

Working through this each time would be quite tedious, so the samtools command contains the subcommand samtools flags, which can translate decimal and hexidecimal flags:

```
$ samtools flags 147
0x93    147    PAIRED,PROPER_PAIR,REVERSE,READ2
$ samtools flags 0x93
0x93    147    PAIRED,PROPER_PAIR,REVERSE,READ2
```

samtools flags can also convert attributes (of the prespecified list given by running samtools flags without arguments) to hexidecimal and decimal flags:

```
$ samtools flags paired,read1,qcfail
0x241   577    PAIRED,READ1,QCFAIL
```

Later on, we'll see how PySAM simplifies this through an interface where properties about an alignment are stored as attributes of an AlignedSegment Python object, which allows for easier checking of bitflag values with a syntax like `aln.is_proper_pair` or `aln.is_reverse`.

CIGAR Strings

Like bitwise flags, SAM's CIGAR strings are another specialized way to encode information about an aligned sequence. While bitwise flags store true/false properties about an alignment, CIGAR strings encode information about which bases of an alignment are matches/mismatches, insertions, deletions, soft or hard clipped, and so on. I'll assume you are familiar with the idea of matches, mismatches, insertions, and deletions, but it's worth describing soft and hard clipping (as SAM uses them).

Soft clipping is when only part of the query sequence is aligned to the reference, leaving some portion of the query sequence unaligned. Soft clipping occurs when an aligner can partially map a read to a location, but the head or tail of the query sequence doesn't match (or the alignment at the end of the sequence is questionable). Hard clipping is similar, but hard-clipped regions are not present in the sequence stored in the SAM field `SEQ`. A basic CIGAR string contains concatenated pairs of integer *lengths* and character *operations* (see Table 11-2 for a table of these operations).

Table 11-2. CIGAR operations

Operation	Value	Description
M	0	Alignment match (note that this could be a sequence match or mismatch!)
I	1	Insertion (to reference)
D	2	Deletion (from reference)
N	3	Skipped region (from reference)
S	4	Soft-clipped region (soft-clipped regions are present in sequence in SEQ field)
H	5	Hard-clipped region (not in sequence in SEQ field)
P	6	Padding (see section 3.1 of the SAM format specification for detail)
=	7	Sequence match

Operation	Value	Description
X	8	Sequence mismatch

For example, a fully aligned 51 base pair read without insertions or deletions would have a CIGAR string containing a single length/operation pair: 51M. By the SAM format specification, M means there's an *alignment match*, not that all bases in the query and reference sequence are identical (it's a common mistake to assume this!).

Sequence Matches and Mismatches, and the NM and MD Tags

It's important to remember that the SAM format specification simply lists what's possible with the format. Along these lines, aligners choose how to output their results in this format, and there are differences among aligners in how they use the CIGAR string (and other parts of the SAM format). It's common for many aligners to forgo using = and X to indicate sequence matches and mismatches, and instead just report these as M.

However, this isn't as bad as it sounds—many aligners have different goals (e.g., general read mapping, splicing-aware aligning, aligning longer reads to find chromosomal breakpoints). These tasks don't require the same level of detail about the alignment, so in some cases explicitly reporting matches and mismatches with = and X would lead to needlessly complicated CIGAR strings.

Also, much of the information that = and X convey can be found in optional SAM tags that many aligners can include in their output. The NM tag is an integer that represents the edit distance between the aligned portion (which excludes clipped regions) and the reference. Additionally, the MD tag encodes mismatching positions from the reference (and the reference's sequence at mismatching positions). See the SAM format specification for more detail. If your BAM file doesn't have the NM and MD tags, samtools calmd can add them for you.

Let's look at a trickier example: 43S6M1I26M. First, let's break this down into pairs: 43S, 6M, 1I, and 26M. Using Table 11-2, we see this CIGAR string tells us that the first 43 bases were soft clipped, the next 6 were matches/mismatches, then a 1 base pair insertion to the reference, and finally, 26 matches. The SAM format specification mandates that all M, I, S, =, and X operations' lengths must add to the length of the sequence. We can validate that's the case here: 43 + 6 + 1 + 26 = 76, which is the length of the sequence in this SAM entry.

Mapping Qualities

Our discussion of the SAM and BAM formats is not complete without mentioning *mapping qualities* (Li et al., 2008). Mapping qualities are one of the most important diagnostics in alignment. All steps downstream of alignment in all bioinformatics projects (e.g., SNP calling and genotyping, RNA-seq, etc.) critically depend on reliable mapping. Mapping qualities quantify mapping reliability by estimating how likely a read is to actually originate from the position the aligner has mapped it to. Similar to base quality, mapping quality is a log probability given by $Q = -10 \log_{10}P(incorrect\ mapping\ position)$. For example, a mapping quality of 20 translates to a $10^{(20/-10)} = 1\%$ chance the alignment is incorrect.

The idea of mapping quality is also related to the idea of *mapping uniqueness*. This is often defined as when a read's second best hit has more mismatches than its first hit. However, this concept of uniqueness doesn't account for the base qualities of mismatches, which carry a lot of information about whether a mismatch is due to a base calling error or a true variant (Li et al., 2008). Mapping quality estimates do account for base qualities of mismatches, which makes them a far better metric for measuring mapping uniqueness (as well as general mapping reliability).

We can use mapping qualities to filter out likely incorrect alignments (which we can do with `samtools view`, which we'll learn about later), find regions where mapping quality is unusually low among most alignments (perhaps in repetitive or paralogous regions), or assess genome-wide mapping quality distributions (which could indicate alignment problems in highly repetitive or polyploid genomes).

Command-Line Tools for Working with Alignments in the SAM Format

In this section, we'll learn about the Samtools suite of tools for manipulating and working with SAM, BAM, and CRAM files. These tools are incredibly powerful, and becoming skilled in working with these tools will allow you to both quickly move forward in file-processing tasks and explore the data in alignment files. All commands are well documented both online (see the Samtools website (*http://www.htslib.org/doc/*)) and in the programs themselves (run a program without arguments or use `--help` for more information).

Using samtools view to Convert between SAM and BAM

Many `samtools` subcommands such as `sort`, `index`, `depth`, and `mpileup` all require input files (or streams) to be in BAM format for efficiency, so we often need to convert between plain-text SAM and binary BAM formats. `samtools view` allows us to convert SAM to BAM with the `-b` option:

```
$ samtools view -b celegans.sam > celegans_copy.bam
```

Similarly, we can go from BAM to SAM:

```
$ samtools view celegans.bam > celegans_copy.sam
$ head -n 3 celegans_copy.sam
I_2011868_2012306_0:0:0_0:0:0_2489       83   I   2012257   40 [...]
I_2011868_2012306_0:0:0_0:0:0_2489       163  I   2011868   60 [...]
I_13330604_13331055_2:0:0_0:0:0_3dd5     83   I   13331006  60 [...]
```

However, `samtools view` will not include the SAM header (see "The SAM Header" on page 356) by default. SAM files without headers cannot be turned back into BAM files:

```
$ samtools view -b celegans_copy.sam > celegans_copy.bam
[E::sam_parse1] missing SAM header
[W::sam_read1] parse error at line 1
[main_samview] truncated file.
```

Converting BAM to SAM loses information when we don't include the header. We can include the header with `-h`:

```
$ samtools view -h celegans.bam > celegans_copy.sam
$ samtools view -b celegans_copy.sam > celegans_copy.bam #now we can convert back
```

Usually we only need to convert BAM to SAM when manually inspecting files. In general, it's better to store files in BAM format, as it's more space efficient, compatible with all `samtools` subcommands, and faster to process (because tools can directly read in binary values rather than require parsing SAM strings).

The CRAM Format

Samtools now supports (after version 1) a new, highly compressed file format known as *CRAM* (see Fritz et al., 2011). Compressing alignments with CRAM can lead to a 10%–30% filesize reduction compared to BAM (and quite remarkably, with no significant increase in compression or decompression time compared to BAM). CRAM is a *reference-based* compression scheme, meaning only the aligned sequence that's different from the reference sequence is recorded. This greatly reduces file size, as many sequences may align with minimal difference from the reference. As a consequence of this reference-based approach, it's imperative that the reference is available and does not change, as this would lead to a loss of data kept in the CRAM format. Because the reference is so important, CRAM files contain an MD5 checksum of the reference file to ensure it has not changed.

CRAM also has support for multiple different *lossy compression* methods. Lossy compression entails some information about an alignment and the original read is lost. For example, it's possible to bin base quality scores using a lower resolution binning scheme to reduce the filesize. CRAM has other lossy compression models; see CRAMTools (*http://github.com/enasequence/cramtools*) for more details.

> Overall, working with CRAM files is not much different than working with SAM or BAM files; CRAM support is integrated into the latest Samtools versions. See the documentation (*http://bit.ly/cram-st*) for more details on CRAM-based Samtools workflows.

Samtools Sort and Index

In "Indexed FASTA Files" on page 352, we saw how we can index a FASTA file to allow for faster random access of the sequence at specific regions in a FASTA file. Similarly, we sort (by alignment position) and index a BAM file to allow for fast random access to reads aligned within a certain region (we'll see how to extract these regions in the next section).

We sort alignments by their alignment position with `samtools sort`:

```
$ samtools sort celegans_unsorted.bam celegans_sorted
```

Here, the second argument is the output filename prefix (`samtools sort` will append the *.bam* extension for you).

Sorting a large number of alignments can be very computationally intensive, so `samtools sort` has options that allow you to increase the memory allocation and parallelize sorting across multiple threads. Very often, large BAM files won't fit entirely in memory, so `samtools sort` will divide the file into chunks, sort each chunk and write to a temporary file on disk, and then merge the results together; in computer science lingo, `samtools sort` uses a *merge sort* (which was also discussed in "Sorting Plain-Text Data with Sort" on page 147). Increasing the amount of memory `samtools sort` can use decreases the number of chunks `samtools sort` needs to divide the file into (because larger chunks can fit in memory), which makes sorting faster. Because merge sort algorithms sort each chunk independently until the final merge step, this can be parallelized. We can use the `samtools sort` option -m to increase the memory, and -@ to specify how many threads to use. For example:

```
$ samtools sort -m 4G -@ 2 celegans_unsorted.bam celegans_sorted
```

`samtools sort`'s -m option supports the suffixes K (kilobytes), M (megabytes), and G (gigabytes) to specify the units of memory. Also, note though that in this example, the toy data file *celegans_unsorted.bam* is far too small for there to be any benefits in increasing the memory or parallelization.

Position-sorted BAM files are the starting point for most later processing steps such as SNP calling and extracting alignments from specific regions. Additionally, sorted BAM files are much more disk-space efficient than unsorted BAM files (and certainly more than plain-text SAM files). Most SAM/BAM processing tasks you'll do in daily bioinformatics work will be to get you to this point.

Often, we want to work with alignments within a particular region in the genome. For example, we may want to extract these reads using `samtools view` or only call SNPs within this region using FreeBayes (Garrison et al., 2012) or `samtools mpileup`. Iterating through an entire BAM file just to work with a subset of reads at a position would be inefficient; consequently, BAM files can be *indexed* like we did in "Indexed FASTA Files" on page 352 with FASTA files. The BAM file must be sorted first, and we cannot index SAM files. To index a position-sorted BAM file, we simply use:

```
$ samtools index celegans_sorted.bam
```

This creates a file named *celegans_sorted.bam.bai*, which contains the index for the *celegans_sorted.bam* file.

Extracting and Filtering Alignments with samtools view

Earlier, we saw how we can use `samtools view` to convert between SAM and BAM, but this is just scratching the surface of `samtools view`'s usefulness in working with alignment data. `samtools view` is a workhorse tool in extracting and filtering alignments in SAM and BAM files, and mastering it will provide you with important skills needed to explore alignments in these formats.

Extracting alignments from a region with samtools view

With a position-sorted and indexed BAM file, we can extract specific regions of an alignment with `samtools view`. To make this example more interesting, let's use a subset of the 1000 Genomes Project data (1000 Genomes Project Consortium, 2012) that's included in this chapter's repository on GitHub. First, let's index it:

```
$ samtools index NA12891_CEU_sample.bam
```

Then, let's take a look at some alignments in the region chromosome 1, 215,906,469-215,906,652:

```
$ samtools view NA12891_CEU_sample.bam 1:215906469-215906652 | head -n 3
SRR003212.5855757   147  1  215906433  60  33S43M  =  215906402 [...]
SRR003206.18432241  163  1  215906434  60  43M8S   =  215906468 [...]
SRR014595.5642583   16   1  215906435  37  8S43M   *  0         [...]
```

We could also write these alignments in BAM format to disk with:

```
$ samtools view -b NA12891_CEU_sample.bam 1:215906469-215906652 >
    USH2A_sample_alns.bam
```

Lastly, note that if you have many regions stored in the BED format, `samtools view` can extract regions from a BED file with the -L option:

```
$ samtools view -L USH2A_exons.bed NA12891_CEU_sample.bam | head -n 3
SRR003214.11652876  163  1  215796180  60  76M  =  215796224  92   [...]
SRR010927.6484300   163  1  215796188  60  51M  =  215796213  76   [...]
SRR005667.2049283   163  1  215796190  60  51M  =  215796340  201  [...]
```

Filtering alignments with samtools view

samtools view also has options for filtering alignments based on bitwise flags, mapping quality, read group. samtools view's filtering features are extremely useful; very often we need to query BAM files for reads that match some criteria such as "all aligned proper-paired end reads with a mapping quality over 30." Using samtools view, we can stream through and filter reads, and either pipe the results directly into another command or write them to a file.

First, note that samtools view (like other samtools subcommands) provides handy documentation within the program for all of its filtering options. You can see these any time by running the command without any arguments (to conserve space, I've included only a subset of the options):

```
$ samtools view

Usage:   samtools view [options] <in.bam>|<in.sam>|<in.cram> [region ...]

Options: -b       output BAM
         -C       output CRAM (requires -T)
         -1       use fast BAM compression (implies -b)
         -u       uncompressed BAM output (implies -b)
         -h       include header in SAM output
         -H       print SAM header only (no alignments)
         -c       print only the count of matching records
[...]
```

Let's first see how we can use samtools view to filter based on bitwise flags. There are two options related to this: -f, which only outputs reads *with* the specified flag(s), and -F, which only outputs reads *without* the specified flag(s). Let's work through an example, using the samtools flags command to assist in figuring out the flags we need. Suppose you want to output all reads that are unmapped. UNMAP is a flag according to samtools flags:

```
$ samtools flags unmap
0x4     4       UNMAP
```

Then, we use samtools view -f 4 to output reads with this flag set:

```
$ samtools view -f 4 NA12891_CEU_sample.bam | head -n 3
SRR003208.1496374   69   1  215623168  0  35M16S  =  215623168  0  [...]
SRR002141.16953736  181  1  215623191  0  40M11S  =  215623191  0  [...]
SRR002143.2512308   181  1  215623216  0  *       =  215623216  0  [...]
```

Note that each of these flags (the second column) have the bit corresponding to unmapped set. We could verify this with samtools flags:

```
$ samtools flags 69
0x45    69      PAIRED,UNMAP,READ1
```

It's also possible to output reads with multiple bitwise flags set. For example, we could find the first reads that aligned in a proper pair alignment. First, we use samtools flags to find out what the decimal representation of these two flags is:

```
$ samtools flags READ1,PROPER_PAIR
0x42    66      PROPER_PAIR,READ1
```

Then, use samtools view's -f option to extract these alignments:

```
$ samtools view -f 66 NA12891_CEU_sample.bam | head -n 3
SRR005672.8895       99  1  215622850  60  51M  =  215623041  227  [...]
SRR005674.4317449    99  1  215622863  37  51M  =  215622987  175  [...]
SRR010927.10846964   83  1  215622892  60  51M  =  215622860  -83  [...]
```

We can use the -F option to extract alignments that *do not* have any of the bits set of the supplied flag argument. For example, suppose we wanted to extract all *aligned* reads. We do this by filtering out all reads with the 0x4 bit (meaning unmapped) set:

```
$ samtools flags UNMAP
0x4     4       UNMAP
$ samtools view -F 4 NA12891_CEU_sample.bam | head -n 3
SRR005672.8895       99   1  215622850  60  51M     =  215623041  227  [...]
SRR010927.10846964   163  1  215622860  60  35M16S  =  215622892  83   [...]
SRR005674.4317449    99   1  215622863  37  51M     =  215622987  175  [...]
```

Be aware that you will likely have to carefully combine bits to build queries that extract the information you want. For example, suppose you wanted to extract all reads that did not align in a proper pair. You might be tempted to approach this by filtering out all alignments that have the proper pair bit (0x2) set using:

```
$ samtools flags PROPER_PAIR
0x2     2       PROPER_PAIR
$ samtools view -F 2 NA12891_CEU_sample.bam | head -n 3
SRR005675.5348609    0  1  215622863  37  51M    [...]
SRR002133.11695147   0  1  215622876  37  48M    [...]
SRR002129.2750778    0  1  215622902  37  35M1S  [...]
```

But beware—this would be incorrect! Both unmapped reads and unpaired reads will also be included in this output. Neither unmapped reads, nor unpaired reads will be in a proper pair and have this bit set.

SAM Bitwise Flags and SAM Fields

It's vital to consider how some bitflags may affect other bitflags (technically speaking, some bitflags are non-orthogonal). Similarly, if some bitflags are set, certain SAM fields may no longer apply.

For example, 0x4 (unmapped) is the only reliable way to tell if an alignment is unaligned. In other words, one cannot tell if a read is aligned by looking at fields such as mapped position (POS and reference RNAME); the SAM format specification does not limit these fields' values if a read is unaligned. If the 0x4 bit is set (meaning the read is unmapped), the fields regarding alignment including position, CIGAR string, mapping quality, and reference name are not relevant and their values cannot be relied upon. Similarly, if the 0x4 bit is set, bits that only apply to *mapped reads* such as 0x2 (proper pair), 0x10 (aligned to reverse strand), and others cannot be relied upon. The primary lesson is you should carefully consider all flags that may apply when working with SAM entries, and start with low-level attributes (whether it's aligned, paired). See the SAM format specification for more detail on bitflags.

Instead, we want to make sure the unmapped (0x4) and proper paired bits are unset (so the read is aligned and paired), and the paired end bit is set (so the read is not in a proper pair). We do this by combining bits:

```
$ samtools flags paired
0x1     1       PAIRED
$ samtools flags unmap,proper_pair
0x6     6       PROPER_PAIR,UNMAP
$ samtools view -F 6 -f 1 NA12891_CEU_sample.bam | head -n 3
SRR003208.1496374   137 1 215623168  0  35M16S  = 215623168 [...]
ERR002297.5178166   177 1 215623174  0  36M     = 215582813 [...]
SRR002141.16953736  121 1 215623191  0  7S44M   = 215623191 [...]
```

One way to verify that these results make sense is to check the counts (note that this may be very time consuming for large files). In this case, our total number of reads that are mapped and paired should be equal to the sum of the number of reads that are mapped, paired, and properly paired, and the number of reads that are mapped, paired, and *not* properly paired:

```
$ samtools view -F 6 NA12891_CEU_sample.bam | wc -l  # total mapped and paired
233628
$ samtools view -F 7 NA12891_CEU_sample.bam | wc -l  # total mapped, paired,
201101                                                # proper paired
$ samtools view -F 6 -f 1 NA12891_CEU_sample.bam | wc -l # total mapped, paired,
32527                                                    # and not proper paired
$ echo "201101 + 32527" | bc
233628
```

Summing these numbers with the command-line bench calculator bc validates that our totals add up. Also, for set operations like this drawing a Venn diagram can help you reason through what's going on.

Visualizing Alignments with samtools tview and the Integrated Genomics Viewer

As we saw in Chapter 8, we can learn a lot about our data through visualization. The same applies with alignment data: one of the best ways to explore alignment data is through visualization. The samtools suite includes the useful tview subcommand for quickly looking at alignments in your terminal. We'll take a brief look at tview first, then look at the Broad Institute's Integrated Genomics Viewer (IGV) application.

samtools tview requires position-sorted and indexed BAM files as input. We already indexed the position-sorted BAM file *NA12891_CEU_sample.bam* (in this chapter's GitHub directory) in "Extracting alignments from a region with samtools view" on page 368, so we're ready to visualize it with samtools tview. samtools tview can also load the reference genome alongside alignments so the reference sequence can be used in comparisons. The reference genome file used to align the reads in *NA12891_CEU_sample.bam* is *human_g1k_v37.fasta*, and although it's too large to include in this chapter's GitHub directory, it can be easily downloaded (see the directory's *README.md* for directions). So to view these alignments with samtools tview, we use:

```
$ samtools tview NA12891_CEU_sample.bam human_g1k_v37.fasta
```

However, this will view the very beginning of a chromosome; because *NA12891_CEU_sample.bam* is a subset of reads, let's go to a specific region with the option -p:

```
$ samtools tview -p 1:215906469-215906652 NA12891_CEU_sample.bam \
    human_g1k_v37.fasta
```

This will open up a terminal-based application. samtools tview has many options to navigate around, jump to particular regions, and change the output format and colors of alignments; press ? to see these options (and press again to close the help screen). Figure 11-1 shows an example of what tview looks like.

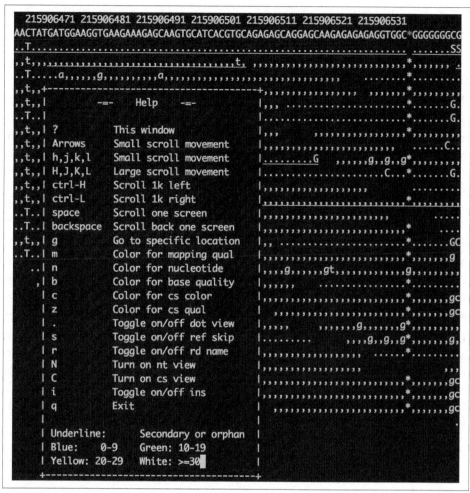

Figure 11-1. A region in samtools tview with the help dialog box open

As a command-line visualization program, samtools tview is great for quickly inspecting a few alignments. However, if you need to spend more time investigating alignments, variants, and insertions/deletions in BAM data, the Integrated Genomics Viewer (IGV) may be more well suited. As an application with a graphical user interface, IGV is easy to use. IGV also has numerous powerful features we won't cover in this brief introduction, so I encourage you to explore IGV's excellent documentation.

First, we need to install IGV. It's distributed as a Java application, so you'll need to have Java installed on your system. After Java is installed, you can install IGV through a package manager such as Homebrew on OS X or apt-get. See the *README.md* file in this chapter's GitHub directory for more detail on installing IGV.

If you've installed IGV through Homebrew or `apt-get`, you can launch the application with:

```
$ igv
```

The command `igv` calls a small shell script wrapper for the Java application and opens IGV.

Once in IGV, we need load our reference genome before loading our alignments. Reference genomes can be loaded from a file by navigating to *Genomes → Load Genome from File*, and then choosing your reference genome through the file browser. IGV also has prepared reference genomes for common species and versions; these can be accessed through *Genomes → Load Genome From Server*. Navigate to this menu and load the "Human (1kg, b37+decoy)" genome. This prepared genome file has some nice additional features used in display such as gene tracks and chromosome ideograms.

Once our reference genome is loaded, we can load in the BAM alignments in *NA12891_CEU_sample.bam* by navigating to File → Load from File and choosing the reference genome through the file browser. Note that you will not see any alignments, as this file contains a subset of alignments in a region IGV's not currently focused on.

IGV's graphical user interface provides many methods for navigation, zooming in and out of a chromosome, and jumping to particular regions. Let's jump to a region that our alignments in *NA12891_CEU_sample.bam* overlap: 1:215,906,528-215,906,567 (there's a copyable version of this region in this chapter's *README.md* for convenience). Enter this region in the text box at the top of the window, and press Enter; this region should look like Figure 11-2.

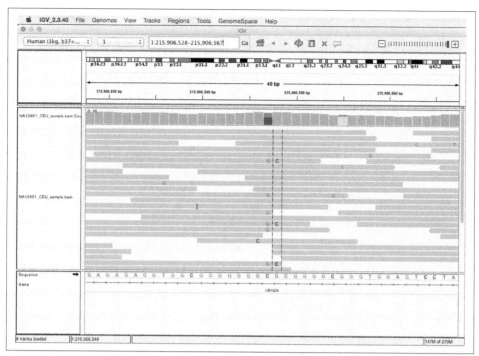

Figure 11-2. A region in IGV that shows possible paralogous alignments

As shown in Figure 11-2, IGV shows an ideogram to indicate where on the chromosome the region is and base pair positions in the top pane, coverage and alignment tracks in the middle pane, and the sequence and gene track information in the bottom pane. The colored letters in alignments indicate bases mismatched between the read sequence and the reference. These mismatching bases can be caused either by sequencing errors, misalignment, errors in library preparation, or an authentic SNP. In this case, we might make a ballpark guess that the stacked mismatches at positions 215,906,547, 215,906,548 and 215,906,555 are true polymorphisms. However, let's use some of IGV's features to take a closer look, specifically at the variants around 215,906,547–215,906,548.

Let's start our exploration by hovering over alignments in this region to reveal IGV's pop-up window full of that alignment's information (see Figure 11-3). This allows you to inspect useful information about each alignment such as the base qualities of mismatches and the alignment's mapping quality. For example, hovering over the alignments in the region from 215,906,547–215,906,548 shows that some aligned with lower mapping quality.

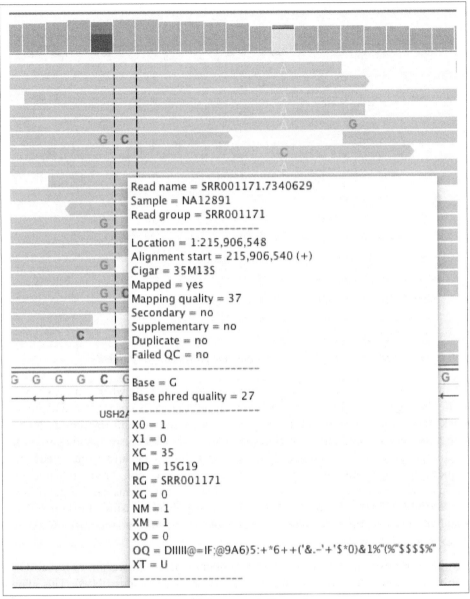

Figure 11-3. IGV's pop-up window of alignment information

We can also learn a lot about our variants by looking at which reads they are carried on. For the potential variants at 215,906,547 and 215,906,548, mismatches are carried on three categories of reads that span these positions:

- Reads with a reference base at 215,906,547 and a reference base at 215,906,548 (e.g., the first alignment from the top)
- Reads with a G and a reference base (e.g., the twelfth alignment from the top)
- Reads with a G and a C (e.g., the sixth alignment from the top)

A single read contains a single continuous stretch of DNA, so the presence of three different combinations of mismatching bases in these reads indicates *three different haplotypes* in this region (ignoring for the moment the possibility that these mismatches might be sequencing errors). Because these sequences come from a single diploid human individual, this indicates a likely problem—probably due to misalignment.

The mismatches in this region could be due to misalignment caused by reads from a different region in the genome aligning to this region due to similar sequence (creating that third haplotype). Misalignments of this nature can be caused by common repeats, paralogous sequences, or similar domains. Misalignments of this type are a major cause of false positive variant calls, and visual inspection with tools like IGV can greatly help in recognizing these issues. While somewhat tedious, manual inspection of alignment data is especially important for entries at the top of lists of significant variants. However, for this particular region, it's more likely a different misalignment mishap is creating these mismatches that look like variants.

Note that IGV also displays the reference sequence track in the bottom pane, which gives us another clue as to what could be causing these variants. This region is composed of *low-complexity sequences*: GGCGGGGGGGCGGGGGGCGGG. Low-complexity sequences are composed of runs of bases or simple repeats, and are a major cause of erroneous variant calls (Li, 2014). In this low-complexity region, reads containing the mismatches G-C might actually contain an upstream G insertion. In low-complexity regions, an aligner may align a read with a single base mismatch rather than an indel, creating a false SNP. Indeed, there's some evidence that this type of misalignment is occurring here: the only reads that carry the G-C mismatches are those that do not span the low-complexity region on the left.

Finally, it's worth noting that in addition to being a low-complexity sequence, the base composition in this region might be a concern as well. GGC sequences are known to generate sequence-specific errors in some Illumina data (see Nakamura et al., 2011). We can inspect the sequencing platform metadata in the BAM header and indeed see that this data comes from an Illumina sequencer:

```
$ samtools view -H NA12891_CEU_sample.bam | grep PL | tail -n 1
@RG  ID:SRR035334  PL:ILLUMINA  LB:Solexa-3628  [...]  SM:NA12891  CN:BI
```

Pileups with samtools pileup, Variant Calling, and Base Alignment Quality

In this section, we'll discuss the *pileup format*, a plain-text format that summarizes reads' bases at each chromosome position by stacking or "piling up" aligned reads. The per-base summary of the alignment data created in a pileup can then be used to identify variants (regions different from the reference), and determine sample individuals' genotypes. `samtools`'s `mpileup` subcommand creates pileups from BAM files, and this tool is the first step in `samtools`-based variant calling pipelines. In this chapter, we'll introduce the pileup format and `samtools` variant calling tools through simple examples that explore how the misalignments we saw in the previous section can lead to erroneous variant calls. We'll also examine how `samtools`'s clever Base Alignment Quality algorithm can prevent erroneous variant calls due to misalignment. Note that this chapter is *not* meant to teach variant calling, as this subject is complex, procedures are project- and organism-specific, and methods will rapidly change with improvements to sequencing technology and advancement in variant calling methods.

To begin, let's look at how a "vanilla" (all extra features turned off) `samtools` variant calling pipeline handles this region. We'll start by creating a pileup to learn more about this format. To do this, we run `samtools mpileup` in the same region we visualized in IGV in "Visualizing Alignments with samtools tview and the Integrated Genomics Viewer" on page 372 (again using the *human_g1k_v37.fasta* file):

```
$ samtools mpileup --no-BAQ --region 1:215906528-215906567 \   ❶
    --fasta-ref human_g1k_v37.fasta NA12891_CEU_sample.bam
[mpileup] 1 samples in 1 input files
<mpileup> Set max per-file depth to 8000
1   215906528   G   21   ,,,,,,,,.,,,,.,,.,,,      ;=?./:?>>;=7?>>@A?==:  ❷
1   215906529   A   18   ,,,,,,,,,,.,,,,,,.,.      D>AA:@A9>?;;?>>@=
[...]
1   215906547   C   15   gGg$,GggGG,,....          <;80;><9=86=C>=       ❸
1   215906548   G   19   c$,ccC.,.,,,.,....,^].    ;58610=7=>75=7<463;
[...]
1   215906555   G   16   .$aaaaaA.AAAaAAA^:A       2@>?8?;<:335?:A>       ❹
[...]
```

❶ First, `samtools mpileup` requires an input BAM file (in this example, we use *NA12891_CEU_sample.bam*). We also supply a reference genome in FASTA format through the `--fasta-ref` or `-f` options (be sure to use the exact same reference used for mapping) so `samtools mpileup` knows each reference base. Additionally, we use the `--region` or `-r` options to limit our pileup to the same region we visualized with IGV. Lastly, we disable Base Alignment Quality (BAQ) with `--no-BAQ` or `-B`, an additional feature of `samtools mpileup` we'll discuss later.

---

❷ This is is a typical line in the pileup format. The columns are:

- Reference sequence name (e.g., chromosome 1 in this entry).
- Position in reference sequence, 1-indexed (position 215,906,528 in this entry).
- Reference sequence base at this position (G in this entry).
- Depth of aligned reads at this position (the depth or coverage, 21 in this entry).
- This column encodes the reference reads bases. Periods (.) indicate a reference sequence match on the forward strand, commas (,) indicate a reference sequence match to the reverse strand, an uppercase base (either A, T, C, G, or N) indicates a mismatch on the forward strand, and a lowercase base (a, t, c, g, or n) indicates a mismatch on the reverse strand. The ^ and $ characters indicate the start and end of reads, respectively. Insertions are denoted with a plus sign (+), followed by the length of the insertion, followed by the sequence of the insertion. An insertion on a line indicates it's between the current line and the next line. Similarly, deletions are encoded with a minus sign (-), followed by the length of the deletion and the deleted sequence. Lastly, the mapping quality of each alignment is specified after the beginning of the alignment character, ^ as the ASCII character value minus 33.
- Finally, the last column indicates the base qualities (the ASCII value minus 33).

❸ On this line and the next line are the stacked mismatches we saw at positions 215,906,547 and 215,906,548 with IGV in Figure 11-2. These lines show the stacked mismatches as nucleotides in the fifth column. Note that variants in this column are both lower- and uppercase, indicating they are supported by both reads aligning to both the forward and reverse strands.

❹ Finally, note at this position most reads disagree with the reference base. Also, this line contains the start of a new read (indicated with the ^ mark), with mapping quality 25 (the character : has ASCII code 58, and 58 - 33 = 25; try ord(:) - 33 in Python).

Altering our mpileup command to output variant calls rather than a pileup is not difficult (we'll see how in a bit). However, while pileups are simply per-position summaries of the data in aligned reads, variant and genotype calls require making inferences from noisy alignment data. Most variant calling approaches utilize probabilistic frameworks to make reliable inferences in spite of low coverage, poor base qualities, possible misalignments, and other issues. Additionally, methods can increase power to detect variants by jointly calling variants on many individuals simultaneously. sam tools mpileup will jointly call SNPs and genotypes for multiple individuals—each individual simply needs to be identified through the SM tags in the @RG lines in the SAM header.

Calling variants with samtools and its companion tool bcftools is a two-step process (excluding the very important steps of validation). In the first step, samtools mpi leup called with the -v or -g arguments will generate genotype likelihoods for every site in the genome (or all sites within a region if one is specified). These results will be returned in either a plain-text tab-delimited *Variant Call Format* (known more commonly by its abbreviation, *VCF*) if -v is used, or *BCF* (the binary analog of VCF) if -g is used. In the second step, bcftools call will filter these results so only variant sites remain, and call genotypes for all individuals at these sites. Let's see how the first step works by calling samtools mpileup with -v in the region we investigated earlier with IGV:

```
$ samtools mpileup -v --no-BAQ --region 1:215906528-215906567 \
    --fasta-ref human_g1k_v37.fasta NA12891_CEU_sample.bam    \
    > NA12891_CEU_sample.vcf.gz
[mpileup] 1 samples in 1 input files
<mpileup> Set max per-file depth to 8000
```

The VCF Format

VCF is a tab-delimited format that has three parts:

- A metadata header consisting of lines that start with ##
- A header line with the eight mandatory fields and if genotypes are called, the individuals' sample names
- The data lines, where each line consists of the information for a variant at a particular position and all individuals' genotypes for this variant

VCF can be a deceivingly complex format that comes in different versions and flavors (and is very likely to change, and may at some point be replaced entirely). For these reasons, the specifics of VCF won't be covered in depth here, but see the format specification (*http://github.com/samtools/hts-specs*) and the supplementary information in this chapter's GitHub directory. I encourage you to become familiar with the VCF format, as it will aid in understanding these examples.

This produces a gzipped VCF file full of intermediate variant and genotype data for every site in the region (or if you haven't specified a region with --region or -r, the entire genome). Note that you can force samtools mpileup to output uncompressed results with -u. These intermediate results are then fed into bcftools call, which uses the estimated genotype likelihoods and other information to call certain sites as variant or not, and infer the genotypes of all individuals (in our example, our file only has one individual, NA12891). The VCF file produced is long and too wide to show in a book page, so I've used zgrep -v to remove the header and awk to select some columns of interest. Because VCF is a tab-delimited file, we can use all of our Unix tricks to manipulate and explore it:

```
$ zgrep "^##" -v NA12891_CEU_sample.vcf.gz | \
  awk 'BEGIN{OFS="\t"} {split($8, a, ";"); print $1,$2,$4,$5,$6,a[1],$9,$10}'
#CHROM  POS         REF  ALT     QUAL  INFO    FORMAT  NA12891
1       215906528   G    <X>     0     DP=21   PL      0,63,236
1       215906529   A    <X>     0     DP=22   PL      0,54,251
[...]
1       215906547   C    G,<X>   0     DP=22   PL      123,0,103,144,127,233
1       215906548   G    C,<X>   0     DP=22   PL      23,0,163,68,175,207
[...]
1       215906555   G    A,<X>   0     DP=19   PL      184,7,0,190,42,204
[...]
```

First, note that some positions such as 215,906,528 and 215,906,528 only contain the alternative allele (in the ALT column) <X>. <X> represents the possibility that an alternative allele has not been observed in the data due to under-sampling (e.g., low coverage). Second, note how the sites 215,906,547, 215,906,548, and 215,906,555 all have an alternative allele other than <X>. These are the same sites we saw earlier in IGV that had multiple mismatches that looked like variants. The alternative alleles in column ALT represent *possible* variants at these positions samtools mpileup has identified. Additional information about these variants is passed in this intermediate VCF file to bcftools call, which uses this information to make an inference whether sites are really variant and what each individuals' genotype is. Let's run this bcftools call step, and then do some exploring of where we lose variants and why:

```
$ bcftools call -v -m NA12891_CEU_sample.vcf.gz > NA12891_CEU_sample_calls.vcf.gz
```

bcftools call run with -m uses the multiallelic caller (the other option is to use the original consensus caller with -c). The -v option only outputs variant sites, which is why our output is much shorter:

```
$ zgrep "^##" -v NA12891_CEU_sample_calls.vcf.gz | \
  awk 'BEGIN{OFS="\t"} {split($8, a, ";"); print $1,$2,$4,$5,$6,a[1],$9,$10}'
#CHROM  POS         REF  ALT  QUAL  INFO    FORMAT  NA12891
1       215906547   C    G    90    DP=22   GT:PL   0/1:123,0,103
1       215906555   G    A    157   DP=19   GT:PL   1/1:184,7,0
```

We see that bcftools calls calls only two variant sites in this region, which is fewer sites than have alternative alleles in the intermediate VCF output from samtools mpileup. Noticeably, the site 215,906,548 is not called as a variant site after being processed through bcftools call's multiallelic caller. This is good—looking at the alignments in IGV ("Visualizing Alignments with samtools tview and the Integrated Genomics Viewer" on page 372), you can see that there are 4 C alleles and 20 reference (G) alleles; hardly convincing evidence that this a true variant. More individuals with this alternative allele in a joint variant calling context may have tipped the balance toward this being called a variant site.

Additionally, bcftools call has estimated a quality score (the QUAL column) for each alternative allele in ALT. These quality scores are Phred-scaled values that estimate the probability that the alternative allele is incorrect (see "Base Qualities" on page 344 for more on Phred-scaled values). Higher QUAL scores indicate the variant caller is *more confident* in a call.

If the alternative allele in column ALT is . (representing no variant), this quality score reflects the probability that the site really does have a variant. We can see how bcftools call is less certain about the nonvariant call at position 215,906,548 by looking at the QUAL value for this variant. Omitting the flag -v with bcftools call to see all sites' information is shown in Example 11-1:

Example 11-1. bcftools call with all sites

```
$ bcftools call -m NA12891_CEU_sample.vcf.gz | grep -v "^##" | \
  awk 'BEGIN{OFS="\t"} {split($8, a, ";"); print $1,$2,$4,$5,$6,a[1],$9,$10}'
#CHROM  POS         REF     ALT   QUAL     INFO    FORMAT  NA12891
1       215906528   G       .     999      DP=21   GT      0/0
1       215906529   A       .     999      DP=22   GT      0/0
[...]
1       215906547   C       G     90       DP=22   GT:PL   0/1:123,0,103
1       215906548   G       .     12.1837  DP=22   GT      0/0
[...]
1       215906555   G       A     157      DP=19   GT:PL   1/1:184,7,0
[...]
```

Compared to other nonvariant sites like 215,906,528 and 215,906,529, and sites with visible stacked mismatches (like 215,906,547), 215,906,548 has a very low QUAL reflecting the variant caller's uncertainty at this site (because there were four mismatches of the same base). Recalculating this Phred-value into a probability using the formula $Q = -10 \log_{10}$ P(alternative call is incorrect), we see that P(alternative call is incorrect) $= 10^{(-12.1837/10)} \approx 0.060$, or about a 6% chance the call that this site is invariant is in incorrect—a fairly large chance. Compare that to the quality at 215,906,555 where the P(alternative call is incorrect) $\approx 2 \times 10^{-16}$.

A lot of information regarding each individual's genotype call can be crammed into the VCF format. The VCF format supports variable numbers of information about genotypes (e.g., genotype call, genotype quality, likelihoods, etc.) by concatenating many values together into a single column per individual. Values are separated by colons (:), and the FORMAT column describes the order of each value in the genotype

---

columns. For example, in Example 11-1, the entry for position 215,906,555 has a FOR MAT entry of GT:PL and the entry for individual NA12891 is 1/1:184,7,0. This means that the key GT has a value 1/1, and the key PL has a value 184,7,0. A really nice feature of the VCF format is that all FORMAT keys are described in the header, so we can use grep to figure out what these keys mean:

```
$ bcftools call -m NA12891_CEU_sample.vcf.gz > NA12891_CEU_sample_calls.vcf.gz

$ grep "FORMAT=<ID=GT" NA12891_CEU_sample_calls.vcf.gz
##FORMAT=<ID=GT,Number=1,Type=String,Description="Genotype">
$ grep "FORMAT=<ID=PL" NA12891_CEU_sample_calls.vcf.gz
##FORMAT=<ID=PL,Number=G,Type=Integer,
    Description="List of Phred-scaled genotype likelihoods">
```

So we see that these values are the genotypes (key GT) and Phred-scaled genotype likelihoods (key PL) of each individual. These are always in the order ref/ref, ref/alt, and alt/alt alleles for biallelic loci (and a similar pattern for multiallelic loci; see the VCF format specification for an example). All genotype likelihoods (PL) are rescaled so the most likely genotype is 1 (so it's Phred-scaled likelihood is 0). Thus, at position 215,906,555 the most likely genotype is alt/alt, the next most likely is ref/alt, and the least likely is ref/ref. There are additional resources on interpreting VCF genotype fields in this chapter's *README* file.

Lastly, let's return to the idea that our reads may have misaligned around the low complexity region near the 215,906,547. Misalignments in low-complexity regions (usually due to indels being aligned as mismatches) are a major cause of erroneous SNP calls. To address this, samtools mpileup enables *Base Alignment Quality* (BAQ), which uses uses a Hidden Markov Model to adjust base qualities to reflect not only the probability of an incorrect base call, but also of a particular base being misaligned. We disabled this algorithm earlier in our simple pipeline, but let's see how it affects our calls:

```
$ samtools mpileup -u -v --region 1:215906528-215906567 \
    --fasta-ref human_g1k_v37.fasta NA12891_CEU_sample.bam > \
    NA12891_CEU_sample_baq.vcf.gz

$ grep -v "^##" NA12891_CEU_sample_baq.vcf.gz | \
    awk 'BEGIN{OFS="\t"} {split($8, a, ";"); print $1,$2,$4,$5,$6,a[1],$9,$10}'
#CHROM  POS        REF  ALT    QUAL  INFO    FORMAT  NA12891
1       215906528  G    <X>    0     DP=21   PL      0,63,236
1       215906529  A    <X>    0     DP=22   PL      0,54,249
[...]
1       215906547  C    <X>    0     DP=22   PL      0,21,141
1       215906548  G    <X>    0     DP=22   PL      0,42,200
[...]
1       215906555  G    A,<X>  0     DP=19   PL      194,36,0,194,36,194
[...]
```

Note how *both* sites 215,906,547 and 215,906,548 are now not considered as possible variant sites—the BAQ algorithm has downweighted the bases around the low complexity region we saw in IGC sufficiently enough that `samtools mpileup` no longer considers these invariant sites.

Creating Your Own SAM/BAM Processing Tools with Pysam

In this section, we'll learn about the basics of Pysam, an *application programming interface* (more commonly known as an *API*). As the name suggests, APIs provide a defined interface to some component (in this case, the data in SAM and BAM files) through a set of well-defined classes and functions. The Pysam API allows for us to quickly work with data in SAM and BAM files without having to implement parsers, write code to read binary BAM files, or parse bitflags. The classes and functions in the `pysam` module take care of all these intermediate steps for you, and provide a consistent interface for you to access the data kept in the lower-level SAM and BAM formats.

Learning to work with software libraries (which implement APIs) is one of the most important steps in becoming a skilled programmer. Often, beginning programmers are eager to implement everything themselves—while this can be a good learning exercise, for production projects it's far better to utilize libraries. Not only does using libraries save a considerable amount of time, libraries are community tested and less likely to contain bugs that can lead to incorrect results. Fundamentally, libraries allow you as the programmer to ignore lower-level details and work with higher-level abstractions.

Let's start by installing Pysam. While the source (*http://github.com/pysam-developers/pysam*) is available on GitHub, the easiest way to install Pysam is through PyPI (the Python Package Index):

```
$ pip install pysam
```

Once Pysam is installed, open its documentation (*http://pysam.readthedocs.org*). Using APIs efficiently is largely about learning how to effectively access information about classes and functions as you need them in the API's documentation. This is one of the most vital skills to develop in working with APIs.

Opening BAM Files, Fetching Alignments from a Region, and Iterating Across Reads

Let's start by opening our *NA12891_CEU_sample.bam* file using Pysam. We'll do this interactively to introduce useful Python functions that help in exploring APIs:

```
>>> import pysam
>>> fname = "NA12891_CEU_sample.bam"
>>> bamfile = pysam.AlignmentFile(filename=fname, mode="rb")
```

```
>>> type(bamfile)
<type 'pysam.calignmentfile.AlignmentFile'>
>>> dir(bamfile)
['__class__', '__delattr__', ..., 'close', 'count', 'fetch', 'filename',
 'getrname', 'gettid', 'head', 'header', 'lengths', 'mapped', 'mate',
 'next', 'nocoordinate', 'nreferences', 'pileup', 'references', 'reset',
 'seek', 'tell', 'text', 'unmapped', 'write']
```

pysam.AlignmentFile opens the specified file and returns an AlignmentFile object.
pysam.AlignmentFile's default mode is to read plain-text SAM files; because we're
reading a BAM file here, we specify the mode as "rb", which opens a binary BAM file
(b) for reading (r). We could also open SAM/BAM files for writing, but we won't
cover that in this quick introduction.

As a well-designed higher-level API, Pysam abstracts away many lower-level technical
details we shouldn't have to worry about when writing higher-level applications. For
example, we use the exact same pysam.AlignmentFile class and its methods when
working with both SAM and BAM files. Similarly, when you open a BAM file, Pysam
will check to see if the file has a corresponding *.bai* index and automatically load this.
This allows for quick random access of position-sorted and indexed BAM files, much
like we achieved earlier with samtools view.

We could use the pysam.AlignmentFile.fetch() method to fetch aligned reads from
a particular region of an indexed BAM file:

```
>>> for read in bamfile.fetch('1', start=215906528, end=215906567):
...     print read.qname, "aligned at position", read.pos
...
SRR005672.5788073 aligned at position 215906479
SRR005666.5830972 aligned at position 215906486
ERR002294.5383813 aligned at position 215906495
[...]
```

Similarly, we can iterate through all reads in the BAM file (aligned and unaligned)
with the code shown in Example 11-2:

Example 11-2. Iterating through all reads with Pysam

```
>>> bamfile = pysam.AlignmentFile(filename=fname, mode="rb")
>>> for read in bamfile:
...     status = "unaligned" if read.is_unmapped else "aligned"
...     print read.qname, "is", status
...
SRR005672.8895 is aligned
SRR010927.10846964 is aligned
SRR005674.4317449 is aligned
SRR005675.5348609 is aligned
[...]
```

Each iteration returns a single `AlignedSegment` object, which contains all information kept in a single SAM/BAM entry. These objects contain attributes that store alignment and read information, as well as many useful methods that simplify common tasks such as counting how many bases an `AlignedSegment` object overlaps a supplied region, or returning the reference positions this aligned read overlaps. We'll cover `AlignedSegment` objects in "Working with AlignedSegment Objects" on page 388.

There's a very important but subtle behavior to remember: when we iterate through an `AlignmentFile` object this way, the current position in the file is remembered. If you iterate through half a file, break out of the loop, and then begin again, you will start *where you left off, not at the beginning of the file*. This should be no surprise if you've used iterators in Python before; state about the current position is also stored when you iterate through the lines of regular files in Python. But if you're still relatively unfamiliar with Python, this is a common gotcha to remember. As a result of the file position being stored, if you iterate through an entire file, you'll need to reset the file position with reset. Let's demonstrate this gotcha and `Alignment File.reset()`:

```
>>> bamfile = pysam.AlignmentFile(filename=fname, mode="rb")
>>> nmapped = 0
>>> for read in bamfile: ❶
...     nmapped += not read.is_unmapped
...
...
>>> bamfile.next() ❷
Traceback (most recent call last):
  File "<stdin>", line 1, in <module>
  File "calignmentfile.pyx", line 1282, [...]
StopIteration
>>> bamfile.reset() ❸
0
>>> read = bamfile.next() ❹
>>> read.qname
'SRR005672.8895'
```

❶ We iterate through each entry in the `bamfile` object in this for loop (which in the body simply counts how many mapped reads there are).

❷ Once the for loop is complete, the `bamfile` object's internal state is pointing at the end of the file (because we looped through everything). If we were to try to grab another entry in `bamfile` using the `AlignmentFile.next()` method, the Python exception `StopIteration` is raised signaling that there's nothing left to iterate over.

❸ Calling the `AlignmentFile.reset()` method resets the file position to the head of the BAM file (below the header).

❹ After resetting the position, calling the `AlignmentFile.next()` method returns the first alignment. Note that the query name, given by the attribute `AlignedSeg ment.qname`, is the same as the first alignment in Example 11-2.

Once you're done working with an open SAM/BAM file, it's a good idea to close it, which can be done with:

```
>>> bamfile.close()
```

Note, though, that all examples in this section will assume that `bamfile` is open (so if you've just executed the preceding line, reopen the file).

Extracting SAM/BAM Header Information from an AlignmentFile Object

`AlignmentFile` objects also contain all information in the SAM header, which can be accessed using a variety of the `AlignmentFile`'s attributes and methods. The entire header is kept in a Python dictionary, which has the same possible keys as a SAM header:

```
>>> bamfile.header.keys()
['SQ', 'RG', 'PG', 'HD']
```

Each of the values corresponding to these keys contains a list of SAM header records. For example, the first record in the read group (`RG`) section and the third record in the sequence (`SQ`) section are:

```
>>> bamfile.header['RG'][0]
{'LB': 'g1k-sc-NA12891-CEU-1', 'CN': 'SC', 'DS': 'SRP000032',
 'SM': 'NA12891', 'PI': '200', 'ID': 'ERR001776', 'PL': 'ILLUMINA'}
>>> bamfile.header['SQ'][0]
{'LN': 249250621, 'M5': '1b22b98cdeb4a9304cb5d48026a85128',
'AS': 'NCBI37', 'SN': '1',
 'UR': 'file:/lustre/scratch102/projects/g1k/ref/main_project/
 human_g1k_v37.fasta'}
```

While the `AlignmentFile.header` dictionary contains all information in the header, there are higher-level and easier ways to access header information. For example, the reference sequence names and their lengths are stored in two tuples, `Alignment File.references` and `AlignmentFile.lengths`:

```
>>> bamfile.references
('1', '2', '3', '4', '5', '6', '7', [...])
>>> bamfile.lengths
(249250621, 243199373, 198022430, 191154276,
 180915260, 171115067, 159138663, [...])
```

When working with Pysam, one of the most common `AlignmentFile` methods you'll use is `AlignmentFile.getrname()`. Whenever you process an `AlignedSegment` object, the reference is stored as a non-negative integer *reference ID* (also known as a *target ID* or *tid* in the documentation). The reference ID corresponds to a particular reference sequence. The `AlignmentFile.getrname()` method is used to retrieve the name of the reference corresponding to this this integer reference ID:

```
>>> # check if aligned (only aligned reads have meaningful reference)
>>> read.is_unmapped
False
>>> read.tid
0
>>> bamfile.getrname(read.reference_id)
'1'
```

It's also possible to go the opposite direction, from name to reference ID:

```
>>> bamfile.gettid('MT')
24
```

Reference IDs can only be non-negative integers; a value of –1 indicates something is wrong (and your code should explicitly test for this possibility!). For example, if we attempted to access a reference named "Mt" by mistake:

```
>>> bamfile.gettid('Mt')
-1
```

There are numerous other useful `AlignmentFile` attributes we don't have the space to cover here; see the Pysam documentation for a full list.

Working with AlignedSegment Objects

Most SAM/BAM processing tasks you'll tackle with Pysam will involve heavy work with `AlignedSegment` objects. These objects contain information about individual alignment records, which is what most SAM/BAM processing is centered around. All information about an alignment record is stored in this `AlignedSegment` object's attributes. We'll explore some of these attributes by taking a closer look at one `AlignedSegment` object from the *NA12891_CEU_sample.bam* file in this section. But note, the best and most up-to-date information will always come from the API documentation itself. This section quickly steps through the basics and highlights a few parts that can be tricky for those just starting out with Pysam.

First, let's load a sample read from this file:

```
>>> bamfile = pysam.AlignmentFile(filename=fname, mode="rb")
>>> read = bamfile.next()
```

Basic information about the read (also known as the *query*), including its name, sequence, base qualities, and length are all stored as attributes:

```
>>> read.query_name ❶
'SRR005672.8895'
>>> read.reference_start ❷
```

```
215622849
>>> read.query_sequence ❸
'GGAATAAATATAGGAAATGTATAATATATAGGAAATATATATATATAGTAA'
>>> read.query_qualities ❹
array('B', [26, 28, 27, 29, 28, 27, 29, 27, 24, 27, [...]])
>>> read.query_length ❺
51
>>> read.query_length == len(read.query_sequence)
True
```

❶ The `AlignedSegment.query_name` attribute returns the read (query) name.

❷ The `AlignedSegment.reference_start` attribute returns the alignment position (the column POS in the SAM/BAM file).

❸ `AlignedSegment.query_sequence` retrieves the read sequence.

❹ `AlignedSegment.query_qualities` retrieves the read qualities, as a Python array. These qualities have already been converted; there's no need to subtract 33 from these values.

❺ The length of the query be accessed with `AlignedSegment.query_length`. This gives the entire query length, which as you see on the next line will be equal to the length of the sequence.

Earlier, in "Bitwise Flags" on page 360, we learned how to extract information from the SAM format's bitwise flags. Pysam makes this much simpler by providing a series of clearly named attributes that get at the values of these bitwise flags. The following are some examples, but see the Pysam documentation for a full list:

```
>>> read.is_unmapped
False
>>> read.is_paired
True
>>> read.is_proper_pair
True
>>> read.mate_is_unmapped
False
>>> read.is_read1
True
>>> read.is_read2
False
>>> read.is_reverse
False
>>> read.mate_is_reverse
True
>>> read.is_qcfail
False
```

There are two attributes to access an alignment's CIGAR alignment details: `Aligned Segment.cigartuples` and `AlignedSegment.cigarstring`. The former returns a list of tuples that use integers to encode the CIGAR operation and its length (see the documentation for these operation codes). The latter returns the CIGAR alignment as a string.

One important consideration when working with Pysam's `AlignedSegment` objects is how certain values handle soft-clipped alignments. With soft clipping, it's possible that the length of the aligned query sequence is less than the original read sequence. Because we often want to work with only the aligned portion of query sequences when processing alignments, `AlignedSegment` has attributes that take soft clipping into account—be sure to use the correct attribute for your task! For example, let's find a soft-clipped read by iterating over all reads until we find one with "S" in the CIGAR string, then break out of the loop:

```
>>> bamfile = pysam.AlignmentFile('NA12891_CEU_sample.bam')
>>> for read in bamfile:
...     if 'S' in read.cigarstring:
...         break
...
>>> read.cigarstring
'35M16S'
```

`AlignedSegment` objects have some attributes with "alignment" in the name to emphasize these exclude soft-clipped bases. For example, note the difference between `AlignedSegment.query_sequence` and `AlignedSegment.query_align ment_sequence`:

```
>>> read.query_sequence
'TAGGAAATGTATAATATATAGGAAATATATATATATAGGAAATATATAATA'
>>> read.query_alignment_sequence
'TAGGAAATGTATAATATATAGGAAATATATATATA'
>>> len(read.query_sequence) - len(read.query_alignment_sequence)
16
```

The difference between these two strings is the soft-clipped bases; `AlignedSeg ment.query_alignment_sequence` doesn't include these soft-clipped bases and is 16 bases shorter. There are similar attributes, including `AlignedSegment.query_align ment_start` and `AlignedSegment.query_alignment_end` (which give the start and end indices in the aligned portion of the query sequence), `AlignedSeg ment.query_alignment_length` (which returns the alignment length), and `Aligned Segment.query_alignment_qualities` (which returns the sequence base qualities, excluding the soft-clipped bases).

Writing a Program to Record Alignment Statistics

We're now ready to employ our knowledge of AlignedSegment objects to write a simple short program that gathers alignment statistics from a SAM or BAM file. We'll then validate this tool by comparing its results to those from samtools flagstat, a tool that reports statistics about SAM bitwise flags:

```python
import sys
import pysam
from collections import Counter

if len(sys.argv) < 2:
    sys.exit("usage: alnstat.py in.bam")

fname = sys.argv[1]
bamfile = pysam.AlignmentFile(fname)  ❶

stats = Counter()  ❷
for read in bamfile:
    stats['total'] += 1
    stats['qcfail'] += int(read.is_qcfail)  ❸

    # record paired end info
    stats['paired'] += int(read.is_paired)  ❹
    stats['read1'] += int(read.is_read1)
    stats['read2'] += int(read.is_read2)

    if read.is_unmapped:  ❺
        stats['unmapped'] += 1
        continue # other flags don't apply

    # record if mapping quality <= 30  ❻
    stats["mapping quality <= 30"] += int(read.mapping_quality <= 30)

    stats['mapped'] += 1  ❼
    stats['proper pair'] += int(read.is_proper_pair)  ❽

# specify the output order, since dicts don't have order  ❾
output_order = ("total", "mapped", "unmapped", "paired",
                "read1", "read2", "proper pair", "qcfail",
                "mapping quality <= 30")

# format output and print to standard out
for key in output_order:
    format_args = (key, stats[key], 100*stats[key]/float(stats["total"]))  ❿
    sys.stdout.write("%s: %d (%0.2f%%)\n" % format_args)  ⓫
```

This is a quick script, without many bells and whistles. In general, when writing scripts for simple tasks like this, it's best to follow the KISS principle. Don't needlessly overcomplicate things or add extra features—these can all come in later revisions after daily use has shown an inadequacy. Let's step through some key parts:

---

❶ After loading the necessary modules and checking there are a sufficient number of arguments to run the script, we open the SAM or BAM file. Here, we don't explicitly specify a mode, because Pysam's `AlignmentFile` can infer this (see the `AlignmentFile` documentation for more information).

❷ Here, we initiate the `Counter` object from the `collections` module (`Counter` is an incredibly useful class in bioinformatics scripts!). `Counter` behaves much like a Python dictionary, which is how we see it used here.

❸ Here's an example of incrementing a key in the `Counter` object depending on what the value of an `AlignedSegment` attribute is. The attribute `AlignedSegment.is_qcfail` returns whether the QC fail bit is set, which is a Boolean value (`True` or `False`). To be explicit, we convert this to an integer, and add this value (either 1 for `True` or 0 for `False`) to the current count kept in the `Counter` object `stats`.

❹ Similar to how we increment the key for QC fail, we increment the counters for whether a read is paired, is read 1, or is read 2.

❺ Some flags only make sense if a read mapped to the reference. At this point, we check if the read is unmapped, and increment the unmapped counter if so. Since the rest of the values we check in the `for` loop block depend on a read being aligned, if a read is unmapped we use `continue` to skip over this code (because none of these later keys should be incremented if a read is unmapped).

❻ At this point, all alignments are mapped. We increment a counter here recording whether the read mapped with mapping quality is less than 30 (any threshold can be used, but this is a common threshold for "low quality" alignments).

❼ ❽Here we increment the `'mapped'` key, since all alignments at this point are mapped. We also increment if the read is in a proper pair (only aligned reads can be a proper pair).

❾ `Counter` objects are based on Python's dictionaries, so keys are not in any particular order. Here, we define the order to output keys (and their values) as a tuple. We'll see how this is used in the next step.

❿ In this loop, we iterate over each of the keys in the tuple `output_order`, which defines the order of output values in the `Counter` object `stats`. Each line of output has three values: the key name, the total counts for that key in the `Counter` object `stats`, and that key's percentage of the total number of SAM/BAM entries.

This line creates a tuple of these values, which are used to propagate the string formatting values in the next line.

⓫ This line writes a formatted string (using the tuple format_args created in the previous line) to standard output.

Let's run this script on our *NA12891_CEU_sample.bam* file:

```
$ python alnstat/alnstat.py NA12891_CEU_sample.bam
total: 636207 (100.00%)
mapped: 630875 (99.16%)
unmapped: 5332 (0.84%)
paired: 435106 (68.39%)
read1: 217619 (34.21%)
read2: 217487 (34.18%)
proper pair: 397247 (62.44%)
qcfail: 0 (0.00%)
mapping quality <= 30: 90982 (14.30%)
```

Even for simple scripts such as this one, it's essential to validate the results. For longer scripts, frequently used programs, or critical parts of a workflow, we might employ *unit testing* (because if something's wrong, everything will be wrong). Unit testing involves breaking code up into separate functions, and writing code that automatically tests that each function is working by checking its values (called *unit tests*). For a small script like this, this might be overkill. Fortunately, there are two simpler (but less robust) alternatives:

- Creating a small test dataset where you've worked out what the results should be by hand and can check your program's results
- Using another program or application to validate the results

We'll use the latter option, by validating our tool with samtools flagstat (which our program emulates):

```
$ samtools flagstat NA12891_CEU_sample.bam
636207 + 0 in total (QC-passed reads + QC-failed reads)
0 + 0 secondary
0 + 0 supplimentary
29826 + 0 duplicates
630875 + 0 mapped (99.16%:nan%)
435106 + 0 paired in sequencing
217619 + 0 read1
217487 + 0 read2
397247 + 0 properly paired (91.30%:nan%)
424442 + 0 with itself and mate mapped
5332 + 0 singletons (1.23%:nan%)
5383 + 0 with mate mapped to a different chr
2190 + 0 with mate mapped to a different chr (mapQ>=5)
```

Looking at the raw counts (in the first column, before +) we can see that each our values check out. samtools flagstat doesn't return a figure for how many reads have a mapping quality less than or equal to 30, so we have to validate our value another way. One way is to use samtools view's -q option, which returns the number of alignments with a mapping quality greater than or equal to the supplied value. We combine this with -c, which returns the count, rather than the alignments themselves:

```
$ samtools view -c -q 31 NA12891_CEU_sample.bam
539893
```

Now, to find the number of reads with mapping quality less than or equal to 30, we need to subtract this value from the total number of mapped reads. We count the total number of mapped reads with samtools flagstat, filtering out alignments with the unmapped bit (0x4) set:

```
samtools view -c -F 4 NA12891_CEU_sample.bam
630875
```

Note that 630,875 – 539,893 = 90,982, which is exactly what we found with our *aln-stat.py* script—this value checks out, too.

Additional Pysam Features and Other SAM/BAM APIs

In this introduction, we've skipped over some additional features of Pysam. For example, Pysam also includes an interface for creating and working with pileup data, through the AlignmentFile.pileup() method. In addition, Pysam implements some common samtools subcommands—for example, sort, view, and calmd are implemented as pysam.sort, pysam.view, pysam.calmd. Pysam also has a Python interface to Tabix files through pysam.TabixFile (we cover Tabix files later in "Fast Access to Indexed Tab-Delimited Files with BGZF and Tabix" on page 425), and FASTQ files through pysam.FastqFile. These are all well-written and easy-to-use interfaces; unfortunately, we don't have the space to cover all of these in this section, so see Pysam's excellent documentation for more details.

Finally, Pysam is just one popular SAM/BAM API—there are numerous others available in different languages. For example, Samtools has its own C API (*http://bit.ly/st-capi*). If you work in Java, Picard (*http://broadinstitute.github.io/picard/*) offers a Java API for working with SAM and BAM files. Finally, Bioconductor has two excellent packages used for working with SAM and BAM files and genome alignments: Rsamtools (*http://bit.ly/rsamtools*) and GenomicAlignment (*http://bit.ly/genom-align*).

## CHAPTER 12

Bioinformatics Shell Scripting, Writing Pipelines, and Parallelizing Tasks

I've waited until the penultimate chapter this book to share a regrettable fact: everyday bioinformatics work often involves a great deal of tedious data processing. Bioinformaticians regularly need to run a sequence of commands on not just one file, but dozens (sometimes even hundreds) of files. Consequently, a large part of bioinformatics is patching together various processing steps into a pipeline, and then repeatedly applying this pipeline to many files. This isn't exciting scientific work, but it's a necessary hurdle before tackling more exciting analyses.

While writing pipelines is a daily burden of bioinformaticians, it's essential that pipelines are written to be robust and reproducible. Pipelines must be robust to problems that might occur during data processing. When we execute a series of commands on data directly into the shell, we usually clearly see if something goes awry—output files are empty when they should contain data or programs exit with an error. However, when we run data through a processing pipeline, we sacrifice the careful attention we paid to each step's output to gain the ability to automate processing of numerous files. The catch is that not only are errors likely to still occur, they're *more likely* to occur because we're automating processing over more data files and using more steps. For these reasons, it's critical to construct robust pipelines.

Similarly, pipelines also play an important role in reproducibility. A well-crafted pipeline can be a perfect record of exactly how data was processed. In the best cases, an individual could download your processing scripts and data, and easily replicate your exact steps. However, it's unfortunately quite easy to write obfuscated or sloppy pipelines that hinder reproducibility. We'll see some principles that can help you avoid these mistakes, leading to more reproducible projects.

In this chapter, we'll learn the essential tools and skills to construct robust and reproducible pipelines. We'll see how to write rerunnable Bash shell scripts, automate file-processing tasks with `find` and `xargs`, run pipelines in parallel, and see a simple makefile. Note that the subset of techniques covered in this chapter do not target a specific cluster or high-performance computing (HPC) architecture—these are general Unix solutions that work well on any machine. For parallelization techniques specific to your HPC system, you will need consult its documentation.

Basic Bash Scripting

> We've found that duct tape is not a perfect solution for anything. But with a little ingenuity, in a pinch, it's an adequate solution for just about everything.
>
> — *Mythbusters'* Jamie Hyneman

Bash, the shell we've used interactively throughout the book, is also a full-fledged scripting language. Like many other tools presented in this book, the trick to using Bash scripts effectively in bioinformatics is knowing when to use them and when not to. Unlike Python, Bash is not a general-purpose language. Bash is explicitly designed to make running and interfacing command-line programs as simple as possible (a good characteristic of a shell!). For these reasons, Bash often takes the role as the duct tape language of bioinformatics (also referred to as a *glue language*), as it's used to tape many commands together into a cohesive workflow.

Before digging into how to create pipelines in Bash, it's important to note that Python may be a more suitable language for commonly reused or advanced pipelines. Python is a more modern, fully featured scripting language than Bash. Compared to Python, Bash lacks several nice features useful for data-processing scripts: better numeric type support, useful data structures, better string processing, refined option parsing, availability of a large number of libraries, and powerful functions that help with structuring your programs. However, there's more overhead when calling command-line programs from a Python script (known as *calling out* or *shelling out*) compared to Bash. Although Bash lacks some of Python's features, Bash is often the best and quickest "duct tape" solution (which we often need in bioinformatics).

Writing and Running Robust Bash Scripts

Most Bash scripts in bioinformatics are simply commands organized into a rerunnable script with some added bells and whistles to check that files exist and ensuring any error causes the script to abort. These types of Bash scripts are quite simple to write: you've already learned important shell features like pipes, redirects, and background processes that play an important role in Bash scripts. In this section, we'll cover the basics of writing and executing Bash scripts, paying particular attention to how create *robust* Bash scripts.

A robust Bash header

By convention, Bash scripts have the extension *.sh*. You can create them in your favorite text editor (and decent text editors will have support for Bash script syntax highlighting). Anytime you write a Bash script, you should use the following Bash script header, which sets some Bash options that lead to more robust scripts (there's also a copy of this header in the *template.sh* file in this chapter's directory on GitHub):

```
#!/bin/bash ❶
set -e ❷
set -u ❸
set -o pipefail ❹
```

❶ This is called the *shebang*, and it indicates the path to the interpreter used to execute this script. This is only necessary when running the script as a program (more on this in a bit). Regardless of how you plan to run your Bash script, it's best to include a shebang line.

❷ By default, a shell script containing a command that fails (exits with a nonzero exit status) will *not* cause the entire shell script to exit—the shell script will just continue on to the next line. This is not a desirable behavior; we always want errors to be loud and noticeable. `set -e` prevents this, by terminating the script if any command exited with a nonzero exit status. Note, however, that `set -e` has complex rules to accommodate cases when a nonzero exit status indicates "false" rather than failure. For example, `test -d file.txt` will return a nonzero exit status if its argument is not a directory, but in this context this isn't meant to represent an error. `set -e` ignores nonzero statuses in `if` conditionals for this reason (we'll discuss this later). Also, `set -e` ignores all exit statuses in Unix pipes except the last one—this relates to `set -o pipefail`, which we discuss later.

❸ `set -u` fixes another unfortunate default behavior of Bash scripts: any command containing a reference to an unset variable name will still run. As a horrifying example of what this can lead to, consider: `rm -rf $TEMP_DIR/*`. If the shell variable `$TEMP_DIR` isn't set, Bash will still substitute its value (which is nothing) in place of it. The end result is `rm -rf /*`! You can see this for yourself:

```
$ echo "rm $NOTSET/blah"
rm /blah
```

`set -u` prevents this type of error by aborting the script if a variable's value is unset.

❹ As just discussed, `set -e` will cause a script to abort if a nonzero exit status is encountered, with some exceptions. One such exception is if a program run in a Unix pipe exited unsuccessfully; unless this program was the last program in the pipe, this would *not* cause the script to abort even with `set -e`. Including `set -o pipefail` will prevent this undesirable behavior—any program that returns a nonzero exit status in the pipe will cause the entire pipe to return a nonzero status. With `set -e` enabled, too, this will lead the script to abort.

The Robust Bash Header in Bash Script Examples in this Chapter

I will omit this header in Bash scripts throughout this chapter for clarity and to save space, but you should always use it in your own work.

These three options are the first layer of protection against Bash scripts with silent errors and unsafe behavior. Unfortunately, Bash is a fragile language, and we need to mind a few other oddities to use it safely in bioinformatics. We'll see these as we learn more about the language.

Running Bash scripts

Running Bash scripts can be done one of two ways: with the `bash` program directly (e.g., `bash script.sh`), or calling your script as a program (`./script.sh`). For our purposes, there's no technical reason to prefer one approach over the other. Practically, it's wise to get in the habit of running scripts you receive with `./script.sh`, as they might use interpreters other than /bin/bash (e.g., `zsh`, `csh`, etc.). But while we can run any script (as long as it has read permissions) with `bash script.sh`, calling the script as an executable requires that it have executable permissions. We can set these using:

```
$ chmod u+x script.sh
```

This adds executable permissions (+x) for the user who owns the file (u). Then, the script can be run with `./script.sh`.

Variables and Command Arguments

Bash variables play an extremely important role in robust, reproducible Bash scripts. Processing pipelines having numerous settings that should be stored in variables (e.g., which directories to store results in, parameter values for commands, input files, etc.). Storing these settings in a variable defined at the top of the file makes adjusting settings and rerunning your pipelines much easier. Rather than having to change numerous hardcoded values in your scripts, using variables to store settings means you only have to change one value—the value you've assigned to the variable. Bash also reads command-line arguments into variables, so you'll need to be familiar with accessing variables' values to work with command-line arguments.

Unlike other programming languages, Bash's variables don't have data types. It's helpful to think of Bash's variables as strings (but that may behave differently depending on context). We can create a variable and assign it a value with (note that spaces matter when setting Bash variables—do not use spaces around the equals sign!):

```
results_dir="results/"
```

To access a variable's value, we use a dollar sign in front of the variable's name (e.g., `$results_dir`). You can experiment with this in a Bash script, or directly on the command line:

```
$ results_dir="results/"
$ echo $results_dir
results/
```

As mentioned in the previous section, you should always set `set -u` to force a Bash script to exit if a variable is not set.

Even though accessing a variable's value using the dollar sign syntax works, it has one disadvantage: in some cases it's not clear where a variable name ends and where an adjacent string begins. For example, suppose a section of your Bash script created a directory for a sample's alignment data, called *<sample>_aln/*, where *<sample>* is replaced by the sample's name. This would look like:

```
sample="CNTRL01A"
mkdir $sample_aln/
```

Although the intention of this code block was to create a directory called *CNTRL01A_aln/*, this would actually fail, because Bash will try to retrieve the value of a variable named `$sample_aln`. To prevent this, wrap the variable name in braces:

```
sample="CNTRL01A"
mkdir ${sample}_aln/
```

Now, a directory named *CNTRL01A_aln/* will be created. While this solves our immediate problem of Bash interpreting `sample_aln` as the variable name, there's one more step we should take to make this more robust: quoting variables. This prevents commands from interpreting any spaces or other special characters that the variable may contain. Our final command would look as follows:

```
sample="CNTRL01A"
mkdir "${sample}_aln/"
```

Command-line arguments

Let's now look at how Bash handles command-line arguments (which are assigned to the value $1, $2, $3, etc.). The variable $0 stores the name of the script. We can see this ourselves with a simple example script:

```
#!/bin/bash
```

```
echo "script name: $0"
echo "first arg: $1"
echo "second arg: $2"
echo "third arg: $3"
```

Running this file prints arguments assigned to $0, $1, $2, and $3:

```
$ bash args.sh arg1 arg2 arg3
script name: args.sh
first arg: arg1
second arg: arg2
third arg: arg3
```

Bash assigns the number of command-line arguments to the variable $# (this does *not* count the script name, $0, as an argument). This is useful for user-friendly messages (this uses a Bash if conditional, which we'll cover in more depth in the next section):

```
#!/bin/bash

if [ "$#" -lt 3 ] # are there less than 3 arguments?
then
    echo "error: too few arguments, you provided $#, 3 required"
    echo "usage: script.sh arg1 arg2 arg3"
    exit 1
fi

echo "script name: $0"
echo "first arg: $1"
echo "second arg: $2"
echo "third arg: $3"
```

Running this with too few arguments gives an error (and leads the process to exit with a nonzero exit status—see "Exit Status: How to Programmatically Tell Whether Your Command Worked" on page 52 if you're rusty on what exit statuses mean):

```
$ ./script.sh some_arg
error: too few arguments, you provided 1, 3 required
usage: script.sh arg1 arg2 arg3
```

It's possible to have more complex options and argument parsing with the Unix tool getopt. This is out of the scope of this book, but the manual entry for getopt is quite thorough. However, if you find your script requires numerous or complicated options, it might be easier to use Python instead of Bash. Python's argparse module is much easier to use than getopt.

Reproducibility and Environmental Variables

Some bioinformaticians make use of *environmental variables* to store settings using the command export, but in general this makes scripts less portable and reproducible. Instead, all important settings should be stored *inside* the script as variables, rather than as external environmental variables. This way, the script is self-contained and reproducible.

Variables created in your Bash script will only be available for the duration of the Bash process running that script. For example, running a script that creates a variable with some_var=3 will not create some_var in your current shell, as the script runs in an entirely separate shell process.

Conditionals in a Bash Script: if Statements

Like other scripting languages, Bash supports the standard if conditional statement. What makes Bash a bit unique is that a command's exit status provides the true and false (remember: contrary to other languages, 0 represents true/success and anything else is false/failure). The basic syntax is:

```
if [commands] ❶
then
  [if-statements] ❷
else
  [else-statements] ❸
fi
```

❶ [commands] is a placeholder for any command, set of commands, pipeline, or test condition (which we'll see later). If the exit status of these commands is 0, execution continues to the block after then; otherwise execution continues to the block after else. The then keyword can be placed on the same line as if, but then a semicolon is required: if [commands]; then.

❷ [if-statements] is a placeholder for all statements executed if [commands] evaluates to true (0).

❸ [else-statements] is a placeholder for all statements executed if [commands] evaluates to false (1). The else block is optional.

Bash's if condition statements may seem a bit odd compared to languages like Python, but remember: Bash is primarily designed to stitch together other commands. This is an advantage Bash has over Python when writing pipelines: Bash allows your scripts to directly work with command-line programs without requiring any overhead to call programs. Although it can be unpleasant to write complicated programs in Bash, writing simple programs is exceedingly easy because Unix tools and Bash harmonize well. For example, suppose we wanted to run a set of commands only if a file contains a certain string. Because grep returns 0 only if it matches a pattern in a file and 1 otherwise, we could use:

```
#!/bin/bash

if grep "pattern" some_file.txt > /dev/null ❶
then
  # commands to run if "pattern" is found
  echo "found 'pattern' in 'some_file.txt"
fi
```

❶ This grep command is our condition statement. The redirection is to tidy the output of this script such that grep's output is redirected to */dev/null* and not to the script's standard out.

The set of commands in an if condition can use all features of Unix we've mastered so far. For example, chaining commands with logical operators like && (logical AND) and || (logical OR):

```
#!/bin/bash

if grep "pattern" file_1.txt > /dev/null &&
    grep "pattern" file_2.txt > /dev/null
then
    echo "found 'pattern' in 'file_1.txt' and in 'file_2.txt'"
fi
```

We can also negate our program's exit status with !:

```
#!/bin/bash

if ! grep "pattern" some_file.txt > /dev/null
then
    echo "did not find 'pattern' in 'some_file.txt"
fi
```

Finally, it's possible to use pipelines in if condition statements. Note, however, that the behavior depends on set -o pipefail. If pipefail is set, any nonzero exit status in a pipe in your condition statement will cause execution to continue on, skipping the if-statements section (and going on to the else block if it exists). However, if pipefail is *not* set, only the exit status of the *last* command is considered. Rather than trying to remember all of these rules, just use the robust header provided earlier —pipefail is a more sensible default.

The final component necessary to understand Bash's if statements is the test command. Like other programs, test exits with either 0 or 1. However test's exit status indicates the return value of the test specified through its arguments, rather than exit success or error. test supports numerous standard comparison operators (whether two strings are equal, whether two integers are equal, whether one integer is greater than or equal to another, etc.). Bash can't rely on familiar syntax such as > for "greater than," as this is used for redirection: instead, `test` has its own syntax (see Table 12-1 for a full list). You can get a sense of how `test` works by playing with it directly on the command line (using `;` `echo` `"$?"` to print the exit status):

```
$ test "ATG" = "ATG" ; echo "$?"
0
$ test "ATG" = "atg" ; echo "$?"
1
$ test 3 -lt 1 ; echo "$?"
1
$ test 3 -le 3 ; echo "$?"
0
```

Table 12-1. String and integer comparison operators

String/integer	Description
-z str	String str is null (empty)
str1 = str2	Strings str1 and str2 are identical
str1 != str2	Strings str1 and str2 are different
int1 -eq int2	Integers int1 and int2 are equal
int1 -ne int2	Integers int1 and int2 are not equal
int1 -lt int2	Integer int1 is less than int2
int1 -gt int2	Integer int1 is greater than int2
int1 -le int2	Integer int1 is less than or equal to int2
int1 -ge int2	Integer int1 is greater than or equal to int2

In practice, the most common conditions you'll be checking aren't whether some integer is less than another, but rather checking to see if files or directories exist and whether you can write to them. `test` supports numerous file- and directory-related test operations (the few that are most useful in bioinformatics are in Table 12-2). Let's look at a few basic command-line examples:

```
$ test -d some_directory ; echo $? # is this a directory?
0
$ test -f some_file.txt ; echo $? # is this a file?
0
$ test -r some_file.txt ; echo $? $ is this file readable?
```

```
0
$ test -w some_file.txt ; echo $? $ is this file writable?
1
```

Table 12-2. File and directory test expressions

File/directory expression	Description
-d dir	dir is a directory
-f file	file is a file
-e file	file exists
-h link	link is a link
-r file	file is readable
-w file	file is writable
-x file	file is executable (or accessible if argument is expression)

Combining test with if statements is simple; test is a command, so we could use:

```
if test -f some_file.txt
then
  [...]
fi
```

However, Bash provides a simpler syntactic alternative to test statements: [-f some_file.txt] . Note the spaces around and within the brackets—these are required. This makes for much simpler if statements involving comparisons:

```
if [ -f some_file.txt ]
then
  [...]
fi
```

When using this syntax, we can chain test expressions with -a as logical AND, -o as logical OR, ! as negation, and parentheses to group statements. Our familiar && and || operators won't work in test, because these are shell operators. As an example, suppose we want to ensure our script has enough arguments and that the input file is readable:

```
#!/bin/bash
set -e
set -u
set -o pipefail

if [ "$#" -ne 1 -o ! -r "$1" ]
```

```
then
    echo "usage: script.sh file_in.txt"
    exit 1
fi
```

As discussed earlier, we quote variables (especially those from human input); this is a good practice and prevents issues with special characters.

When chained together with `-a` or `-e`, `test`'s syntax uses *short-circuit evaluation*. This means that `test` will only evaluate as many expressions as needed to determine whether the entire statement is true or false. In this example, `test` won't check if the file argument `$1` is readable if there's not exactly one argument provided (the first condition is true). These two expressions are combined with a logical OR, which only requires *one* expression to be true for the entire condition to be true.

Processing Files with Bash Using for Loops and Globbing

In bioinformatics, most of our data is split across multiple files (e.g., different treatments, replicates, genotypes, species, etc.). At the heart of any processing pipeline is some way to apply the same workflow to each of these files, taking care to keep track of sample names. Looping over files with Bash's `for` loop is the simplest way to accomplish this. This is such an important part of bioinformatics processing pipelines that we'll cover additional useful tools and methods in the next section of this chapter.

There are three essential parts to creating a pipeline to process a set of files:

- Selecting which files to apply the commands to
- Looping over the data and applying the commands
- Keeping track of the names of any output files created

There are different computational tricks to achieve each of these tasks. Let's first look at the simple ways to select which files to apply commands to.

There are two common ways to select which files to apply a bioinformatics workflow to: approaches that start with a text file containing information about samples (their sample names, file path, etc.), and approaches that select files in directories using some criteria. Either approach is fine—it mostly comes down to what's most efficient for a given task. We'll first look at an approach that starts with sample names and return to how to look for files later. Suppose you have a file called *samples.txt* that tells you basic information about your raw data: sample name, read pair, and where the file is. Here's an example (which is also in this chapter's directory on GitHub):

```
$ cat samples.txt
zmaysA  R1  seq/zmaysA_R1.fastq
zmaysA  R2  seq/zmaysA_R2.fastq
```

```
zmaysB  R1  seq/zmaysB_R1.fastq
zmaysB  R2  seq/zmaysB_R2.fastq
zmaysC  R1  seq/zmaysC_R1.fastq
zmaysC  R2  seq/zmaysC_R2.fastq
```

The first column gives sample names, the second column contains read pairs, and the last column contains the path to the FASTQ file for this sample/read pair combination. The first two columns are called *metadata* (data about data), which is vital to relating sample information to their physical files. Note that the metadata is also in the filename itself, which is useful because it allows us to extract it from the filename if we need to.

With this *samples.txt* file, the first step of creating the pipeline is complete: all information about our files to be processed, including their path, is available. The second and third steps are to loop over this data, and do so in a way that keeps the samples straight. How we accomplish this depends on the specific task. If your command takes a single file and returns a single file, the solution is trivial: files are the unit we are processing. We simply loop over each file and use a modified version of that file's name for the output.

Let's look at an example: suppose that we want to loop over every file, gather quality statistics on each and every file (using the imaginary program fastq_stat), and save this information to an output file. Each output file should have a name based on the input file, so if a summary file indicates something is wrong we know which file was affected. There's a lot of little parts to this, so we're going to step through how to do this a piece at a time learning about Bash arrays, basename, and a few other shell tricks along the way.

First, we load our filenames into a *Bash array*, which we can then loop over. Bash arrays can be created manually using:

```
$ sample_names=(zmaysA zmaysB zmaysC)
```

And specific elements can be extracted with (note Bash arrays are 0-indexed):

```
$ echo ${sample_names[0]}
zmaysA
$ echo ${sample_names[2]}
zmaysC
```

All elements are extracted with the cryptic-looking ${sample_files[@]}:

```
$ echo ${sample_names[@]}
zmaysA zmaysB zmaysC
```

You can also access how many elements are in the array (and each element's index) with the following:

```
$ echo ${#sample_names[@]}
3
```

```
$ echo ${!sample_names[@]}
0 1 2
```

But creating Bash arrays by hand is tedious and error prone, especially because we already have our filenames in our *sample.txt* file. The beauty of Bash is that we can use a *command substitution* (discussed in "Command Substitution" on page 54) to construct Bash arrays (though this can be dangerous; see the following warning). Because we want to loop over each file, we need to extract the third column using cut -f 3 from *samples.txt*. Demonstrating this in the shell:

```
$ sample_files=($(cut -f 3 samples.txt))
$ echo ${sample_files[@]}
seq/zmaysA_R1.fastq seq/zmaysA_R2.fastq seq/zmaysB_R1.fastq
seq/zmaysB_R2.fastq seq/zmaysC_R1.fastq seq/zmaysC_R2.fastq
```

Again, this only works if you can make strong assumptions about your filenames—namely that they only contain alphanumeric characters, (_), and (-)! If spaces, tabs, newlines, or special characters like * end up in filenames, it will break this approach.

The Internal Field Separator, Word Splitting, and Filenames

When creating a Bash array through command substitution with sample_files=($(cut -f 3 samples.txt)), Bash uses *word splitting* to split fields into array elements by splitting on the characters in the *Internal Field Separator* (IFS). The Internal Field Separator is stored in the Bash variable IFS, and by default includes spaces, tabs, and newlines. You can inspect the value of IFS with:

```
$ printf %q "$IFS"
$' \t\n'
```

Note that space is included in IFS (the first character). This can introduce problems when filenames contain spaces, as Bash will split on space characters breaking the filename into parts. Again, the best way to avoid issues is to not use spaces, tabs, newlines, or special characters (e.g., *) in filenames—only use alphanumeric characters, (-), and (_). The techniques taught in this section assume files are properly named and are not robust against improperly named files. If spaces are present in filenames, you can set the value of IFS to just tabs and newlines; see this chapter's *README* file on GitHub for additional details on this topic.

With our filenames in a Bash array, we're almost ready to loop over them. The last component is to strip the path and extension from each filename, leaving us with the most basic filename we can use to create an output filename. The Unix program base name strips paths from filenames:

```
$ basename seqs/zmaysA_R1.fastq
zmaysA_R1.fastq
```

`basename` can also strip a suffix (e.g., extension) provided as the second argument from a filename (or alternatively using the argument `-s`):

```
$ basename seqs/zmaysA_R1.fastq .fastq
zmaysA_R1
$ basename -s .fastq seqs/zmaysA_R1.fastq
zmaysA_R1
```

We use `basename` to return the essential part of each filename, which is then used to create output filenames for `fastq_stat`'s results.

Now, all the pieces are ready to construct our processing script:

```
#!/bin/bash

set -e
set -u
set -o pipefail

# specify the input samples file, where the third
# column is the path to each sample FASTQ file
sample_info=samples.txt

# create a Bash array from the third column of $sample_info
sample_files=($(cut -f 3 "$sample_info")) ❶

for fastq_file in ${sample_files[@]} ❷
do
    # strip .fastq from each FASTQ file, and add suffix
    # "-stats.txt" to create an output filename for each FASTQ file
    results_file="$(basename $fastq_file .fastq)-stats.txt" ❸

    # run fastq_stat on a file, writing results to the filename we've
    # above
    fastq_stat $fastq_file > stats/$results_file ❹
done
```

❶ This line uses command substitution to create a Bash array containing all FASTQ files. Note that this uses the filename contained in the variable `$sample_info`, which can later be easily be changed if the pipeline is to be run on a different set of samples.

❷ Next, we loop through each sample filename using a Bash for loop. The expression `${sample_files[@]}` returns all items in the Bash array.

❸ This important line creates the output filename, using the input file's filename. The command substitution `$(basename $fastq_file .fastq)` takes the current filename of the iteration, stored in `$fastq_file`, and strips the *fastq* extension off. What's left is the portion of the sample name that identifies the original file, to which the suffix `-stats.txt` is added.

❹ Finally, the command `fastq_stat` is run, using the filename of the current iteration as input, and writing results to `stats/$results_file`.

That's all there is to it. A more refined script might add a few extra features, such as using an `if` statement to provide a friendly error if a FASTQ file does not exist or a call to `echo` to report which sample is currently being processed.

This script was easy to write because our processing steps took a single file as input, and created a single file as output. In this case, simply adding a suffix to each filename was enough to keep our samples straight. However, many bioinformatics pipelines *combine* two or more input files into a single output file. Aligning paired-end reads is a prime example: most aligners take two input FASTQ files and return one output alignment file. When writing scripts to align paired-end reads, we can't loop over each file like we did earlier. Instead, each *sample*, rather than each file, is the processing unit. Our alignment step takes both FASTQ files for each sample, and turns this into a single alignment file for this sample. Consequently, our loop must iterate over unique sample names, and we use these sample names to re-create the input FASTQ files used in alignment. This will be clearer in an example; suppose that we use the aligner BWA and our genome reference is named *zmays_AGPv3.20.fa*:

```
#!/bin/bash
set -e
set -u
set -o pipefail

# specify the input samples file, where the third
# column is the path to each sample FASTQ file
sample_info=samples.txt

# our reference
reference=zmays_AGPv3.20.fa

# create a Bash array from the first column, which are
# sample names. Because there are duplicate sample names
# (one for each read pair), we call uniq
sample_names=($(cut -f 1 "$sample_info" | uniq)) ❶

for sample in ${sample_names[@]}
do
    # create an output file from the sample name
    results_file="${sample}.sam" ❷
    bwa mem $reference ${sample}_R1.fastq ${sample}_R2.fastq \ ❸
      > $results_file
done
```

❶ This is much like the previous example, except now we use cut to grab the first column (corresponding to sample names), and (most importantly) pipe these sample names to uniq so duplicates of the same sample name are removed. This is necessary because our first column repeats each sample name twice, once for each paired-end file.

❷ As before, we create an output filename for the current sample being iterated over. In this case, all that's needed is the sample name stored in $sample.

❸ Our call to bwa provides the reference, and the two paired-end FASTQ files for this sample as input. Note how we can re-create the two input FASTQ files for a given sample name, as the naming of these files across samples is consistent. In practice, this is possible for a large amount of bioinformatics data, which often comes from machines that name files consistently. Finally, the output of bwa is redirected to $results_file. For clarity, I've omitted quoting variables in this command, but you may wish to add this.

Finally, in some cases it might be easier to directly loop over files, rather than working a file containing sample information like *samples.txt*. The easiest (and safest) way to do this is to use Bash's wildcards to glob files to loop over (recall we covered globbing in "Organizing Data to Automate File Processing Tasks" on page 26). The syntax of this is quite easy:

```
#!/bin/bash
set -e
set -u
set -o pipefail

for file in *.fastq
do
    echo "$file: " $(bioawk -c fastx 'END {print NR}' $file)
done
```

This simple script uses bioawk to count how many FASTQ records are in a file, for each file in the current directory ending in *.fastq*.

Bash's loops are a handy way of applying commands to numerous files, but have a few downsides. First, compared to the Unix tool find (which we see in the next section), globbing is not a very powerful way to select certain files. Second, Bash's loop syntax is lengthy for simple operations, and a bit archaic. Finally, there's no easy way to parallelize Bash loops in a way that constrains the number of subprocesses used. We'll see a powerful file-processing Unix idiom in the next section that works better for some tasks where Bash scripts may not be optimal.

Automating File-Processing with find and xargs

In this section, we'll learn about a more powerful way to specify files matching some criteria using Unix find. We'll also see how files printed by find can be passed to another tool called xargs to create powerful Unix-based processing workflows.

Using find and xargs

First, let's look at some common shell problems that find and xargs solve. Suppose you have a program named process_fq that takes multiple filenames through standard in to process. If you wanted to run this program on all files with the suffix *.fq*, you might run:

```
$ ls *.fq
treatment-01.fq treatment 02.fq treatment-03.fq
$ ls *.fq | process_fq
```

Your shell expands this wildcard to all matching files in the current directory, and ls prints these filenames. Unfortunately, this leads to a common complication that makes ls and wildcards a fragile solution. Suppose your directory contains a filename called *treatment 02.fq*. In this case, ls returns *treatment 02.fq* along with other files. However, because files are separated by spaces, and this file *contains* a space, process_fq will interpret *treatment 02.fq* as two separate files, named *treatment* and *02.fq*. This problem crops up periodically in different ways, and it's necessary to be aware of when writing file-processing pipelines. Note that this does *not* occur with file globbing in arguments—if process_fq takes multiple files as arguments, your shell handles this properly:

```
$ process_fq *.fq
```

Here, your shell automatically escapes the space in the filename *treatment 02.fq*, so process_fq will correctly receive the arguments treatment-01.fq, treatment 02.fq, treatment-03.fq. So why not use this approach? Alas, there's a limit to the number of files that can be specified as arguments. It's not unlikely to run into this limit when processing numerous files. As an example, suppose that you have a *tmp/* directory with thousands and thousands of temporary files you want to remove before rerunning a script. You might try rm tmp/*, but you'll run into a problem:

```
$ rm tmp/*
/bin/rm: cannot execute [Argument list too long]
```

New bioinformaticians regularly encounter these two problems (personally, I am asked how to resolve these issues at least once every other month by various colleagues). The solution to both of these problems is through find and xargs, as we will see in the following sections.

Finding Files with find

Unlike `ls`, `find` is recursive (it will search for matching files in subdirectories, and subdirectories of subdirectories, etc.). This makes `find` useful if your project directory is deeply nested and you wish to search the entire project for a file. In fact, running `find` on a directory (without other arguments) can be a quick way to see a project directory's structure. Again, using the *zmays-snps/* toy directory we created in "Organizing Data to Automate File Processing Tasks" on page 26:

```
$ find zmays-snps
zmays-snps
zmays-snps/analysis
zmays-snps/data
zmays-snps/data/seqs
zmays-snps/data/seqs/zmaysA_R1.fastq
zmays-snps/data/seqs/zmaysA_R2.fastq
zmays-snps/data/seqs/zmaysB_R1.fastq
zmays-snps/data/seqs/zmaysB_R2.fastq
zmays-snps/data/seqs/zmaysC_R1.fastq
zmays-snps/data/seqs/zmaysC_R2.fastq
zmays-snps/scripts
```

`find`'s recursive searching can be limited to search only a few directories deep with the argument `-maxdepth`. For example, to search only within the current directory, use `-maxdepth 1`; to search within the current directory and its subdirectories (but not within those subdirectories), use `-maxdepth 2`.

The basic syntax for `find` is `find path expression`. Here, `path` specifies which directory `find` is to search for files in (if you're currently in this directory, it's simply `find .`). The `expression` part of `find`'s syntax is where `find`'s real power comes in. Expressions are how we describe which files we want to `find` return. Expressions are built from predicates, which are chained together by logical AND and OR operators. `find` only returns files when the expression evaluates to true. Through expressions, `find` can match files based on conditions such as creation time or the permissions of the file, as well as advanced combinations of these conditions, such as "find all files created after last week that have read-only permissions."

To get a taste of how simple predicates work, let's see how to use `find` to match files by filename using the `-name` predicate. Earlier we used unquoted wildcards with `ls`, which are expanded by the shell to all matching filenames. With `find`, we quote patterns (much like we did with `grep`) to avoid our shells from interpreting characters like `*`. For example, suppose we want to find all files matching the pattern `"zmaysB*fastq"` (e.g., FASTQ files from sample "B", both read pairs) to pass to a pipeline. We would use the command shown in Example 12-1:

Example 12-1. Find through filename matching

```
$ find data/seqs -name "zmaysB*fastq"
data/seqs/zmaysB_R1.fastq
data/seqs/zmaysB_R2.fastq
```

This gives similar results to `ls zmaysB*fastq`, as we'd expect. The primary difference is that `find` reports results separated by newlines and, by default, `find` is recursive.

find's Expressions

`find`'s expressions allow you to narrow down on specific files using a simple syntax. In the previous example, the `find` command (Example 12-1) would return directories also matching the pattern `"zmaysB*fastq"`. Because we only want to return FASTQ files (and not directories with that matching name), we might want to limit our results using the `-type` option:

```
$ find data/seqs -name "zmaysB*fastq" -type f
data/seqs/zmaysB_R1.fastq
data/seqs/zmaysB_R2.fastq
```

There are numerous different types you can search for (e.g., files, directories, named pipes, links, etc.), but the most commonly used are f for files, d for directories, and l for links.

By default, `find` connects different parts of an expression with logical AND. The `find` command in this case returns results where the name matches `"zmaysB*fastq"` *and* is a file (type "f"). `find` also allows explicitly connecting different parts of an expression with different operators. The preceding command is equivalent to:

```
$ find data/seqs -name "zmaysB*fastq" -and -type f
data/seqs/zmaysB_R1.fastq
data/seqs/zmaysB_R2.fastq
```

We might also want all FASTQ files from samples A *or* C. In this case, we'd want to chain expressions with another operator, `-or` (see Table 12-3 for a full list):

```
$ find data/seqs -name "zmaysA*fastq" -or -name "zmaysC*fastq" -type f
data/seqs/zmaysA_R1.fastq
data/seqs/zmaysA_R2.fastq
data/seqs/zmaysC_R1.fastq
data/seqs/zmaysC_R2.fastq
```

Table 12-3. Common find expressions and operators

Operator/expression	Description
-name <pattern>	Match a filename to <pattern>, using the same special characters (*, ?, and [...] as Bash)

Operator/expression	Description
`-iname <pattern>`	Identical to `-name`, but is case insensitive
`-empty`	Matches empty files or directories
`-type <x>`	Matches types x (f for files, d for directories, l for links)
`-size <size>`	Matches files where the `<size>` is the file size (shortcuts for kilobytes (k), megabytes (M), gigabytes (G), and terabytes (T) can be used); sizes preceded with + (e.g., `+50M`) match files *at least* this size; sizes preceded with - (e.g., `-50M`) match files *at most* this size
`-regex`	Match by regular expression (extended regular expressions can be enabled with `-E`)
`-iregex`	Identical to `-regex`, but is case insensitive
`-print0`	Separate results with null byte, not newline
`expr -and expr`	Logical "and"
`expr -or expr`	Logical "or"
`-not / "!" expr`	Negation
`(expr)`	Group a set of expressions

An identical way to select these files is with negation:

```
$ find seqs -type f "!" -name "zmaysC*fastq"
seqs/zmaysA_R1.fastq
seqs/zmaysA_R2.fastq
seqs/zmaysB_R1.fastq
seqs/zmaysB_R2.fastq
```

Let's see one more advanced example. Suppose that a messy collaborator decided to create a file named *zmaysB_R1-temp.fastq* in *seqs/*. You notice this file because now your find command is matching it (we are still in the *zmays/data* directory):

```
$ touch seqs/zmaysB_R1-temp.fastq
$ find seqs -type f "!" -name "zmaysC*fastq"
seqs/zmaysB_R1-temp.fastq
seqs/zmaysA_R1.fastq
seqs/zmaysA_R2.fastq
seqs/zmaysB_R1.fastq
seqs/zmaysB_R2.fastq
```

You don't want to delete his file or rename it, because your collaborator may need that file and/or rely on it having that specific name. So, the best way to deal with it seems to be to change your find command and talk to your collaborator about this mystery file later. Luckily, find allows this sort of advanced file querying:

```
$ find seqs -type f "!" -name "zmaysC*fastq" -and "!" -name "*-temp*"
seqs/zmaysA_R1.fastq
seqs/zmaysA_R2.fastq
seqs/zmaysB_R1.fastq
seqs/zmaysB_R2.fastq
```

Note that find's operators like !, (, and) should be quoted so as to avoid your shell from interpreting these.

find's -exec: Running Commands on find's Results

While find is useful for locating a file, its real strength in bioinformatics is as a tool to programmatically access certain files you want to run a command on. In the previous section, we saw how find's expressions allow you to select distinct files that match certain conditions. In this section, we'll see how find allows you to run commands on each of the files find returns, using find's -exec option.

Let's look at a simple example to understand how -exec works. Continuing from our last example, suppose that a messy collaborator created numerous temporary files. Let's emulate this (in the *zmays-snps/data/seqs* directory):

```
$ touch zmays{A,C}_R{1,2}-temp.fastq
$ ls
zmaysA_R1-temp.fastq zmaysB_R1-temp.fastq zmaysC_R1.fastq
zmaysA_R1.fastq      zmaysB_R1.fastq      zmaysC_R2-temp.fastq
zmaysA_R2-temp.fastq zmaysB_R2.fastq      zmaysC_R2.fastq
zmaysA_R2.fastq      zmaysC_R1-temp.fastq
```

Suppose your collaborator allows you to delete all of these temporary files. One way to delete these files is with rm *-temp.fastq. However, rm with a wildcard in a directory filled with important data files is too risky. If you've accidentally left a space between * and -temp.fastq, the wildcard * would match *all* files in the current directory and pass them to rm, leading to *everything* in the directory being accidentally deleted. Using find's -exec is a much safer way to delete these files.

find's -exec works by passing each file that matches find's expressions to the command specified by -exec. With -exec, it's necessary to use a semicolon at the end of the command to indicate the end of your command. For example, let's use find - exec and rm -i to delete these temporary files. rm's -i forces rm to be interactive, prompting you to confirm that you want to delete a file. Our find and remove command is:

```
$ find . -name "*-temp.fastq" -exec rm -i {} \;
remove ./zmaysA_R1-temp.fastq? y
remove ./zmaysA_R2-temp.fastq? y
remove ./zmaysB_R1-temp.fastq? y
remove ./zmaysC_R1-temp.fastq? y
remove ./zmaysC_R2-temp.fastq? y
```

In one line, we're able to pragmatically identify and execute a command on files that match a certain pattern. Our command was rm but can just as easily be a bioinformatics program. Using this approach allows you to call a program on any number of files in a directory. With find and -exec, a daunting task like processing a directory of 100,000 text files with a program is simple.

Deleting Files with find -exec

Deleting files with find -exec is a such a common operation that find also has a -delete option you can use in place of -exec -rm {} (but it will not be interactive, unlike rm with -i).

When using -exec, always write your expression first and check that the files returned are those you want to apply the command to. Then, once your find command is returning the proper subset of files, add in your -exec statement. find -exec is most appropriate for quick, simple tasks (like deleting files, changing permissions, etc.). For larger tasks, xargs (which we see in the next section) is a better choice.

xargs: A Unix Powertool

If there were one Unix tool that introduced me to the incredible raw power of Unix, it is xargs. xargs allows us to take input passed to it from standard in, and use this input as *arguments* to another program, which allows us to build commands programmatically from values received through standard in (in this way, it's somewhat similar to R's do.call()). Using find with xargs is much like find with -exec, but with some added advantages that make xargs a better choice for larger bioinformatics file-processing tasks.

Let's re-create our messy temporary file directory example again (from the *zmays-snps/data/seqs* directory):

```
$ touch zmays{A,C}_R{1,2}-temp.fastq   # create the test files
```

xargs works by taking input from standard in and splitting it by spaces, tabs, and newlines into arguments. Then, these arguments are passed to the command supplied. For example, to emulate the behavior of find -exec with rm, we use xargs with rm:

```
$ find . -name "*-temp.fastq"
./zmaysA_R1-temp.fastq
./zmaysA_R2-temp.fastq
./zmaysC_R1-temp.fastq
./zmaysC_R2-temp.fastq
$ find . -name "*-temp.fastq" | xargs rm
```

Playing It Safe with find and xargs

There's one important gotcha with find and xargs: spaces in file-names can break things, because spaces are considered argument separators by xargs. This would lead to a filename like *treatment 02.fq* being interpreted as two separate arguments, *treatment* and *02.fq*. The find and xargs developers created a clever solution: both allow for the option to use the null byte as a separator. Here is an example of how to run find and xargs using the null byte delimiter:

```
$ find . -name "samples [AB].txt" -print0 | xargs -0 rm
```

In addition to this precaution, it's also wise to simply not use file-names that contain spaces or other strange characters. Simple alphanumeric names with either dashes or underscores are best. To simplify examples, I will omit -print0 and -0, but these should always be used in practice.

Essentially, xargs is splitting the output from find into arguments, and running:

```
$ rm ./zmaysA_R1-temp.fastq ./zmaysA_R2-temp.fastq \
    ./zmaysC_R1-temp.fastq ./zmaysC_R2-temp.fastq
```

xargs passes *all* arguments received through standard in to the supplied program (rm in this example). This works well for programs like rm, touch, mkdir, and others that take multiple arguments. However, other programs only take a single argument at a time. We can set how many arguments are passed to each command call with xargs's -n argument. For example, we could call rm four separate times (each on one file) with:

```
$ find . -name "*-temp.fastq" | xargs -n 1 rm
```

One big benefit of xargs is that it separates the process that specifies the files to operate on (find) from applying a command to these files (through xargs). If we wanted to inspect a long list of files find returns before running rm on all files in this list, we could use:

```
$ find . -name "*-temp.fastq" > files-to-delete.txt
$ cat files-to-delete.txt
./zmaysA_R1-temp.fastq
./zmaysA_R2-temp.fastq
./zmaysC_R1-temp.fastq
```

```
./zmaysC_R2-temp.fastq
$ cat files-to-delete.txt | xargs rm
```

Another common trick is to use xargs to build commands that are written to a simple Bash script. For example, rather than running rm directly, we could call echo on rm, and then allow xargs to place arguments after this command (remember, xargs's behavior is very simple: it just places arguments after the command you provide). For example:

```
$ find . -name "*-temp.fastq" | xargs -n 1 echo "rm -i" > delete-temp.sh
$ cat delete-temp.sh
rm -i ./zmaysA_R1-temp.fastq
rm -i ./zmaysA_R2-temp.fastq
rm -i ./zmaysC_R1-temp.fastq
rm -i ./zmaysC_R2-temp.fastq
```

Breaking up the task in this way allows us to inspect the commands we've built using xargs (because the command xargs runs is echo, which just prints everything). Then, we could run this simple script with:

```
$ bash delete-temp.sh
remove ./zmaysA_R1-temp.fastq? y
remove ./zmaysA_R2-temp.fastq? y
remove ./zmaysC_R1-temp.fastq? y
remove ./zmaysC_R2-temp.fastq? y
```

Using xargs with Replacement Strings to Apply Commands to Files

So far, xargs builds commands purely by adding arguments to the end of the command you've supplied. However, some programs take arguments through options, like program --in file.txt --out-file out.txt; others have many positional arguments like program arg1 arg2. xargs's -I option allows more fine-grained placement of arguments into a command by replacing all instances of a placeholder string with a single argument. By convention, the placeholder string we use with -I is {}.

Let's look at an example. Suppose the imaginary program fastq_stat takes an input file through the option --in, gathers FASTQ statistics information, and then writes a summary to the file specified by the --out option. As in our Bash loop example ("Processing Files with Bash Using for Loops and Globbing" on page 405), we want our output filenames to be paired with our input filenames and have corresponding names. We can tackle this with find, xargs, and basename. The first step is to use find to grab the files you want to process, and then use xargs and basename to extract the sample name. basename allows us to remove suffixes through the argument -s:

```
$ find . -name "*.fastq" | xargs basename -s ".fastq"
zmaysA_R1
```

```
zmaysA_R2
zmaysB_R1
zmaysB_R2
zmaysC_R1
zmaysC_R2
```

Then, we want to run the command `fastq_stat --in file.fastq --out ../`
`summaries/file.txt`, but with file replaced with the file's base name. We do this by
piping the sample names we've created earlier to another `xargs` command that runs
`fastq_stat`:

```
$ find . -name "*.fastq" | xargs basename -s ".fastq" | \
    xargs -I{} fastq_stat --in {}.fastq --out ../summaries/{}.txt
```

BSD and GNU xargs

Unfortunately, the behavior of `-I` differs across BSD `xargs` (which
is what OS X uses) and GNU `xargs`. BSD `xargs` will only replace
up to five instances of the string specified by `-I` by default, unless
more are set with `-R`. In general, it's better to work with GNU
`xargs`. If you're on a Mac, you can install GNU Coreutils with
Homebrew. To prevent a clash with your system's `xargs` (the BSD
version), Homebrew prefixes its version with g so the GNU version
of `xargs` would be `gxargs`.

Combining `xargs` with `basename` is a powerful idiom used to apply commands to
many files in a way that keeps track of which output file was created by a particular
input file. While we could accomplish this other ways (e.g., through Bash `for` loops or
custom Python scripts), `xargs` allows for very quick and incremental command
building. However, as we'll see in the next section, `xargs` has another very large
advantage over `for` loops: it allows parallelization over a prespecified number of processes. Overall, it may take some practice to get these `xargs` tricks under your fingers,
but they will serve you well for decades of bioinformatics work.

xargs and Parallelization

An incredibly powerful feature of `xargs` is that it can launch *a limited number* of processes in parallel. I emphasize limited number, because this is one of `xargs`'s strengths
over Bash's `for` loops. We could launch numerous multiple background processes
with Bash `for` loops, which on a system with multiple processors would run in parallel (depending on other running tasks):

```
for filename in *.fastq; do
  program "$filename" &
done
```

But this launches however many background processes there are files in *.fastq! This is certainly not good computing etiquette on a shared server, and even if you were the only user of this server, this might lead to bottlenecks as all processes start reading from and writing to the disk. Consequently, when running multiple process in parallel, we want to explicitly limit how many processes run simultaneously. xargs allows us to do this with the option -P <num> where <num> is the number of processes to run simultaneously.

Let's look at a simple example—running our imaginary program fastq_stat in parallel, using at most six processes. We accomplish this by adding -P 6 to our second xargs call (there's no point in parallelizing the basename command, as this will be very fast):

```
$ find . -name "*.fastq" | xargs basename -s ".fastq" | \
    xargs -P 6 -I{} fastq_stat --in {}.fastq --out ../summaries/{}.txt
```

Generally, fastq_stat could be any program or even a shell script that performs many tasks *per sample*. The key point is that we provide all information the program or script needs to run through the sample name, which is what replaces the string {}.

xargs, Pipes, and Redirects

One stumbling block beginners frequently encounter is trying to use pipes and redirects with xargs. This won't work, as the shell process that reads your xargs command will interpret pipes and redirects as what to do with xarg's output, not as part of the command run by xargs. The simplest and cleanest trick to get around this limitation is to create a small Bash script containing the commands to process a single sample, and have xargs run this script in many parallel Bash processes. For example:

```
#!/bin/bash
set -e
set -u
set -o pipefail

sample_name=$(basename -s ".fastq" "$1")

some_program ${sample_name}.fastq | another_program >
    ${sample_name}-results.txt
```

Then, run this with:

```
$ find . -name "*.fastq" | xargs -n 1 -P 4 bash script.sh
```

Where -n 1 forces xargs to process one input argument at a time. This could be easily parallelized by specifying how many processes to run with -P.

Admittedly, the price of some powerful `xargs` workflows is complexity. If you find yourself using `xargs` mostly to parallelize tasks or you're writing complicated `xargs` commands that use `basename`, it may be worthwhile to learn GNU Parallel (*http://www.gnu.org/software/parallel/*). GNU Parallel extends and enhances `xargs`'s functionality, and fixes several limitations of `xargs`. For example, GNU parallel can handle redirects in commands, and has a shortcut (`{/.}`) to extract the base filename without `basename`. This allows for very short, powerful commands:

```
$ find . -name "*.fastq" | parallel --max-procs=6 'program {/.} > {/.}-out.txt'
```

GNU Parallel has numerous other options and features. If you find yourself using `xargs` frequently for complicated workflows, I'd recommend learning more about GNU Parallel. The GNU Parallel website has numerous examples and a detailed tutorial (*http://www.gnu.org/software/parallel*).

Make and Makefiles: Another Option for Pipelines

Although this chapter has predominantly focused on building pipelines from Bash, and using `find` and `xargs` to apply commands to certain files, I can't neglect to quickly introduce another very powerful tool used to create bioinformatics pipelines. This tool is Make, which interprets *makefiles* (written in their own makefile language). Make was intended to compile software, which is a complex process as each step that compiles a file needs to ensure every dependency is already compiled or available. Like SQL (which we cover in Chapter 13), the makefile language is *declarative*—unlike Bash scripts, makefiles don't run from top to bottom. Rather, makefiles are constructed as a set of *rules*, where each rule has three parts: the *target*, the *prerequisites*, and the *recipe*. Each recipe is a set of commands used to build a target, which is a file. The prerequisites specify which files the recipe needs to build the target file (the dependencies). The amazing ingenuity of Make is that the program figures out how to use all rules to build files for you from the prerequisites and targets. Let's look at a simple example—we want to write a simple pipeline that downloads a file from the Internet and creates a summary of it:

```
FASTA_FILE_LINK=http://bit.ly/egfr_flank ❶

.PHONY: all clean ❷

all: egfr_comp.txt ❸

egfr_flank.fa: ❹
        curl -L $(FASTA_FILE_LINK) > egfr_flank.fa

egfr_comp.txt: egfr_flank.fa ❺
        seqtk comp egfr_flank.fa > egfr_comp.txt
```

```
    clean: ❻
            rm -f egfr_comp.txt egfr_flank.fa
```

❶ We define a variable in a makefile much like we do in a Bash script. We keep this link to this FASTA file at the top so it is noticeable and can be adjusted easily.

❷ The targets all and clean in this makefile aren't the names of files, but rather are just names of targets we can refer to. We indicate these targets aren't files by specifying them as prerequisites in the special target .PHONY.

❸ all is the conventional name of the target used to build all files this makefile is meant to build. Here, the end goal of this simple example makefile is to download a FASTA file from the Internet and run seqtk comp on it, returning the sequence composition of this FASTA file. The final file we're writing sequence composition to is *egfr_comp.txt*, so this is the prerequisite for the all target.

❹ This rule creates the file *egfr_flank.fa*. There are no prerequisites in this rule because there are no local files needed for the creation of *egfr_flank.fa* (as we're downloading this file). Our recipe uses curl to download the link stored in the variable FASTA_FILE_LINK. Since this is a shortened link, we use curl's -L option to follow redirects. Finally, note that to reference a variable's value in a makefile, we use the syntax $(VARIABLE).

❺ This rule creates the file containing the sequencing composition data, *egfr_comp.txt*. Because we need the FASTA file *egfr_flank.fa* to create this file, we specify *egfr_flank.fa* as a prerequisite. The recipe runs seqtk comp on the prerequisite, and redirects the output to the target file, *egfr_comp.txt*.

❻ Finally, it's common to have a target called clean, which contains a recipe for cleaning up all files this makefile produces. This allows us to run make clean and return the directory to the state before the makefile was run.

We run makefiles using the command make. For the preceding makefile, we'd run it using make all, where the all argument specifies that make should start with the all target. Then, the program make will first search for a file named *Makefile* in the current directory, load it, and start at the target all. This would look like the following:

```
$ make all
curl -L http://bit.ly/egfr_flank > egfr_flank.fa
  % Total    % Received % Xferd  Average Speed   Time    Time     Time  Current
                                 Dload  Upload   Total   Spent    Left  Speed
100   190  100   190    0     0    566      0 --:--:-- --:--:-- --:--:--   567
100  1215  100  1215    0     0    529      0  0:00:02  0:00:02 --:--:--   713
seqtk comp egfr_flank.fa > egfr_comp.txt
```

An especially powerful feature of make is that it only generates targets when they don't exist or when their prerequisites have been modified. This is very powerful: if you have a long and complex makefile, and you modified one file, *make will only rerun the recipes for the targets that depend on this modified file* (assuming you fully specified dependencies). Note what happens if we run make all again:

```
$ make all
make: Nothing to be done for `all'.
```

Because all targets have been created and no input files have been modified, there's nothing to do. Now, look what happens if we use touch to change the modification time of the *egfr_flank.fa* file:

```
$ touch egfr_flank.fa
$ make all
seqtk comp egfr_flank.fa > egfr_comp.txt
```

Because *egfr_flank.fa* is a prerequisite to create the *egfr_comp.txt* file, make reruns this rule to update *egfr_comp.txt* using the newest version of *egfr_flank.txt*.

Finally, we can remove all files created with our clean target:

```
$ make clean
rm -f egfr_comp.txt egfr_flank.fa
```

We're just scratching the surface of Make's capabilities in this example; a full tutorial of this language is outside the scope of this book. Unfortunately, like Bash, Make's syntax can be exceedingly cryptic and complicated for some more advanced tasks. Additionally, because makefiles are written in a declarative way (and executed in a nonlinear fashion), debugging makefiles can be exceptionally tricky. Still, Make is a useful tool that you should be aware of in case you need an option for simple tasks and workflows. For more information, see the GNU Make documentation (*http://bit.ly/gnu-make*).

## CHAPTER 13

Out-of-Memory Approaches:
Tabix and SQLite

In this chapter, we'll look at *out-of-memory approaches*—computational strategies built around storing and working with data kept out of memory on the disk. Reading data from a disk is much, much slower than working with data in memory (see "The Almighty Unix Pipe: Speed and Beauty in One" on page 45), but in many cases this is the approach we have to take when in-memory (e.g., loading the entire dataset into R) or streaming approaches (e.g., using Unix pipes, as we did in Chapter 7) aren't appropriate. Specifically, we'll look at two tools to work with data out of memory: Tabix and SQLite databases.

Fast Access to Indexed Tab-Delimited Files with BGZF and Tabix

BGZF and Tabix solve a really important problem in genomics: we often need fast read-only random access to data linked to a genomic location or range. For the scale of data we encounter in genomics, retrieving this type of data is not trivial for a few reasons. First, the data may not fit entirely in memory, requiring an approach where data is kept out of memory (in other words, on a slow disk). Second, even powerful relational database systems can be sluggish when querying out millions of entries that overlap a specific region—an incredibly common operation in genomics. The tools we'll see in this section are specially designed to get around these limitations, allowing fast random-access of tab-delimited genome position data.

In chapter on alignment, we saw how sorted and indexed BAM files allow us to quickly access alignments from within a particular region. The technology that allows fast random access of alignments within a region is based on an ingenious compression format called *BGZF* (Blocked GNU Zip Format), which uses the GNU Zip (gzip) compression format we first saw in "gzip" on page 119. However, while gzip compresses the *entire* file, BGZF files are compressed in *blocks*. These blocks provide BGZF files with a useful feature that gzip-compressed files don't have: we can jump to and decompress a specific block without having to decompress the entire file. Block compression combined with file indexing is what enables fast random access of alignments from large BAM files with samtools view. In this section, we'll utilize BGZF compression and a command-line tool called Tabix to provide fast random access to a variety of tab-delimited genomic formats, including GFF, BED, and VCF.

To use Tabix to quickly retrieve rows that overlap a query region, we first must prepare our file containing the genomic regions. We can prepare a file using the following steps:

1. Sort the file by chromosome and start position, using Unix sort.

2. Use the bgzip command-line tool to compress the file using BGZF.

3. Index the file with tabix, specifying which format the data is in (or by manually specifying which columns contain the chromosome, start, and end positions).

Before getting started, check that you have the bgzip and tabix programs installed. These are both included with Samtools, and should already be installed if you've worked through Chapter 11. We'll use both bgzip and tabix throughout this section.

Compressing Files for Tabix with Bgzip

Let's use the gzipped *Mus_musculus.GRCm38.75.gtf.gz* file from this chapter's directory in the GitHub repository in our Tabix examples. First, note that this file is compressed using gzip, *not* BGZF. To prepare this file, we first need to unzip it, sort by chromosome and start position, and then compress it with bgzip. We can transform this gzipped GTF file into a sorted BGZF-compressed file using a single line built from piped commands.

Unfortunately, we have to deal with one minor complication before calling sort—this GTF file has a metadata header at the top of the file:

```
$ gzcat Mus_musculus.GRCm38.75.gtf.gz | head -n5
#!genome-build GRCm38.p2
#!genome-version GRCm38
#!genome-date 2012-01
#!genome-build-accession NCBI:GCA_000001635.4
#!genebuild-last-updated 2013-09
```

We'll get around this using the subshell trick we learned in "Subshells" on page 169, substituting the gzip step with bgzip:

```
$ (zgrep "^#" Mus_musculus.GRCm38.75.gtf.gz; \
  zgrep -v "^#" Mus_musculus.GRCm38.75.gtf.gz | \
  sort -k1,1 -k4,4n) | bgzip > Mus_musculus.GRCm38.75.gtf.bgz
```

Remember, subshells are interpreted in a separate shell. All standard output produced from each sequential command is sent to bgzip, in the order it's produced. In the preceding example, this has the effect of outputting the metadata header (lines that start with #), and then outputting all nonheader lines sorted by the first and fourth columns to bgzip. The end result is a bzip-compressed, sorted GTF file we can now index with tabix. Subshells are indeed a bit tricky—if you forget the specifics of how to use them to bgzip files with headers, the tabix man page (man tabix) has an example.

Indexing Files with Tabix

Once our tab-delimited data file (with data linked to genomic ranges) is compressed with BGZF, we can use the tabix command-line tool to index it. Indexing files with tabix is simple for files in standard bioinformatics formats—tabix has preset options for GTF/GFF, BED, SAM, VCF, and PSL (a tab-delimited format usually generated from BLAT). We can index a file in these formats using the -p argument. So, we index a GTF file that's been compressed with bgzip (remember, files must be sorted and compressed with BGZF!) with:

```
$ tabix -p gff Mus_musculus.GRCm38.75.gtf.bgz
```

Note that Tabix created a new index file, ending with the suffix *.tbi*:

```
$ ls *tbi
Mus_musculus.GRCm38.75.gtf.bgz.tbi
```

Tabix will also work with custom tab-delimited formats—we just need to specify the columns that contain the chromosome, start, and end positions. Run tabix without any arguments to see its help page for more information (but in general, it's better to stick to an existing bioinformatics format whenever you can).

Using Tabix

Once our tab-delimited file is indexed with Tabix, we can make fast random queries with the same tabix command we used to index the file. For example, to access features in *Mus_musculus.GRCm38.75.gtf.bgz* on chromosome 16, from 23,146,536 to 23,158,028, we use:

```
$ tabix Mus_musculus.GRCm38.75.gtf.bgz 16:23146536-23158028 | head -n3
16   protein_coding   UTR      23146536   23146641   [...]
16   protein_coding   exon     23146536   23146641   [...]
16   protein_coding   gene     23146536   23158028   [...]
```

From here, we could redirect these results to a file or use them directly in a stream. For example, we might want to pipe these results to awk and extract rows with the feature column "exon" in this region:

```
$ tabix Mus_musculus.GRCm38.75.gtf.bgz 16:23146536-23158028  | \
awk '$3 ~ /exon/ {print}'
16      protein_coding  exon    23146536        23146641 [...]
16      protein_coding  exon    23146646        23146734 [...]
16      protein_coding  exon    23155217        23155447 [...]
16      protein_coding  exon    23155217        23155447 [...]
16      protein_coding  exon    23157074        23157292 [...]
16      protein_coding  exon    23157074        23158028 [...]
```

A nice feature of Tabix is that it works across an HTTP or FTP server. Once you've sorted, bgzipped, and indexed a file with tabix, you can host the file on a shared server so others in your group can work with it remotely. Sharing files this way is outside the scope of this book, but it's worth noting Tabix supports this possibility.

Introducing Relational Databases Through SQLite

Many standard bioinformatics data formats (GTF/GFF, BED, VCF/BCF, SAM/BAM) we've encountered so far store tabular data in single *flat file* formats. Flat file formats don't have any internal hierarchy or structured relationship with other tables. While we're able to join tables using Unix tools like join, and R's match() and merge() functions, the flat files themselves do not encode any relationships between tables. Flat file formats are widely used in bioinformatics because they're simple, flexible, and portable, but occasionally we do need to store and manipulate data that is best represented in many related tables—this is where *relational databases* are useful.

Unlike flat files, relational databases can contain multiple tables. Relational databases also support methods that use relationships between tables to join and extract specific records using specialized queries. The most common language for querying relational databases is *SQL* (Structured Query Language), which we'll use throughout the remainder of this chapter.

In this section, we'll learn about a *relational database management system* (the software that implements a relational database, also known as RDBMS) called *SQLite*. SQLite is an entirely self-contained database system that runs as a single process on your machine. We'll use SQLite in this section because it doesn't require any setup—you can create a database and start making queries with minimal time spent on configuring and administrating your database. In contrast, other database systems like MySQL and PostgreSQL require extensive configuration just to get them up and running. While SQLite is not as powerful as these larger database systems, it works surprisingly well for databases in the gigabytes range. In cases when we do need a relational database management system in bioinformatics (less often than you'd think, as we'll see in the next section), SQLite performs well. I like to describe SQLite as having the highest ratio of power and usability to setup cost of any relational database system.

When to Use Relational Databases in Bioinformatics

As it turns out, you'll need to directly interact with relational databases less often than you might think, for two reasons. First, many large bioinformatics databases provide user-friendly *application programming interfaces* (known more commonly by their acronym, *API*) for accessing data. It's often more convenient to access data through an API than by interacting with databases directly. This is because APIs allow you to interact with higher-level applications so you don't have to worry about the gritty details of how the database is structured to get information out. For example, both Ensembl (*http://ensembl.org*) and Entrez Global Query (*http://www.ncbi.nlm.nih.gov/ gquery/*) have APIs that simplify accessing data from these resources.

Second, for many bioinformatics tasks, using a relational database often isn't the best solution. Relational databases are designed for storing and managing multiple records in many tables, where users will often need to add new records, update existing records, and extract specific records using queries. To get an idea of when relational databases are a suitable choice, let's compare working with two different types of data: a curated set of gene models and a set of alignment data from a sequencing experiment.

Adding and Merging Data

Suppose we need to merge a set of new gene models with an existing database of gene models. Simply concatenating the new gene models to the database of existing gene models isn't appropriate, as we could end up with duplicate exons, gene identifiers, gene names, and the like. These duplicate entries would lead to cluttered queries in the future (e.g., fetching all exons from a certain gene would return duplicate exons). With datasets like these, we need a structured way to query existing records, and only add gene models that aren't already in the dataset. Relational databases provide data insertion methods that simplify keeping consistent relationships among data.

In contrast, when we merge two alignment files, we don't have to maintain any consistency or structure—each alignment is essentially an observation in an experiment. To merge two alignment files, we can simply concatenate two alignment files together (though with proper tools like `samtools merge` and attention to issues like the `RG` dictionary in the header). Additionally, we don't have to pay attention to details like duplicates during the merge step. Duplicate alignments created by technical steps like PCR are best removed using specialized programs designed to remove duplicates. Overall, the nature of alignment data obviates the need for advanced data merging operations.

Updating Data

Suppose now that a collaborator has run a new version of gene finding software. After some data processing, you produce a list of gene models that differ from your original gene models. After some careful checking, it's decided that the new gene models are superior, and you need to update all original gene models with the new models' coordinates. As with merging data, these types of update operations are much simpler to do using a relational database system. SQL provides an expressive syntax for updating existing data. Additionally, relational databases simplify updating relationships between data. If an updated gene model includes two new exons, these can be added to a relational database without modifying additional data. Relational databases also allow for different versions of databases and tables, which allows you to safely archive past gene models (good for reproducibility!).

In contrast, consider that any updates we make to alignment data:

- Are never made in place (changing the original dataset)
- Usually affect *all* alignments, not a specific subset
- Do not require sophisticated queries

As a result, any updates we need to make to alignment data are made with specialized tools that usually operate on streams of data (which allows them to be included in processing pipelines). Using tools designed specifically to update certain alignment attributes will always be faster than trying to store alignments in a database and update all entries.

Querying data

Querying information from a set of gene models can become quite involved. For example, imagine you had to write a script to find the first exon for all transcripts for all genes in a given set of gene identifiers. This wouldn't be too difficult, but writing separate scripts for additional queries would be tedious. If you were working with a large set of gene models, searching each gene model for a matching identifier would be unnecessarily slow. Again, relational databases streamline these types of queries. First, SQL acts as a language you can use to specify *any* type of query. Second, unlike Python and R, SQL is a *declarative language*, meaning you state what you want, not how to get it (the RDBMS implements this). This means you don't have to implement the specific code that retrieves gene models matching your query. Last, RDBMS allow you to index certain columns, which can substantially speed up querying data.

On the other hand, the queries made to alignment data are often much simpler. The most common query is to extract alignments from a specific region, and as we saw in Chapter 11, this can be easily accomplished with `samtools view`. Querying alignments within a region in a position-sorted and indexed BAM file is much faster than loading all alignments into a database and extracting alignments with SQL. Additionally, sorted BAM files are much more space efficient and other than a small index file, there's no additional overhead in indexing alignments for fast retrieval. For other types of queries, we can use algorithms that stream through an entire BAM file.

This comparison illustrates the types of applications where a relational database would be most suitable. Overall, databases are not appropriate for raw experimental data, which rarely needs to be merged or modified in a way not possible with a streaming algorithm. For these activities, specialized bioinformatics formats (such as SAM/BAM, VCF/VCF) and tools to work with these formats will be more computationally efficient. Relational databases are better suited for smaller, refined datasets where relationships between records are easily represented and can be utilized while querying out data.

Installing SQLite

Unlike MySQL and PostgreSQL, SQLite doesn't require separate server and client programs. You can install on OS X with Homebrew, using `brew install sqlite`, or on an Ubuntu machine with `apt-get` using `apt-get install sqlite3`. The SQLite website (*http://www.sqlite.org/*) also has source code you can download, compile, and install yourself (but it's preferable to just use your ports/packages manager).

Exploring SQLite Databases with the Command-Line Interface

We'll learn the basics of SQL through SQLite, focusing on how to build powerful queries and use joins that take advantage of the relational structure of databases. We'll start off with the *gwascat.db* example database included in this chapter's directory in the book's GitHub repository. This is a SQLite database containing a table of the National Human Genome Research Institute's catalog of published Genome-Wide Association Studies (GWAS) with some modifications (Welter et al., 2014). In later examples, we'll use another database using a different relational structure of this data. The code and documentation on how these databases were created is included in the same directory in the GitHub repository. The process of structuring, creating, and managing a database is outside the scope of this chapter, but I encourage you to refer to Jay A. Kreibich's book *Using SQLite* and Anthony Molinaro's *SQL Cookbook*.

Each SQLite database is stored in its own file (as explained on the SQLite website (*http://www.sqlite.org/onefile.html*)). We can connect to our example database, the *gwascat.db* file, using SQLite's command-line tool `sqlite3`. This leaves you with an interactive SQLite prompt:

```
$ sqlite3 gwascat.db
SQLite version 3.7.13 2012-07-17 17:46:21
Enter ".help" for instructions
```

```
Enter SQL statements terminated with a ";"
sqlite>
```

We can issue commands to the SQLite program (not the databases it contains) using *dot commands*, which start with a . (and must not begin with whitespace). For example, to list the tables in our SQLite database, we use .tables (additional dot commands are listed in Table 13-1):

```
sqlite> .tables
gwascat
```

Table 13-1. Useful SQLite3 dot commands

Command	Description
.exit, .quit	Exit SQLite
.help	List all SQLite dot commands
.tables	Print a list of all tables
.schema	Print the table schema (the SQL statement used to create a table)
.headers <on,off>	Turn column headers on, off
.import	Import a file into SQLite (see SQLite's documentation for more information)
.indices	Show column indices (see "Indexing" on page 457)
.mode	Set output mode (e.g., csv, column, tabs, etc.; see .help for a full list)
.read <filename>	Execute SQL from file

From .tables, we know this database contains one table: gwascat. Our next step in working with this table is understanding its columns and the recommended type used to store data in that column. In database lingo, we call the structure of a database and its tables (including their columns, preferred types, and relationships between tables) the *schema*. Note that it's vital to understand a database's schema before querying data (and even more so when inserting data). In bioinformatics, we often run into the situation where we need to interact with a remote public SQL database, often with very complex schema. Missing important schema details can lead to incorrectly structured queries and erroneous results.

In SQLite, the .schema dot command prints the original SQL CREATE statements used to create the tables:

```
sqlite> .schema
CREATE TABLE gwascat(
```

```
id integer PRIMARY KEY NOT NULL,
dbdate text,
pubmedid integer,
author text,
date text,
journal text,
link text,
study text,
trait text,
[...]
```

Don't worry too much about the specifics of the CREATE statement—we'll cover this later. More importantly, .schema provides a list of all columns and their preferred type. Before going any further, it's important to note an important feature of SQLite: *columns do not have types, data values have types.* This is a bit like a spreadsheet: while a column of height measurements should all be real-valued numbers (e.g., 63.4, 59.4, etc.), there's no strict rule mandating this is the case. In SQLite, data values are one of five types:

- *text*
- *integer*
- *real*
- *NULL*, used for missing data, or no value
- *BLOB*, which stands for *binary large object*, and stores any type of object as bytes

But again, SQLite's table columns do not enforce all data must be the same type. This makes SQLite unlike other database systems such as MySQL and PostgreSQL, which have strict column types. Instead, each column of a SQLite table has a preferred type called a *type affinity*. When data is inserted into a SQLite table, SQLite will try to coerce this data into the preferred type. Much like R, SQLite follows a set of coercion rules that don't lead to a loss of information. We'll talk more about the different SQLite type affinities when we discuss creating tables in "Creating tables" on page 455.

Orderly Columns

Despite SQLite's column type leniency, it's best to try to keep data stored in the same column all the same type. Keeping orderly columns makes downstream processing much easier because you won't need to worry about whether the data in a column is all different types.

With the skills to list tables with `.tables` and peek at their schema using `.schema`, we're ready to start interacting with data held in tables using the SQL `SELECT` command.

Querying Out Data: The Almighty SELECT Command

Querying out data using the SQL `SELECT` command is one of the most important data skills you can learn. Even if you seldom use relational database systems, the reasoning behind querying data using SQL is applicable to many other problems. Also, while we'll use our example *gwascat.db* database in these examples, the `SELECT` syntax is fairly consistent among other relational database management systems. Consequently, you'll be able to apply these same querying techniques when working with public MySQL biological databases like UCSC's Genome Browser and Ensembl databases.

In its most basic form, the `SELECT` statement simply fetches rows from a table. The most basic `SELECT` query grabs all rows from all columns from a table:

```
sqlite> SELECT * FROM gwascat;
id|dbdate|pubmedid|author|date|journal| [...]
1|08/02/2014|24388013|Ferreira MA|12/30/2013|J Allergy Clin Immunol| [...]
2|08/02/2014|24388013|Ferreira MA|12/30/2013|J Allergy Clin Immunol| [...]
3|08/02/2014|24388013|Ferreira MA|12/30/2013|J Allergy Clin Immunol| [...]
4|08/02/2014|24388013|Ferreira MA|12/30/2013|J Allergy Clin Immunol| [...]
[...]
```

The syntax of this simple `SELECT` statement is `SELECT <columns> FROM <tablename>`. Here, the asterisk (*) denotes that we want to select all columns in the table specified by `FROM` (in this case, `gwascat`). Note that SQL statements *must* end with a semicolon.

Working with the SQLite Command-Line Tool

If you make a mistake entering a SQLite command, it can be frustrating to cancel it and start again. With text entered, `sqlite3` won't obey exiting with Control-*c*;—the `sqlite3` command-line client isn't user-friendly in this regard. The best way around this is to use Control-*u*, which clears all input to the beginning of the line. If your input has already carried over to the next line, the only solution is a hack—you'll need to close any open quotes, enter a syntax error so your command won't run, finish your statement with `;`, and run it so `sqlite3` errors out.

The `sqlite3` command-line tool can also take queries directly from the command line (rather than in the interactive SQLite shell). This is especially useful when writing processing pipelines that need to execute a SQL statement to retrieve data. For example, if we wanted to retrieve all data in the `gwascat` table:

```
$ sqlite3 gwascat.db "SELECT * FROM gwascat" > results.txt
$ head -n 1 results.txt
1|2014-08-02|24388013|Ferreira MA|2013-12-30|J Allergy Clin Immunol| [...]
```

The `sqlite3` program also has options that allow us to change how these results are returned. The option `-separator` can be used to specify how columns are separated (e.g., "," for CSV or "\t" for tab) and `-header` and `-noheader` can be used to display or omit the column headers.

Limiting results with LIMIT

To avoid printing all 17,290 rows of the `gwascat` table, we can add a `LIMIT` clause to this `SELECT` statement. `LIMIT` is an optional clause that limits the number of rows that are fetched. Like the Unix command `head`, `LIMIT` can be used to take a peek at a small subset of the data:

```
sqlite> SELECT * FROM gwascat LIMIT 2;
id|dbdate|pubmedid|author|date|journal| [...]
1|08/02/2014|24388013|Ferreira MA|12/30/2013|J Allergy Clin Immunol| [...]
2|08/02/2014|24388013|Ferreira MA|12/30/2013|J Allergy Clin Immunol| [...]
```

Unlike the rows in a file, the order in which rows of a table are stored on disk is not guaranteed, and could change even if the data stays the same. You could execute the same `SELECT` query, but the order of the resulting rows may be entirely different. This is an important characteristic of relational databases, and it's important that how you process the results from a `SELECT` query doesn't depend on row ordering. Later on, we'll see how to use an `ORDER BY` clause to order by one or more columns.

Selecting columns with SELECT

There are 36 columns in the `gwascat` table, far too many to print on a page without wrapping (which is why I've cropped the previous example). Rather than selecting all columns, we can specify a comma-separated subset we care about:

```
sqlite> SELECT trait, chrom, position, strongest_risk_snp, pvalue
   ...> FROM gwascat LIMIT 5;
trait|chrom|position|strongest_risk_snp|pvalue
Asthma and hay fever|6|32658824|rs9273373|4.0e-14
Asthma and hay fever|4|38798089|rs4833095|5.0e-12
Asthma and hay fever|5|111131801|rs1438673|3.0e-11
Asthma and hay fever|2|102350089|rs10197862|4.0e-11
Asthma and hay fever|17|39966427|rs7212938|4.0e-10
```

When we're selecting only a few columns, SQLite's default list output mode isn't very clear. We can adjust SQLite's settings so results are a bit clearer:

```
sqlite> .header on
sqlite> .mode column
```

Now, displaying the results is a bit easier:

```
sqlite> SELECT trait, chrom, position, strongest_risk_snp, pvalue
   ...> FROM gwascat LIMIT 5;
trait                    chrom       position    strongest_risk_snp  pvalue
-----------------------  ----------  ----------  ------------------  ----------
Asthma and hay fever     6           32658824    rs9273373           4.0e-14
Asthma and hay fever     4           38798089    rs4833095           5.0e-12
Asthma and hay fever     5           111131801   rs1438673           3.0e-11
Asthma and hay fever     2           102350089   rs10197862          4.0e-11
Asthma and hay fever     17          39966427    rs7212938           4.0e-10
```

In cases where there are many columns, SQLite's list mode is usually clearer (you can turn this back on with .mode list).

Ordering rows with ORDER BY

As mentioned earlier, the rows returned by SELECT are not ordered. Often we want to get order by a particular column. For example, we could look at columns related to the study, ordering by author's last name:

```
sqlite> SELECT author, trait, journal
   ...> FROM gwascat ORDER BY author LIMIT 5;
author      trait                                              journal
----------  -------------------------------------------------  ------------------
Aberg K     Antipsychotic-induced QTc interval prolongation    Pharmacogenomics J
Aberg K     Antipsychotic-induced QTc interval prolongation    Pharmacogenomics J
Aberg K     Antipsychotic-induced QTc interval prolongation    Pharmacogenomics J
Aberg K     Response to antipsychotic therapy (extrapyramid    Biol Psychiatry
Aberg K     Response to antipsychotic therapy (extrapyramid    Biol Psychiatry
```

To return results in descending order, add DESC to the ORDER BY clause:

```
sqlite> SELECT author, trait, journal
   ...> FROM gwascat ORDER BY author DESC LIMIT 5;
author              trait         journal
------------------  ------------  ----------
van der Zanden LF   Hypospadias   Nat Genet
van der Valk RJ     Fractional    J Allergy
van der Valk RJ     Fractional    J Allergy
van der Valk RJ     Fractional    J Allergy
van der Valk RJ     Fractional    J Allergy
```

Remember, SQLite's columns do not have strict types. Using ORDER BY on a column that contains a mix of data value types will follow the order: NULL values, integer and real values (sorted numerically), text values, and finally blob values. It's especially important to note that NULL values will always be first when sorting by ascending or descending order. We can see this in action by ordering by ascending *p*-value:

```
sqlite> SELECT trait, chrom, position, strongest_risk_snp, pvalue
   ...> FROM gwascat ORDER BY pvalue LIMIT 5;
trait           chrom       position    strongest_risk_snp   pvalue
-------------   ----------  ----------  -------------------  ----------
Brain imaging                           rs10932886
```

```
Brain imaging                      rs429358
Brain imaging                      rs7610017
Brain imaging                      rs6463843
Brain imaging                      rs2075650
```

In the next section, we'll see how we can filter out rows with NULL values with WHICH clauses.

As discussed in Chapter 8, ordering data to look for suspicious outliers is a great way to look for problems in data. For example, consider what happens when we sort p-values (which as a probability, must mathematically be between 0 and 1, inclusive) in descending order:

```
sqlite> SELECT trait, strongest_risk_snp, pvalue
   ...> FROM gwascat ORDER BY pvalue DESC LIMIT 5;
trait                        strongest_risk_snp  pvalue
--------------------------   ------------------  ----------
Serum protein levels (sST2)  rs206548            90000000.0
Periodontitis (Mean PAL)     rs12050161          4000000.0
Coronary artery disease or   rs964184            2.0e-05
Lipid traits                 rs10158897          9.0e-06
Rheumatoid arthritis         rs8032939           9.0e-06
```

Yikes! These two erroneous p-values likely occurred as a data entry mistake. It's very easy to miss the negative when entering p-values in scientific notation (e.g., 9e-7 versus 9e7). Problems like these once again illustrate why it's essential to not blindly trust your data.

Filtering which rows with WHERE

Up until now, we've been selecting all rows from a database table and adding ORDER BY and LIMIT clauses to sort and limit the results. But the strength of a relational database isn't in selecting all data from a table, it's in using queries to select out specific informative subsets of data. We filter out particular rows using WHERE clauses, which is the heart of making queries with SELECT statements.

Let's look at a simple example—suppose we wanted to find rows where the strongest risk SNP is rs429358:

```
sqlite> SELECT chrom, position, trait, strongest_risk_snp, pvalue
   ...> FROM gwascat WHERE strongest_risk_snp = "rs429358";

chrom   position    trait                          strongest_risk_snp  pvalue
------  ----------  -----------------------------  ------------------  -------
19      44908684    Alzheimer's disease biomarkers  rs429358            5.0e-14
19      44908684    Alzheimer's disease biomarkers  rs429358            1.0e-06
                    Brain imaging                   rs429358
```

SQLite uses = or == to compare values. Note that all entries are case sensitive, so if your values have inconsistent case, you can use the `lower()` function to convert values to lowercase before comparison:

```
sqlite> SELECT chrom, position, trait, strongest_risk_snp, pvalue
   ...> FROM gwascat WHERE lower(strongest_risk_snp) = "rs429358";
chrom   position    trait                            strongest_risk_snp pvalue
------  ----------  -------------------------------  ------------------  -------
19      44908684    Alzheimer's disease biomarkers   rs429358            5.0e-14
19      44908684    Alzheimer's disease biomarkers   rs429358            1.0e-06
                    Brain imaging                    rs429358
```

In our case, this doesn't lead to a difference because the RS identifiers all have "rs" in lowercase. It's best to try to use consistent naming, but if you're not 100% sure that all RS identifiers start with a lowercase "rs," converting everything to the same case is the robust solution.

The equals operator (=) is just one of many useful SQLite comparison operators; see Table 13-2 for a table of some common operators.

Table 13-2. Common operators used in WHERE statements

Operator	Description
=, ==	Equals
!=, <>	Not equals
IS, IS NOT	Identical to = and !=, except that IS and IS NOT can be used to check for NULL values
<, <=	Less than, less than or equal to
>, >=	Greater than, greater than or equal to
x IN (a, b, ...)	Returns whether x is in the list of items (a, b, ...)
x NOT IN (a, b, ...)	Returns whether x is not in the list of items (a, b, ...)
NOT	Logical negation
LIKE	Pattern matching (use % as a wildcard, e.g. like Bash's *)
BETWEEN	x BETWEEN start AND END is a shortcut for x >= start AND x <= end
-	Negative

Operator	Description
+	Positive

Additionally, we can build larger expressions by using AND and OR. For example, to retrieve all rows with *p*-value less than 10 x 10^{-15} and on chromosome 22:

```
sqlite> SELECT chrom, position, trait, strongest_risk_snp, pvalue
   ...> FROM gwascat WHERE chrom = "22" AND pvalue < 10e-15;
chrom       position    trait                strongest_risk_snp  pvalue
----------  ----------  -------------------  ------------------  ----------
22          39351666    Rheumatoid arthritis rs909685            1.0e-16
22          21577779    HDL cholesterol      rs181362            4.0e-18
22          39146287    Multiple myeloma     rs877529            8.0e-16
22          37185445    Graves' disease      rs229527            5.0e-20
22          40480230    Breast cancer        rs6001930           9.0e-19
[...]
```

It's important to note that using colname = NULL will not work, as NULL is not equal to anything by definition; IS needs to be used in this case. This is useful when we combine create SELECT statements that use ORDER BY to order by a column that contains NULLs (which show up at the top). Compare:

```
sqlite> SELECT chrom, position, trait, strongest_risk_snp, pvalue
   ...> FROM gwascat ORDER BY pvalue LIMIT 5;
chrom       position    trait           strongest_risk_snp  pvalue
----------  ----------  -------------   ------------------  ----------
                        Brain imaging   rs10932886
                        Brain imaging   rs429358
                        Brain imaging   rs7610017
                        Brain imaging   rs6463843
```

to using WHERE pvalue IS NOT NULL to eliminate NULL values before ordering:

```
sqlite> SELECT chrom, position, trait, strongest_risk_snp, pvalue
   ...> FROM gwascat WHERE pvalue IS NOT NULL ORDER BY pvalue LIMIT 5;
chrom       position    trait           strongest_risk_snp  pvalue
----------  ----------  -------------   ------------------  ----------
16          56959412    HDL cholesterol rs3764261           0.0
10          122454932   Age-related mac rs10490924          0.0
1           196710325   Age-related mac rs10737680          0.0
4           9942428     Urate levels    rs12498742          0.0
6           43957789    Vascular endoth rs4513773           0.0
```

OR can be used to select rows that satisfy either condition:

```
sqlite> SELECT chrom, position, strongest_risk_snp, pvalue FROM gwascat
   ...> WHERE (chrom = "1" OR chrom = "2" OR chrom = "3")
   ...> AND pvalue < 10e-11 ORDER BY pvalue LIMIT 5;
chrom       position    strongest_risk_snp  pvalue
----------  ----------  ------------------  ----------
1           196710325   rs10737680          0.0
```

```
2            233763993   rs6742078          4.94065645
3            165773492   rs1803274          6.0e-262
1            196690107   rs1061170          1.0e-261
2            73591809    rs13391552         5.0e-252
```

This approach works, but is complex. It's helpful to group complex statements with parentheses, both to improve readability and ensure that SQLite is interpreting your statement correctly.

Rather than listing out all possible values a column can take with OR and =, it's often simpler to use IN (or take the complement with NOT IN):

```
sqlite> SELECT chrom, position, strongest_risk_snp, pvalue FROM gwascat
   ...> WHERE chrom IN ("1", "2", "3") AND pvalue < 10e-11
   ...> ORDER BY pvalue LIMIT 5;
chrom       position    strongest_risk_snp  pvalue
----------  ----------  ------------------  ----------
1           196710325   rs10737680          0.0
2           233763993   rs6742078           4.94065645
3           165773492   rs1803274           6.0e-262
1           196690107   rs1061170           1.0e-261
2           73591809    rs13391552          5.0e-252
```

Finally, the BETWEEN ... AND ... operator is useful for retrieving entries between specific values. x BETWEEN start AND end works like x > = start AND x < = end:

```
sqlite> SELECT chrom, position, strongest_risk_snp, pvalue
   ...> FROM gwascat WHERE chrom = "22"
   ...> AND position BETWEEN 24000000 AND 25000000
   ...> AND pvalue IS NOT NULL ORDER BY pvalue LIMIT 5;
chrom       position    strongest_risk_snp  pvalue
----------  ----------  ------------------  ----------
22          24603137    rs2073398           1.0e-109
22          24594246    rs4820599           7.0e-53
22          24600663    rs5751902           8.0e-20
22          24594246    rs4820599           4.0e-11
22          24186073    rs8141797           2.0e-09
```

However, note that this approach is not as efficient as other methods that use specialized data structures to handle ranges, such as Tabix, BedTools, or the interval trees in GenomicRanges. If you find many of your queries require finding rows that overlap regions and you're running into bottlenecks, using one of these tools might be a better choice. Another option is to implement *binning* scheme, which assigns features to specific bins. To find features within a particular range, one can precalculate which bins these features would fall in and include only these bins in the WHERE clause. The UCSC Genome Browser uses this scheme (originally suggested by Lincoln Stein and Richard Durbin): see Kent et al., (2002) *The Human Genome Browser at UCSC* (*http://bit.ly/hgb-ucsc*) for more information. We'll discuss more efficiency tricks later on when we learn about table indexing.

SQLite Functions

So far, we've just accessed column data exactly as they are in the table using SELECT. But it's also possible to use expressions to create new columns from existing columns. We use SQLite functions and operators to do this, and AS to give each new column a descriptive name. For example, we could create a region string in the format "chr4:165773492" for each row and convert all traits to lowercase:

```
sqlite> SELECT lower(trait) AS trait,
   ...> "chr" || chrom || ":" || position AS region FROM gwascat LIMIT 5;
trait                   region
--------------------    -------------
asthma and hay fever    chr6:32658824
asthma and hay fever    chr4:38798089
asthma and hay fever    chr5:11113180
asthma and hay fever    chr2:10235008
asthma and hay fever    chr17:3996642
```

Here, || is the concatenation operator, which concatenates two strings. As another example, we may want to replace all NULL values with "NA" if we needed to write results to a file that may be fed into R. We can do this using the ifnull() function:

```
sqlite> SELECT ifnull(chrom, "NA") AS chrom, ifnull(position, "NA") AS position,
   ...> strongest_risk_snp, ifnull(pvalue, "NA") AS pvalue FROM gwascat
   ...> WHERE strongest_risk_snp = "rs429358";
chrom       position    strongest_risk_snp  pvalue
----------  ----------  ------------------  ----------
19          44908684    rs429358            5.0e-14
19          44908684    rs429358            1.0e-06
NA          NA          rs429358            NA
```

Later, we'll see a brief example of how we can interact with SQLite databases directly in R (which automatically converts NULLs to R's NAs). Still, if you're interfacing with a SQLite database at a different stage in your work, ifnull() can be useful.

We're just scratching the surface of SQLite's capabilities here; see Table 13-3 for other useful common functions.

Table 13-3. Common SQLite functions

Function	Description
ifnull(x, val)	If x is NULL, return with val, otherwise return x; shorthand for coalesce() with two arguments
min(a, b, c, ...)	Return minimum in a, b, c, ...
max(a, b, c, ...)	Return maximum in a, b, c, ...
abs(x)	Absolute value

Function	Description
coalesce(a, b, c, ...)	Return first non-NULL value in a, b, c, ... or NULL if all values are NULL
length(x)	Returns number of characters in x
lower(x)	Return x in lowercase
upper(x)	Return x in uppercase
replace(x, str, repl)	Return x with all occurrences of str replaced with repl
round(x, digits)	Round x to digits (default 0)
trim(x, chars), ltrim(x, chars), rtrim(x, chars)	Trim off chars (spaces if chars is not specified) from both sides, left side, and right side of x, respectively.
substr(x, start, length)	Extract a substring from x starting from character start and is length characters long

SQLite Aggregate Functions

Another type of SQLite function takes a column retrieved from a query as input and returns a single value. For example, consider counting all values in a column: a count() function would take all values in the column and return a single value (how many non-NULL values there are in the column). SQLite's count() function given the argument * (and without a filtering WHERE clause) will return how many rows are in a database:

```
sqlite> SELECT count(*) FROM gwascat;
count(*)
----------
17290
```

Using count(*) will always count the rows, regardless of whether there are NULLs. In contrast, calling count(colname) where colname is a particular column will return the number of non-NULL values:

```
sqlite> SELECT count(pvalue) FROM gwascat;
count(pvalue)
-------------
17279
```

We can use expressions and AS to print the number of NULL *p*-values, by subtracting the total number non-NULL *p*-values from the total number of rows to return the number of NULL *p*-values:

```
sqlite> SELECT count(*) - count(pvalue) AS number_of_null_pvalues FROM gwascat;
number_of_null_pvalues
----------------------
11
```

Other aggregate functions can be used to find the average, minimum, maximum, or sum of a column (see Table 13-4). Note that all of these aggregating functions work with WHERE clauses, where the aggregate function is only applied to the filtered results. For example, we could find out how many entries have publication dates in 2007 (i.e., after December 31, 2006, but before January 1, 2008):

```
sqlite> select "2007" AS year, count(*) AS number_entries
   ...> from gwascat WHERE date BETWEEN "2007-01-01" AND "2008-01-01";
year        number_entries
----------  --------------
2007        435
```

Table 13-4. Common SQLite aggregate functions

Function	Description
count(x), count(*)	Return the number of non-NULL values in column x and the total number of rows, respectively
avg(x)	Return the average of column x
max(x), min(x)	Return the maximum and minimum values in x, respectively
sum(x), total(x)	Return the sum of column x; if all values of x are NULL, sum(x) will return NULL and total(x) will return 0. sum() will return integers if all data are integers; total() will always return a real value

Note that unlike other relational databases like MySQL and PostgreSQL, SQLite does not have a dedicated date type. Although the date is text, this works because dates in this table are stored in ISO 8601 format (introduced briefly in "Command Substitution" on page 54), which uses leading zeros and is in the form: YYYY-MM-DD. Because this format cleverly arranges the date from largest period (year) to smallest (day), sorting and comparing this date as text is equivalent to comparing the actual dates. This is the benefit of storing dates in a tidy format (see XKCD's ISO 8601 (*http://xkcd.com/1179/*)).

A nice feature of aggregate functions is that they allow prefiltering of duplicated values through the DISTINCT keyword. For example, to count the number of unique non-NULL RS IDs in the strongest_risk_snp column:

```
sqlite> SELECT count(DISTINCT strongest_risk_snp) AS unique_rs FROM gwascat;
unique_rs
----------
13619
```

Grouping rows with GROUP BY

If you're familiar with R (covered in Chapter 8), you've probably noticed that many of SQLite's querying capabilities overlap R's. We extract specific columns from a SQLite table using SELECT a, b FROM tbl;, which is similar to tbl[, c("a", "b")] to access columns a and b in an R dataframe. Likewise, SQL's WHERE clauses filter rows much like R's subsetting. Consider the similarities between SELECT a, b FROM tbl WHERE a < 4 AND b = 1; and tbl[tbl$a < 4 & b == 1, c("a", "b")]. In this section, we'll learn about the SQL GROUP BY statement, which is similar to the split-apply-combine strategy covered in "Working with the Split-Apply-Combine Pattern" on page 239.

As the name suggests, GROUP BY gathers rows into groups by some column's value. Rather importantly, the aggregate functions we saw in the previous section are applied to *each group separately*. In this way, GROUP BY is analogous to using R's split() function to split a dataframe by a column and using lapply() to apply an aggregating function to each split list item. Let's look at an example—we'll count how many associations there are in the GWAS catalog per chromosome:

```
sqlite> SELECT chrom, count(*) FROM gwascat GROUP BY chrom;
chrom       count(*)
----------  ----------
            70
1           1458
10          930
11          988
12          858
13          432
[...]
```

We can order our GROUP BY results from most hits to least with ORDER BY. To tidy up our columns, we'll also use AS to name the count(*) nhits:

```
sqlite> SELECT chrom, count(*) as nhits FROM gwascat GROUP BY chrom
   ...> ORDER BY nhits DESC;
chrom       nhits
----------  ----------
6           1658
1           1458
2           1432
```

```
3           1033
11          988
10          930
[...]
```

As a more interesting example, let's look at the top five most frequent strongest risk SNPs in the table:

```
sqlite> select strongest_risk_snp, count(*) AS count
   ...> FROM gwascat GROUP BY strongest_risk_snp
   ...> ORDER BY count DESC LIMIT 5;
strongest_risk_snp  count
------------------  ----------
rs1260326           36
rs4420638           30
rs1800562           28
rs7903146           27
rs964184            25
```

It's also possible to group by multiple columns. Suppose, for example, that you weren't just interested in the number of associations per strongest risk SNP, but also the number of associations for each allele in these SNPs. In this case, we want to compute count(*) grouping by both SNP and allele:

```
sqlite> select strongest_risk_snp, strongest_risk_allele, count(*) AS count
   ...> FROM gwascat GROUP BY strongest_risk_snp, strongest_risk_allele
   ...> ORDER BY count DESC LIMIT 10;
strongest_risk_snp  strongest_risk_allele  count
------------------  ---------------------  ----------
rs1260326           T                      22
rs2186369           G                      22
rs1800562           A                      20
rs909674            C                      20
rs11710456          G                      19
rs7903146           T                      19
rs4420638           G                      18
rs964184            G                      15
rs11847263          G                      14
rs3184504           T                      12
```

All other aggregate functions similarly work on grouped data. As an example of how we can combine aggregate functions with GROUP BY, let's look at the average log10 *p*-value for all association studies grouped by year. The gwascat table doesn't have a year column, but we can easily extract year from the date column using the substr() function. Let's build up the query incrementally, first ensuring we're extracting year correctly:

```
sqlite> SELECT substr(date, 1, 4) AS year FROM gwascat GROUP BY year;
year
----------
2005
2006
```

```
2007
2008
2009
2010
2011
2012
2013
```

Now that this is working, let's add in a call to avg() to find the average -log10 *p*-values per year (using round() to simplify the output) and count() to get the number of cases in each group:

```
sqlite> SELECT substr(date, 1, 4) AS year,
   ...> round(avg(pvalue_mlog), 4) AS mean_log_pvalue,
   ...> count(pvalue_mlog) AS n
   ...> FROM gwascat GROUP BY year;
year        mean_log_pvalue  n
----------  ---------------  ----------
2005        6.2474           2
2006        7.234            8
2007        11.0973          434
2008        11.5054          971
2009        12.6279          1323
2010        13.0641          2528
2011        13.3437          2349
2012        9.6976           4197
2013        10.3643          5406
```

The primary trend is an increase in -log10 *p*-values over time. This is likely caused by larger study sample sizes; confirming this will take a bit more sleuthing.

It's important to note that filtering with WHERE applies to rows *before* grouping. If you want to filter groups themselves on some condition, you need to use the HAVING clause. For example, maybe we only want to report per-group averages when we have more than a certain number of cases (because our -log10 *p*-value averages may not be reliable with too few cases). You can do that with:

```
sqlite> SELECT substr(date, 1, 4) AS year,
   ...> round(avg(pvalue_mlog), 4) AS mean_log_pvalue,
   ...> count(pvalue_mlog) AS n
   ...> FROM gwascat GROUP BY year
   ...> HAVING count(pvalue_mlog) > 10;
year        mean_log_pvalue  n
----------  ---------------  ----------
2007        11.0973          434
2008        11.5054          971
2009        12.6279          1323
2010        13.0641          2528
2011        13.3437          2349
2012        9.6976           4197
2013        10.3643          5406
```

We've filtered here on the total per-group number of cases, but in general HAVING works with any aggregate functions (e.g., avg(), sum(), max(), etc.).

Subqueries

As a teaser of the power of SQL queries, let's look at a more advanced example that uses aggregation. The structure of the gwascat table is that a single study can have multiple rows, each row corresponding to a different significant association in the study. As study sample sizes grow larger (and have increased statistical power), intuitively it would make sense if later publications find more significant trait associations. Let's build up a query to summarize the data to look for this pattern. The first step is to count how many associations each study has. Because the PubMed ID (column pubmedid) is unique *per* publication, this is an ideal grouping factor (see Example 13-1).

Example 13-1. An example subquery

```
sqlite> SELECT substr(date, 1, 4) AS year, author, pubmedid,
   ...> count(*) AS num_assoc FROM gwascat GROUP BY pubmedid
   ...> LIMIT 5;
year        author       pubmedid    num_assoc
----------  ----------   ----------  ----------
2005        Klein RJ     15761122    1
2005        Maraganore   16252231    1
2006        Arking DE    16648850    1
2006        Fung HC      17052657    3
2006        Dewan A      17053108    1
```

I've included the author and pubmedid columns in output so it's easier to see what's going on here. These results are useful, but what we really want is *another* level of aggregation: we want to find the average number of associations *across* publications, *per year*. In SQL lingo, this means we need to take the preceding results, group by year, and use the aggregating function avg() to calculate the average number of associations per publication, per year. SQLite (and other SQL databases too) have a method for nesting queries this way called *subqueries*. We can simply wrap the preceding query in parentheses, and treat it like a database we're specifying with FROM:

```
sqlite> SELECT year, avg(num_assoc)
   ...> FROM (SELECT substr(date, 1, 4) AS year,
   ...>       author,
   ...>       count(*) AS num_assoc
   ...>       FROM gwascat GROUP BY pubmedid)
   ...> GROUP BY year;
year        avg(num_assoc)
----------  ----------
2005        1.0
2006        1.6
```

```
2007        5.87837837
2008        7.64566929
2009        6.90625
2010        9.21660649
2011        7.49681528
2012        13.4536741
2013        16.6055045
```

Indeed, it looks like there are more associations being found per study in more recent studies. While this subquery looks complex, structurally it's identical to SELECT year, avg(num_assoc) FROM tbl GROUP BY year; where tbl is replaced by the SELECT query Example 13-1. While we're still only scratching the surface of SQL's capabilities, subqueries are good illustration of the advanced queries that can be done entirely within SQL.

Organizing Relational Databases and Joins

The solution to this redundancy is to organize data into multiple tables and use *joins* in queries to tie the data back together. If WHERE statements are the heart of making queries with SELECT statements, joins are the soul.

Organizing relational databases

If you poke around in the gwascat table, you'll notice there's a great deal of redundancy. Each row of the table contains a single trait association from a study. Many studies included in the table find multiple associations, meaning that there's a lot of duplicate information related to the study. Consider a few rows from the study with PubMed ID "24388013":

```
sqlite> SELECT date, pubmedid, author, strongest_risk_snp
   ...> FROM gwascat WHERE pubmedid = "24388013" LIMIT 5;
date        pubmedid    author        strongest_risk_snp
----------  ----------  -----------   ------------------
2013-12-30  24388013    Ferreira MA   rs9273373
2013-12-30  24388013    Ferreira MA   rs4833095
2013-12-30  24388013    Ferreira MA   rs1438673
2013-12-30  24388013    Ferreira MA   rs10197862
2013-12-30  24388013    Ferreira MA   rs7212938
```

I've only selected a few columns to include here, but all rows from this particular study have duplicated author, date, journal, link, study, initial_samplesize, and replication_samplesize column values. This redundancy is not surprising given that this data originates from a spreadsheet, where data is always organized in flat tables. The database community often refers to unnecessary redundancy in database tables as *spreadsheet syndrome*. With large databases, this redundancy can unnecessarily increase required disk space.

Before learning about the different join operations and how they work, it's important to understand the philosophy of organizing data in a relational database so as to avoid spreadsheet syndrome. This philosophy is *database normalization,* and it's an extensive topic in its own right. Database normalization is a hierarchy of increasingly normalized levels called *normal forms.* Though the technical definitions of the normal forms are quite complicated, the big picture is most important when starting to organize your own relational databases. I'll step through a few examples that illustrate how reorganizing data in a table can reduce redundancy and simplify queries. I'll mention the normal forms in passing, but I encourage you to further explore database normalization, especially if you're creating and working with a complex relational database.

First, it's best to organize data such that every column of a row only contains *one* data value. Let's look at an imaginary table called `assocs` that breaks this rule:

```
id  pubmedid  year  journal  trait            strongest_risk_snp
--  --------  ----  -------  -----            ------------------
1   24388013  2013  J Allergy  Asthma, hay fever  rs9273373,rs4833095,rs1438673
2   17554300  2007  Nature   Hypertension     rs2820037
3   17554300  2007  Nature   Crohn's disease  rs6596075
```

From what we've learned so far about SQL queries, you should be able to see why the organization of this table is going to cause headaches. Suppose a researcher asks you if rs4833095 is an identifier in the strongest risk SNP column of this table. Using the SQL querying techniques you learned earlier in this chapter, you might use `SELECT * FROM assoc WHERE strongest_risk_snp = "rs4833095"` to find matching records. However, this would not work, as multiple values are combined in a single column of the row containing rs4833095!

A better way to organize this data is to create a separate row for each data value. We can do this by splitting the records with multiple values in a column into multiple rows (in database jargon, this is putting the table in first normal form):

```
id  pubmedid  year  journal  trait            strongest_risk_snp
--  --------  ----  -------  -----            ------------------
1   24388013  2013  J Allergy  Asthma, hay fever  rs9273373
2   24388013  2013  J Allergy  Asthma, hay fever  rs4833095
3   24388013  2013  J Allergy  Asthma, hay fever  rs1438673
4   17554300  2007  Nature   Hypertension     rs2820037
5   17554300  2007  Nature   Crohn's disease  rs6596075
```

You might notice something about this table now though: there's now duplication in `journal`, `year`, and `pubmedid` column values. This duplication is avoidable, and arises because the `year` and `journal` columns are directly dependent on the `pubmedid` column (and no other columns).

A better way of organizing this data is to split the table into two tables (one for association results, and one for studies):

```
assocs:

id  pubmedid  trait              strongest_risk_snp
--  --------  -----              ------------------
1   24388013  Asthma, hay fever  rs9273373
2   24388013  Asthma, hay fever  rs4833095
3   24388013  Asthma, hay fever  rs1438673
4   17554300  Hypertension       rs2820037
5   17554300  Crohn's disease    rs6596075


studies:

pubmedid  year  journal
--------  ----  -------
24388013  2013  J Allergy
17554300  2007  Nature
```

Organizing data this way considerably reduces redundancy (in database jargon, this scheme is related to the second and third normal forms). Now, our link between these two different tables are the pubmedid identifiers. Because pubmedid is a primary key that uniquely identifies each study, we can use these keys to reference studies from the assocs table. The pubmedid of the assocs table is called a *foreign key*, as it uniquely identifies a record in a different table, the studies table.

While we used pubmedid as a foreign key in this example, in practice it's also common to give records in the studies column an arbitrary primary key (like id in assocs) and store pubmedid as an additional column. This would look like:

```
assocs:

id  study_id  trait              strongest_risk_snp
--  --------  -----              ------------------
1   1         Asthma, hay fever  rs9273373
2   1         Asthma, hay fever  rs4833095
3   1         Asthma, hay fever  rs1438673
4   2         Hypertension       rs2820037
5   2         Crohn's disease    rs6596075


studies:

id  pubmedid  year  journal
--  --------  ----  -------
1   24388013  2013  J Allergy
2   17554300  2007  Nature
```

Now that we've organized our data into two tables, we're ready to use joins to unite these two tables in queries.

Inner joins

Once data is neatly organized in different tables linked with foreign keys, we're ready to use SQLite queries to join results back together. In these examples, we'll use the *joins.db* database from this chapter's GitHub directory. This small database contains the two tables we just organized, `assocs` and `studies`, with a few additional rows to illustrate some intricacies with different types of joins.

Let's load the *joins.db* database:

```
$ sqlite3 joins.db
SQLite version 3.7.13 2012-07-17 17:46:21
Enter ".help" for instructions
Enter SQL statements terminated with a ";"
sqlite> .mode columns
sqlite> .header on
```

Let's take a peek at these example data:

```
sqlite> SELECT * FROM assocs;
id          study_id    trait              strongest_risk_snp
----------  ----------  -----------------  ------------------
1           1           Asthma, hay fever  rs9273373
2           1           Asthma, hay fever  rs4833095
3           1           Asthma, hay fever  rs1438673
4           2           Hypertension       rs2820037
5           2           Crohn's disease    rs6596075
6                       Urate levels       rs12498742

sqlite> SELECT * FROM studies;
id          pubmedid    year        journal
----------  ----------  ----------  ----------
1           24388013    2013        J Allergy
2           17554300    2007        Nature
3           16252231    2005        Am J Hum G
```

By far, the most frequently used type of join used is an *inner join*. We would use an inner join to reunite the association results in the `assocs` table with the studies in the `studies` table. Here's an example of the inner join notation:

```
sqlite> SELECT * FROM assocs INNER JOIN studies ON assocs.study_id = studies.id;
id  study_id  trait              strongest_risk_snp  id  pubmedid    year   journal
---  --------  -----------------  ------------------  ---  --------  -----  ----------
1    1         Asthma, hay fever  rs9273373           1   24388013    2013   J Allergy
2    1         Asthma, hay fever  rs4833095           1   24388013    2013   J Allergy
3    1         Asthma, hay fever  rs1438673           1   24388013    2013   J Allergy
4    2         Hypertension       rs2820037           2   17554300    2007   Nature
5    2         Crohn's disease    rs6596075           2   17554300    2007   Nature
```

Notice three important parts of this join syntax:

- The type of join (INNER JOIN). The table on the left side of the INNER JOIN statement is known as the *left table* and the table on the right side is known as the *right table*.

- The *join predicate*, which follows ON (in this case, it's `assocs.study_id = studies.id`).

- The new notation `table.column` we use to specify a column from a specific table. This is necessary because there may be duplicate column names in either table, so to select a particular column we need a way to identify which table it's in.

Inner joins only select records where the join predicate is true. In this case, this is all rows where the id column in the `studies` table is equal to the `study_id` column of the assocs table. Note that we must use the syntax `studies.id` to specify the column id from the table `studies`. Using id alone will not work, as id is a column in both assocs and studies. This is also the case if you select a subset of columns where some names are shared:

```
sqlite> SELECT studies.id, assocs.id, trait, year FROM assocs
   ...> INNER JOIN studies ON assocs.study_id = studies.id;
id          id          trait             year
----------  ----------  ----------------  ----------
1           1           Asthma, hay fever 2013
1           2           Asthma, hay fever 2013
1           3           Asthma, hay fever 2013
2           4           Hypertension      2007
2           5           Crohn's disease   2007
```

To make the results table even clearer, use AS to rename columns:

```
sqlite> SELECT studies.id AS study_id, assocs.id AS assoc_id, trait, year
   ...> FROM assocs INNER JOIN studies ON assocs.study_id = studies.id;
study_id    assoc_id    trait             year
----------  ----------  ----------------  ----------
1           1           Asthma, hay fever 2013
1           2           Asthma, hay fever 2013
1           3           Asthma, hay fever 2013
2           4           Hypertension      2007
2           5           Crohn's disease   2007
```

In both cases, our join predicate links the `studies` table's primary key, id, with the study_id foreign key in assocs. But note that our inner join only returns five columns, when there are six records in assocs:

```
sqlite> SELECT count(*) FROM assocs INNER JOIN studies
...> ON assocs.study_id = studies.id;
count(*)
----------
5
sqlite> SELECT count(*) FROM assocs;
```

```
count(*)
----------
6
```

What's going on here? There's one record in `assocs` with a `study_id` not in the `studies.id` column (in this case, because it's NULL). We can find such rows using subqueries:

```
sqlite> SELECT * FROM assocs WHERE study_id NOT IN (SELECT id FROM studies);
id          study_id    trait         strongest_risk_snp
----------  ----------  ------------  -------------------
6                       Urate levels  rs12498742
```

Likewise, there is also a record in the `studies` table that is not linked to any association results in the `assocs` table:

```
sqlite> SELECT * FROM studies WHERE id NOT IN (SELECT study_id FROM assocs);
id          pubmedid    year        journal
----------  ----------  ----------  ---------------
3           16252231    2005        Am J Hum Genet
```

When using inner joins, it's crucial to remember that such cases will be ignored! All records where the join predicate `assocs.study_id = studies.id` is not true are excluded from the results of an inner join. In this example, some records from both the left table (`assocs`) and right table (`studies`) are excluded.

Left outer joins

In some circumstances, we want to join in such a way that includes all records from one table, even if the join predicate is false. These types of joins will leave some column values NULL (because there's no corresponding row in the other table) but pair up corresponding records where appropriate. These types of joins are known as *outer joins*. SQLite only supports a type of outer join known as a *left outer join*, which we'll see in this section.

Other Types of Outer Joins

In addition to the left outer joins we'll discuss in this section, there are two other types of outer joins: *right outer joins* and *full outer joins*. Right outer joins are like left outer joins that include all rows from the right table. Right outer joins aren't supported in SQLite, but we can easily emulate their behavior using left outer joins by swapping the left and right tables (we'll see this in this section).

Full outer joins return *all* rows from both the left and right table, uniting records from both tables where the join predicate is true. SQLite doesn't support full outer joins, but it's possible to emulate full outer joins in SQLite (though this is outside of the scope of this book). In cases where I've needed to make use of extensive outer joins, I've found it easiest to switch to PostgreSQL, which supports a variety of different joins.

Left outer joins include all records from the left table (remember, this is the table on the left of the JOIN keyword). Cases where the join predicate are true are still linked in the results. For example, if we wanted to print all association records in the assocs table (regardless of whether they come from a study in the studies table), we would use:

```
sqlite> SELECT * FROM assocs LEFT OUTER JOIN studies
   ...> ON assocs.study_id = studies.id;
id  study_id  trait            [...]risk_snp   id   pubmedid    year   journal
--- --------- ---------------- --------------- ---  ----------  -----  ---------
1   1         Asthma, hay fever rs9273373       1    24388013    2013   J Allergy
2   1         Asthma, hay fever rs4833095       1    24388013    2013   J Allergy
3   1         Asthma, hay fever rs1438673       1    24388013    2013   J Allergy
4   2         Hypertension      rs2820037       2    17554300    2007   Nature
5   2         Crohn's disease   rs6596075       2    17554300    2007   Nature
6             Urate levels      rs12498742
```

As this example shows (note the last record), SQLite left outer joins cover a very important use case: we need all records from the left table, but want to join on data from the right table *whenever it's possible to do so*. In contrast, remember that an inner join will only return records in which the join predicate is true.

While SQLite doesn't support right outer joins, we can emulate their behavior by swapping the left and right columns. For example, suppose rather than fetching all association results and joining on a study where a corresponding study exists, we wanted to fetch all studies, and join on an association where one exists. This can easily be done by making studies the left table, and assocs the right table in a left outer join:

```
sqlite> SELECT * FROM studies LEFT OUTER JOIN assocs
   ...> ON assocs.study_id = studies.id;
id  pubmedid   year   journal      id  study_id   trait                [...]_risk_snp
--  --------   ----   ----------   --  ---------   ------------------   --------------
1   24388013   2013   J Allergy    3   1           Asthma, hay fever    rs1438673
1   24388013   2013   J Allergy    2   1           Asthma, hay fever    rs4833095
1   24388013   2013   J Allergy    1   1           Asthma, hay fever    rs9273373
2   17554300   2007   Nature       5   2           Crohn's disease      rs6596075
2   17554300   2007   Nature       4   2           Hypertension         rs2820037
3   16252231   2005   Am J Hum G
```

Again, note that the assocs's columns id, study_id, trait, and strongest_risk_snp have some values that are NULL, for the single record (with PubMed ID 16252231) without any corresponding association results in assocs.

Finally, while our example joins all use join predicates that are simply connecting assocs's study_id foreign key with studies's primary key, it's important to recognize that join predicates can be quite advanced if necessary. It's easy to build more complex join predicates that join on multiple columns using AND and ON to link statements.

Writing to Databases

Most bioinformatics databases are primarily read-only: we read data more often than we add new data or modify existing data. This is because data we load into a database is usually generated by a pipeline (e.g., gene annotation software) that takes input data and creates results we load in bulk to a database. Consequently, bioinformaticians need to be primarily familiar with bulk loading data into a database rather than making incremental updates and modifications. In this section, we'll learn the basic SQL syntax to create tables and insert records into tables. Then in the next section, we'll see how to load data into SQLite using Python's sqlite3 module.

Because we mainly load data in bulk into bioinformatics databases once, we won't cover tools common SQL modification and deletion operations like DROP, DELETE, UPDATE, and ALTER used to delete and modify tables and rows.

Creating tables

The SQL syntax to create a table is very simple. Using the .schema dot command, let's look at the statement that was used to create the table study in *gwascat2table.db* (execute this after connecting to the *gwascat2table.db* database with *sqlite3 gwascat2table.db*):

```
sqlite> .schema study
CREATE TABLE study(
  id integer primary key,
  dbdate text,
  pubmedid integer,
```

```
    author text,
    date text,
    journal text,
    link text,
    study text,
    trait text,
    initial_samplesize text,
    replication_samplesize text
);
```

This shows the basic syntax of the CREATE TABLE command. The basic format is:

```
CREATE TABLE tablename(
    id integer primary key,
    column1 column1_type,
    column2 column2_type,
    ...
);
```

Each column can have one of the basic SQLite data types (text, integer, numeric, real, or blob) or none (for no type affinity). In the preceding example (and all table definitions used throughout this chapter), you'll notice that the first column always has the definition id integer primary key. *Primary keys* are unique integers that are used to identify a record in a table. In general, *every table you create should have primary key column*, so you can unambiguously and exactly refer to any particular record. The guaranteed uniqueness means that no two rows can have the same primary key—you'll get an error from SQLite if you try to insert a row with a duplicate primary key. Primary keys are one type of table *constraint*; others like UNIQUE, NOT NULL, CHECK and FOREIGN KEY are also useful in some situations but are outside of the scope of this chapter. If you're creating multiple tables to build a complex relational database, I'd recommend getting the basic idea in this section, and then consulting a book dedicated to SQL. Thoughtfulness and planning definitely pay off when it comes to organizing a database.

Let's create a toy SQLite database, and create a new table we'll use in a later example:

```
$ sqlite3 practice.db

sqlite> CREATE TABLE variants(
   ...>    id integer primary key,
   ...>    chrom text,
   ...>    start integer,
   ...>    end integer,
   ...>    strand text,
   ...>    name text);
```

Then, using the dot command .tables shows this table now exists (and you can check its schema with .schema variants if you like):

```
sqlite> .tables
variants
```

Inserting records into tables

Like creating new tables, inserting records into tables is simple. The basic syntax is:

```
INSERT INTO tablename(column1, column2)
VALUES (value1, value2);
```

It's also possible to omit the column names (`column1, column2`), but I wouldn't recommend this—specifying the column names is more explicit and improves readability.

In the previous section, we learned how all tables should absolutely have a primary key. Primary keys aren't something we need to create manually ourselves—we can insert the value `NULL` and SQLite will automatically increment the last primary key (as long as it's an integer primary key) and use that for the row we're inserting. With this in mind, let's insert a record into the `variants` table we created in the previous section:

```
sqlite> INSERT INTO variants(id, chrom, start, end, strand, name)
   ...> VALUES(NULL, "16", 48224287, 48224287, "+", "rs17822931");
```

Then, let's select all columns and rows from this table (which comprises only the record we just created) to ensure this worked as we expect:

```
sqlite> SELECT * FROM variants;
id          chrom       start       end         strand      name
----------  ----------  ----------  ----------  ----------  ----------
1           16          48224287    48224287    +           rs17822931
```

Indexing

While querying records in our example databases takes less than a second, complex queries (especially those involving joins) on large databases can be quite slow. Under the hood, SQLite needs to search every record to see if it matches the query. For large queries, these *full table scans* are time consuming, as potentially gigabytes of data need to be read from your disk. Fortunately, there's a computational trick SQLite can use: it can *index* a column of a table.

A database index works much like the index in a book. Database indexes contain an sorted listing of all entries found in a particular row, alongside which row these entries can be found. Because the indexed column's entries are sorted, it's much faster to search for particular entries (compare searching for a word in an entire book, versus looking it up in an alphabetized index). Then, once the entry is found in the database index, it's easy to find matching records in the table (similar to turning directly to a page where a word occurs in a book index). Indexes do come at a cost: just as book indexes add additional pages to a text, table indexes take up additional disk space. Indexes for very large tables can be quite large (something to be aware of when indexing bioinformatics data).

Creating an index for a particular table's column is quite simple. Here, we create an index on the `strongest_risk_snp` column of the table `assocs` in the *gwascat2table.db* database:

```
sqlite> CREATE INDEX snp_idx ON assocs(strongest_risk_snp);
```

We can use the SQLite dot command `.indices` to look at all table indexes:

```
sqlite> .indices
snp_idx
```

This index will automatically be updated as new records are added to the table. Behind the scenes, SQLite will utilize this indexed column to more efficiently query records. The databases used in this chapter's examples are quite small, and queries on indexed tables are unlikely to be noticeably faster.

Note that SQLite automatically indexes the primary key for each table—you won't have to index this yourself. In addition, SQLite will not index foreign keys for you, and you should generally index foreign keys to improve the performance joins. For example, we would index the foreign key column `study_id` in `assocs` with:

```
sqlite> CREATE INDEX study_id_idx ON assocs(study_id);
sqlite> .indices
snp_idx
study_id_idx
```

If, for some reason, you need to delete an index, you can use `DROP INDEX`:

```
sqlite> DROP INDEX snp_idx;
sqlite> .indices
study_id_idx
```

Dropping Tables and Deleting Databases

Occasionally, you'll create a table incorrectly and need to delete it. We can delete a table `old_table` using:

```
sqlite> DROP TABLE old_table;
```

It's also possible to modify existing tables with `ALTER TABLE`, but this is outside of the scope of this chapter.

Unlike SQLite and PostgreSQL, which support multiple databases, each database in SQLite is a single file. The best way to delete a database is just to delete the entire SQLite database file.

Interacting with SQLite from Python

The SQLite command-line tool `sqlite3` we've used in examples so far is just one way to interact with SQLite databases. The `sqlite3` tool is primarily useful for extracting data from SQLite databases from a script, or quick exploration and interaction with a SQLite database. For more involved tasks such as loading numerous records into a database or executing complex queries as part of data analysis, it's often preferable to interact with a SQLite database through an API. APIs allow you to interact with a database through an interface in your language of choice (as long as there's an API for that language). In this section, we'll take a quick look at Python's excellent API.

Connecting to SQLite databases and creating tables from Python

Python's standard library includes the module `sqlite3` for working with SQLite databases. As with all of Python's standard library modules, the documentation (*http:// bit.ly/sqlite3-doc*) is thorough and clear, and should be your primary reference (using an API is largely about mastering its documentation!).

Let's take a quick look at the basics of using this Python module. Because Python is well suited to processing data in chunks, it's a good language to reach for when bulk-loading data into a SQLite database. In contrast, while it's certainly possible to use R to bulk-load into a SQLite database, it's a bit trickier if the data is too large to load into the database all at once. The following script is a simple example of how we initialize a connection to a SQLite database and execute a SQL statement to create a simple table:

```
import sqlite3

# the filename of this SQLite database
db_filename = "variants.db"

# initialize database connection
conn = sqlite3.connect(db_filename) ❶

c = conn.cursor() ❷

table_def = """\ ❸
CREATE TABLE variants(
  id integer primary key,
  chrom test,
  start integer,
  end integer,
  strand text,
  rsid text);
"""

c.execute(table_def) ❹
```

```
conn.commit()  ❺
conn.close()  ❻
```

❶ First, we use the `connect()` function to establish a connection with the database (provided by the file in `db_filename`; in this case, we're using *variants.db*). con nect() returns an object with class `Connection`, which here we assign to `conn`. `Connection` objects have numerous methods for interacting with a database connection (the exhaustive list is presented in the Python documentation).

❷ When interacting with a SQLite database through Python's API, we use a `Cursor` object. Cursors allow us to retrieve results from queries we've made. In this script, we're just executing a single SQL statement to create a table (and thus there are no results to fetch), but we'll see more on how cursors are used later on.

❸ This block of text is the SQL statement to create a table. For readability, I've formatted it across multiple lines.

❹ To execute SQL statements, we use the `execute()` method of the `Cursor` object `c`. In cases where there are results returned from the database after executing a SQL statement (as with a `SELECT` statement that has matching records), we'll use this cursor object to retrieve them—(we'll see an example of this later).

❺ We need to commit our statement, using the `Connection.commit()` method. This writes the changes to the SQL database.

❻ Finally, we close our connection to the database using `Connection.close()`.

We'll save this script to *create_table.py*. Before proceeding, let's check that this worked. First, we run this script:

```
$ python create_table.py
```

This creates the *variants.db* SQLite database in the current directory. Let's check that this has the correct schema:

```
$ sqlite3 variants.db
SQLite version 3.7.13 2012-07-17 17:46:21
Enter ".help" for instructions
Enter SQL statements terminated with a ";"
sqlite> .tables
variants
sqlite> .schema variants
CREATE TABLE variants(
  id integer primary key,
  chrom test,
  start integer,
  end integer,
```

```
    strand text,
    rsid text);
```

Loading data into a table from Python

Next, we need to write code to load some data into the `variants` table. Included in this chapter's directory in the GitHub repository is a tab-delimited example data file, *variants.txt*:

```
$ cat variants.txt
chr10    114808901      114808902      +        rs12255372
chr9     22125502       22125503       +        rs1333049
chr3     46414946       46414978       +        rs333
chr2     136608645      136608646      -        rs4988235
```

There are a couple important considerations when bulk-loading data into a SQLite table. First, it's important to make sure data loaded into the database is clean, has the correct data type, and missing values are converted to NULL (which are represented in Python by None). Note that while SQLite is very permissive with data types, it's usually best to try to stick with one type per column—as mentioned earlier, this makes downstream work that utilizes data from a table much simpler.

Let's now write a quick script to load data from the file *variants.txt* into our newly created table. Normally, you might fold this code into the same script that created the tables, but to prevent redundancy and simplify discussion, I'll keep them separate. A simple script that reads a tab-delimited file, coerces each column's data to the appropriate type, and inserts these into the `variants` table would look as follows:

```
import sys
import sqlite3
from collections import OrderedDict

# the filename of this SQLite database
db_filename = "variants.db"

# initialize database connection
conn = sqlite3.connect(db_filename)
c = conn.cursor()

## Load Data
# columns (other than id, which is automatically incremented
tbl_cols = OrderedDict([("chrom", str), ("start", int),  ❶
                        ("end", int), ("strand", str),
                        ("rsid", str)])

with open(sys.argv[1]) as input_file:
    for line in input_file:
        # split a tab-delimited line
        values = line.strip().split("\t")

        # pair each value with its column name
```

```
        cols_values = zip(tbl_cols.keys(), values) ❷

        # use the column name to lookup an appropriate function to coerce each
        # value to the appropriate type
        coerced_values = [tbl_cols[col](value) for col, value in cols_values] ❸

        # create an empty list of placeholders
        placeholders = ["?"] * len(tbl_cols) ❹

        # create the query by joining column names and placeholders quotation
        # marks into comma-separated strings
        colnames = ", ".join(tbl_cols.keys())
        placeholders = ", ".join(placeholders)
        query = "INSERT INTO variants(%s) VALUES (%s);"%(colnames, placeholders)

        # execute query
        c.execute(query, coerced_values) ❺

conn.commit() # commit these inserts
conn.close()
```

❶ First, we use an OrderedDict from the collections module to store each column in the table (and our *variants.txt* file) with its appropriate type. Functions str() and int() coerce their input to strings and integers, respectively. We use these functions to coerce data from the input data into its appropriate table type. Additionally, if the data cannot be coerced to the appropriate type, these functions will raise a loud error and stop the program.

❷ Using the zip() function, we take a list of column names and a list of values from a single line of the tab-delimited input file, and combine them into a list of tuples. Pairing this data allows for easier processing in the next step.

❸ Here, we use a Python list comprehension to extract the appropriate coercion function from tbl_cols for each column. We then call this function on the value: this is what tbl_cols[col](value) does. While storing functions in lists, dictionaries, or OrderedDicts may seem foreign at first, this strategy can drastically simplify code. The end result of this list comprehension is a list of values (still in the order as they appear in the input data), coerced to the appropriate type. To reduce the complexity of this example, I've not handled missing values (because our test data does not contain them). See the *code-examples/README.md* file in this chapter's GitHub repository for more information on this.

❹ While we could directly insert our values into a SQL query statement, this is a bad practice and should be avoided. At the very least, inserting values directly into a SQL query statement string can lead to a host of problems with quotations in the data being interpreted as valid SQL quotations. The solution is to use parameter substitution, or parameterization. Python's `sqlite3` module supports two methods to parameterize a SQL statement. In this case, we replace the values in an INSERT statement with ?, and then pass our values directly as a list to the `Cursor.execute()` method. Note too that we ignore the primary key column `id`, as SQLite will automatically increment this for us.

❺ Finally, we use the `Cursor.execute()` method to execute our SQL INSERT statement.

Let's run this script, and verify our data was loaded:

```
$ python load_variants.py variants.txt
```

Then, in SQLite:

```
sqlite> .header on
sqlite> .mode column
sqlite> select * from variants;
id          chrom       start       end         strand      rsid
----------  ----------  ----------  ----------  ----------  ----------
1           chr10       114808901   114808902   +           rs12255372
2           chr9        22125502    22125503    +           rs1333049
3           chr3        46414946    46414978    +           rs333
4           chr2        136608645   136608646   -           rs4988235
```

In this example, we've manually parsed and loaded multiple lines of a tab-delimited file into a table using `Cursor.execute()` and `Connection.commit()`. However, note that we're only committing all of these INSERT statements *at the end* of the for loop. While this wouldn't be a problem with small datasets that fit entirely in memory, for larger datasets (that may not fit entirely in memory) we need to pay attention to these technical details.

One solution is to commit after each INSERT statement, which we could do explicitly by pushing conn.commit() inside the loop. `sqlite3` also supports *autocommit mode* (which can be enabled in `sqlite.connect()`), which automatically commits SQL statements. Unfortunately, using either conn.commit() or autocommit mode leads to each record being committed one at a time, which can be inefficient. To get around this limitation, Python's `sqlite3` Cursor objects have an executemany() method that can take any Python sequence or iterator of values to fill placeholders in a query. Cursor.executemany() is the preferable way to bulk-load large quantities of data into a SQLite database. Loading data with this approach is a bit more advanced (because it relies on using generator functions), but I've included an example script in this chapter's directory in the GitHub repository. Note that it's also possible to interface Python's csv module's reader objects with executemany().

R's RSQLite Package

Like Python, R has a library to interface with SQLite databases: RSQLite. Like sqlite3, RSQLite is very easy to use. The primary difference is that RSQLite connects well with R's dataframes. As a simple demonstration of the interface, let's connect with the *variants.db* database and execute a query:

```
> library(RSQLite)
> sqlite <- dbDriver("SQLite")
> variants_db <- dbConnect(sqlite, "variants.db")

> dbListTables(variants_db)
[1] "variants"
> dbListFields(variants_db, "variants")
[1] "id"     "chrom"  "start"  "end"    "strand" "rsid"

> d <- dbGetQuery(variants_db, "SELECT * FROM variants;")
> head(d)
  id chrom     start       end strand       rsid
1  1 chr10 114808901 114808902      + rs12255372
2  2  chr9  22125502  22125503      +  rs1333049
3  3  chr3  46414946  46414978      +      rs333
4  4  chr2 136608645 136608646      -  rs4988235
> class(d)
[1] "data.frame"
```

dbSendQuery() grabs all results from a SQL statement at once—much like using fetchall() in sqlite3. RSQLite also has methods that support incrementally accessing records resulting from a SQL statement, through dbSendQuery() and fetch(). Overall, the RSQLite's functions are similar to those in Python's sqlite3; for further information, consult the manual.

Finally, let's look at how to work with Python's Cursor objects to retrieve data. We'll step through this interactively in the Python shell. First, let's connect to the database and initialize a Cursor:

```
>>> import sqlite3
>>> conn = sqlite3.connect("variants.db")
>>> c = conn.cursor()
```

Next, let's use Cursor.execute() to execute a SQL statement:

```
>>> statement = """\
... SELECT chrom, start, end FROM variants WHERE rsid IN ('rs12255372', 'rs333')
... """

>>> c.execute(statement)
<sqlite3.Cursor object at 0x10e249f80>
```

Finally, we can fetch data from this query using the `Cursor.fetchone()`, `Cursor.fetchmany()` (which takes an integer argument of how many records to fetch), and `Cursor.fetchall()` methods. The benefit of the `Cursor` object is that it keeps track of which row's you've fetched and which rows you haven't, so you won't accidentally double process a row. Let's look at `Cursor.fetchone()`:

```
>>> c.fetchone()
(u'chr10', 114808901, 114808902)
>>> c.fetchone()
(u'chr3', 46414946, 46414978)
>>> c.fetchone() # nothing left
>>>
```

Dumping Databases

Finally, let's talk about *database dumps*. A database dump is all SQL commands necessary to entirely reproduce a database. Database dumps are useful to back up and duplicate databases. Dumps can also be useful in sharing databases, though in SQLite it's easier to simply share the database file (but this isn't possible with other database engines like MySQL and PostgreSQL). SQLite makes it very easy to dump a database. We can use the `sqlite3` command-line tool:

```
$ sqlite3 variants.db ".dump"
PRAGMA foreign_keys=OFF;
BEGIN TRANSACTION;
CREATE TABLE variants(
  id integer primary key,
  chrom test,
  start integer,
  end integer,
  strand text,
  rsid text);
INSERT INTO "variants" VALUES(1,'chr10',114808901,114808902,'+','rs12255372');
INSERT INTO "variants" VALUES(2,'chr9',22125502,22125503,'+','rs1333049');
INSERT INTO "variants" VALUES(3,'chr3',46414946,46414978,'+','rs333');
INSERT INTO "variants" VALUES(4,'chr2',136608645,136608646,'-','rs4988235');
COMMIT;
```

The `.dump` dot command also takes an optional table name argument, if you wish to dump a single table and not the entire database. We can use database dumps to create databases:

```
$ sqlite3 variants.db ".dump" > dump.sql
$ sqlite3 variants-duplicate.db < dump.sql
```

This series of commands dumps all tables in the *variants.db* database to a SQL file *dump.sql*. Then, this SQL file is loaded into a new empty database *variants-duplicate.db*, creating all tables and inserting all data in the original *variants.db* database.

Conclusion

When I set out to write *Bioinformatics Data Skills*, I initially struggled with how I could present intermediate-level bioinformatics in book format in a way that wouldn't quickly become obsolete in the fast-moving field of bioinformatics. Even in the time it has taken to complete my book, new shiny algorithms, statistical methods, and bioinformatics software have been released and adopted by the bioinformatics community. It's possible (perhaps even likely) that new sequencing technology will again revolutionize biology and bioinformaticians will need to adapt their approaches and tools. How can a print book be a valuable learning asset in this changing environment?

I found the solution to this problem by looking at the tools I use most in my everyday bioinformatics work: Unix, Python, and R. Unix dates back to the early 1970s, making it over 40 years old. The initial release of Python was in 1991 and R was born soon after in 1993, making both of these languages over 20 years old. These tools have all stood the test of time and are the foundation of modern data processing and statistical computing. Bioinformatics and Unix have a nearly inseparable history—the necessary first step of learning bioinformatics skills is to learn Unix. While genomics is rapidly evolving, bioinformaticians continue to reach for same standard tools to tackle new problems and analyze new datasets. Furthermore, Unix, Python, and R are all extensible tools. Nearly every new bioinformatics program is designed to be used on the Unix command line. Working with the newest bioinformatics and statistical methods often boils down to just downloading and installing to new Python and R packages. All other tools in this book—(from `GenomicRanges`, to `sqlite3`, to `sam tools`) are designed to work with our Unix, Python, and R data toolkits. Furthermore, these tools work together, allowing us to mix and combine them to leverage each of their comparative advantages in our daily work—creating a foundational bioinformatics computing environment. While we can't be certain what future sequencing technology will allow us to do, we can be confident that Unix, Python, and R will continue to be the foundation of modern bioinformatics.

However, powerful tools alone don't create a proficient bioinformatician. Using powerful tools to adeptly solve real problems requires an advanced set of skills. With bioinformatics, this set of skills is only fully developed after years of working with real data. *Bioinformatics Data Skills* focuses on a robust and reproducible approach because this is the best context in which to develop your bioinformatics skills. Distrust of one's tools and data and awareness of the numerous pitfalls that can occur during analysis is one of the most important skills to develop. However, you'll only fully develop these skills when you've encountered and been surprised by serious issues in your own research.

Where to Go From Here?

Bioinformatics Data Skills was designed for readers familiar with scripting and a bit of Unix, but less so with how to apply these skills to everyday bioinformatics problems. Throughout the book, we've seen many other tools and learned important skills to solve nearly any problem in bioinformatics (alongside running tools like aligners, assemblers, etc.). Where do you go from here?

First, I'd recommend you *learn more statistics and probability*. It's impossible to emphasize this point too much. After learning the skills in this book and honing them on real-world data, the next step to becoming a masterful bioinformatician is learning statistics and probability. The practical skills are primarily computational; to turn a glut of genomics data into meaningful biological knowledge depends critically on your ability to use statistics. Similarly, understanding probability theory and having the skills to apply it to biological problems grants you with an entirely new way to approach problems. Even applying simple probabilistic approaches in your own work can free you from unpleasant heuristic methods and often work much, much better. Furthermore, many new innovative bioinformatics methods are built on probabilistic models—being familiar with the underlying probabilistic mechanics is crucial to understanding why these methods work and under which conditions they might not.

Second, I would recommend learning some basic topics in computer science—especially *algorithms and data structures*. We continuously need to process large amounts of data in bioinformatics, and it's far too easy to take an approach that's needlessly computationally inefficient. All too often researchers reach for more computational power to parallelize code that could easily run on their desktop machines if it were written more efficiently. Fortunately, it only takes some basic understanding of algorithms and data structures to design efficient software and scripts. I'd also recommend learning about specialized bioinformatics algorithms used in aligners and assemblers; having an in-depth knowledge of these algorithms can help you understand the limitation of bioinformatics software and choose the right tool (and develop your own if you like!).

For more direction into these topics, see this chapter's *README* file on GitHub. I've included my favorite books on these subjects there—and will continue to add others as I discover them. Finally, the last piece of advice I can give you in your path toward becoming a skilled bioinformatician is to *use the source*. In other words, read code, and read lots of code—(especially from programmers who are more skilled than you). Developing programming skills is 90% about experience—writing, debugging, and wrestling with code for years and years. But reading and learning from others' code is like a secret shortcut in this process. While it can be daunting at first to stare at hundreds of thousands of lines of someone else's complex code, you simultaneously strengthen your ability to quickly understand code while learning the tricks a programmer more clever than you uses. Over time, the exercise of reading others' code will reward you with better programming skills.

Glossary

alignment

(1) The process of ordering a sequence such as DNA, protein, or RNA to another sequence that can be used to infer evolutionary relationships (e.g., homology) or sequence origin (e.g., a sequencing read aligned to a particular region of a chromosome). (2) A single aligned pair of sequences.

allele

An alternative form of a gene at a particular *locus*. For example, SNP rs17822931 has two possible alleles (C and T) that determine earwax type. Individuals that have a C allele (e.g., their genotype is either CC or CT) have wet earwax, while individuals with two T alleles (e.g., their genotype is TT) have dry earwax.

AND

AND is a logical operator commonly used in programming languages. x AND y has the value true if and only if x and y are both true. In Python, the logical AND operator is and; in R, it is either && (for AND on an entire vector) or & (for element-wise AND).

anonymous function

A temporary function (used only once) without a name. Anonymous functions are commonly used in R sapply() or lapply() statements. Python also supports anonymous functions through its lambda expressions (e.g., lambda x: 2*x).

application programming interface (API)

An API or application programming interface is a defined interface to some software component, such as a database or file format (e.g., SAM/BAM files). APIs are often modules or libraries that you can load in and utilize in your software projects, allowing you to use a preexisting set of routines that work with low-level details rather than writing your own.

ASCII (pronounced ask-ee)

A character encoding format that encodes for 128 different characters. The acronym stands for American Standard Code for Information Interchange. ASCII characters take up 7 bits and are usually stored in a single byte in memory (one byte is 8 bits). In addition to the common letters and punctuation characters used in English, the ASCII scheme also supports 33 nonprinting characters. See man ascii for reference.

bacterial artificial chromosomes (BACs)

A type of DNA construct used to clone DNA (up to 350,000 base pairs).

base calling

The process of inferring a nucleotide base from its raw signal (e.g., light intensity data) from a sequencer.

batch effects

Undesirable technical effects that can confound a sequencing experiment. For example, if two different sequencing libraries are prepared using different reagents, observed expression differences between the two libraries could be due to batch effects (and is completely confounded by the reagent used). See Leek et al. (2010) for a good review on batch effects.

Binary Call Format (BCF)

The binary version of *Variant Call Format* (*VCF*); see Variant Call Format for more information.

BEDTools

A software suite of command-line tools for manipulating genomic range data in the BED, GFF, VCF, and BAM formats.

Blocked GNU Zip Format (BGZF)

A variant of GNU zip (or gzip) compression that compresses a file in *blocks* rather than in its entirety. Block compression is often used with tools like Tabix that can *seek* to a specific block and uncompress it, rather than requiring the entire file be uncompressed. This allows for fast random access of large compressed files.

binary

(1) The base-2 numeric system that underlies computing, where values take only 0 (true) or 1 (false). (2) A file is said to be in *binary* if it's not human-readable plain text.

BioMart (*http://www.biomart.org*)

A software project that develops tools and databases to organize diverse types of biological data, and simplifies querying information out of these databases. Large genomic databases like Ensembl use BioMart tools for data querying and retrieval.

bit

Short for *binary digit*, a bit is the smallest value represented on computer hardware; a 0 or 1 value.

bit field or bitwise flag

A technique used to store multiple true/false values using bits. Bit fields or bitwise flags are used in SAM and BAM files to store information about alignments such as "is paired" or "is unmapped."

binary large object (BLOB)

A data type used in database systems for storing *binary* data.

brace expansion

A type of shell expansion in Unix that expands out comma-separated values in braces to all combinations. For example, in the shell {dog,cat,rat}-food expands to dog-food cat-food rat-food. Unlike wildcard expansion, there's no guarantee that expanded results have corresponding files.

branch

In Git, a branch is a path of development in a Git repository; alternative paths can be created by creating new branches. By default, Git commits are made to the *master* branch. Alternative branches are often used to separate new features or bug fixes from the main working version of your code on the master branch.

breakpoint

A point in code where execution is temporarily paused so the developer can debug code at that point. A breakpoint can be inserted in R code by calling the function browser() and in Python code using pdb.set_trace() (after importing the Python Debugger with import pdb).

byte

A common unit in computing equal to eight *bits*.

call

A function evaluated with supplied arguments. For example, in Python sum is a function we can call on a list like sum([1, 2, 3, 4]) to sum the list values.

call stack

A data structure used in programming languages to store data used in open function calls. The call stack can be inspected when debugging code in Python using `where` in the Python debugger and in R using the `traceback()` function.

calling out or shelling out

Executing another program from the Unix shell from within a programming language. For example, in R: `system` (*echo "this is a shellout"*). Python's `subprocess` module is used for calling processes from Python.

capturing groups

A grouped regular expression used to capture matching text. For example, capturing groups could be used to capture the chromosome name in a string like `chr3:831991,832018` with a regular expression like `(.*):`. For instance, with Python: `re.match(r'(.*):', "chr3:831991,832018").groups()` (after the `re` module has been imported with `import re`).

character class

A regular expression component that matches any single character specified between square brackets—for example, `[ATCG]` would match any single A, T, C, or G.

checksums

A special summary of digital data used to ensure data integrity and warns against data corruption. Checksums (such as SHA and MD5) are small summaries of data that change when the data changes even the slightest amount.

CIGAR

A format used to store alignment details in SAM/BAM and other bioinformatics formats. Short for the "Compact Idiosyncratic Gapped Alignment Report."

coercion

In the context of programming languages, coercion is the process of changing one data type to another. For example, numeric data like "54.21" can be stored in a string, and needs to be coerced to a floating-point data type before being manipulated numerically. In R, this would be accomplished with `as.numeric("54.21")`, and in Python with `float("54.21")`.

command substitution

A technique used to create anonymous *named pipes*.

commit

In Git, a commit is a snapshot of your project's current state.

constructor

A function used to create and initialize (also known as instantiate) a new object.

contig

Short for contiguous sequence of DNA; often used to describe an assembled DNA sequence (often which isn't a full chromosome). Consecutive contigs can be *scaffolded* together into a longer sequence, which can contain gaps of unknown sequence and length.

coordinate system

How coordinates are represented in data. For example, some genomic range coordinate systems give the first base in a sequence the position 0 while others give it the position 1. Knowing which coordinate system is used is necessary to prevent errors; see Chapter 9 for details.

corner case

Roughly, an unusual case that occurs when data or parameters take extreme or unexpected values. Usually, a corner case refers to cases where a program may function improperly, unexpectedly, or return incorrect results. Also sometimes referred to as an edge case (which is technically different, but very similar).

Coverage

Short for *depth of coverage*; in bioinformatics, this refers to the average depth of sequencing reads across a genome. For example, "10x coverage" means on average, each base pair of a sequence is covered by 10 reads.

CRAM

A compressed format for storing alignments similar to BAM, which compresses data by only storing differences from the alignment reference (which must be stored too to uncompress files).

CRAN or Comprehensive R Archive Network

A collection of R packages for a variety of statistical and data analysis methods. Packages on CRAN can be installed with `install.packages()`.

data integrity

Data integrity is the state of data being free from corruption.

database dumps

An export of the table schemas and data in a database used to back up or duplicate a database and its tables. Database dumps contain the necessary SQL commands to entirely re-create the database and propagate it with its contents.

declarative language

A programming language paradigm where the user writes code that specifies the result user wants, but not how to perform the computation to arrive at that result. For example, SQL is a declarative language because the relational database management system translates a query describing to the necessary computational steps to produce that result.

dependencies

Dependencies are required data, code, or other external information required by a program to function properly. For example, software may have other software programs it requires to run—these are called dependencies.

diff

(1) A Unix tool (`diff`) for calculating the difference between files. (2) A file produced by `diff` or other programs (e.g., `git diff`) that represents the difference between two file versions. Diff files are also sometimes called *patches*, as they contain the necessary information to turn the original file into the modified file (in other words, to "patch" the other file).

Exploratory Data Analaysis or EDA

A statistical technique pioneered by John Tukey that relies on learning about data not through explicit statistical modeling, but rather exploring the data using summary statistics and especially visualization.

environmental variables

In the Unix shell, environmental variables are global variables shared between all shells and applications. These can contain configurations like your terminal pager (`$PAGER`) or your terminal prompt (`$PS1`). Contrast to *local variables*, which are only available to the current shell.

exit status

An integer value returned by a Unix program to indicate whether the program completed successfully or with an error. Exit statuses with the value zero indicate success, while any other value indicates an error.

foreign key

A key used in one database table that uniquely refers to the entries of another table.

fork, forking

In GitHub lingo, forking refers to cloning a GitHub user's repository to your own GitHub account. This allows you to develop your own version of the repository, and then submit *pull requests* to share your changes with the original project.

FTP

An acronym for *File Transfer Protocol*, a protocol used to transfer files over the Internet by connecting an FTP client to an FTP server.

GC content

The percentage of guanine and cytosine (nitrogenous) bases in a DNA sequence.

genomic range

A region on a genomic sequence, specified by chromosome or sequence name, start position, end position, and strand.

Git

A distributed version control system.

globbing

A type of Unix shell expansion that expands a pattern containing the Unix wildcard character * to all existing and matching files.

greedy

In regular expressions, a greedy pattern is one that matches as many characters as possible. For example, the regular expression (.*): applied to string "1:2:3" would not match just "1", but "1:2" because it is greedy.

hangup

A Unix signal (SIGHUP) that indicates a terminal has been closed. Hangups cause most applications to exit, which can be a problem for long-running bioinformatics applications. A terminal multiplexer like tmux or the command-line tool nohup are used to prevent hangups from terminating programs (see Chapter 4).

hard masked

A hard masked sequence contains certain bases (e.g., low-complexity repeat sequences) that are masked out using Ns. In contrast, *soft masked* sequences are masked using lowercase bases (e.g., a, t, c, and g).

HEAD

In Git, HEAD is a pointer to the current branch's latest commit.

HTTP

An acronym for *Hypertext Transfer Protocol*; HTTP is the protocol used to transfer web content across the World Wide Web.

hunks

Sections of a *diff* file that represent discrete blocks of changes.

indexed, indexing

A computational technique used to speed up lookup operations, often used to decrease the time needed for access to various random (e.g., nonsequential) parts of a file. Indexing is used by many bioinformatics tools such as aligners, Samtools, and Tabix.

inner join

The most commonly used database join used to join two tables on one or more keys. The result of an inner join only contains rows with matching non-null keys in both tables; contrast with *left outer join* and *right outer join*.

interval tree

A data structure used to store range data in a way that optimizes finding overlapping ranges. Interval trees are used in some genomic range libraries like Bioconductor's GenomicRanges and the command-line BEDTools Suite.

IUPAC or International Union of Pure and Applied Chemistry

An organization known for standardizing symbols and names in chemistry; known in bioinformatics for the IUPAC nucleotide base codes.

key

In sorting, a *sort key* is the field or column used to sort data on. In databases, a key is a unique identifier for each entry in a table.

LaTeX

A document markup language and typesetting system often used for technical scientific and mathematical documents.

leading zeros

A number with leading zeros is printed with the same number of digits—for example, 00002412 and 00000337. Encoding identifiers with leading zeros is a useful trick, as these identifiers automatically sort lexicographically.

left outer join

A type of database join in which all rows of the left table are included, but only matching rows of the right table are kept. Note that the left and right tables are the tables left and right of the SQL join statement, (e.g., with x LEFT OUTER JOIN y, x is the left table and y is the right table). A *right outer join* is the same as a left outer join with right and left tables switched.

literate programming

A style of programming pioneered by Donald Knuth that intermixes natural language with code. Code in the literate program can be separated out and executed.

local variables

See *environmental variables*.

locus, loci (plural)

The position of a gene.

lossless compression

A data compression scheme where no information is lost in the compression process.

lossy compression

A data compression scheme where data is intentionally lost in the compression scheme to save space.

master branch

See *branch*.

mapping quality

A measure of how likely a read is to actually originate from the position it maps to.

Markdown

A lightweight document markup language that extends plain-text formats. Mark-down documents can be rendered in HTML and many other languages.

metadata

Data or information about other data or a dataset. For example, a reference genome's version, creation data, etc. are all metadata about the reference genome.

named pipes

A type of Unix pipe that's represented as a special type of file on the filesystem. Data can be written to and read from a named pipe as if it were a regular file, but does not write to the disk, instead connecting the reading and writing processes with a stream.

NOT

A logical operator for negation; returns the logical opposite of the value it's called on. For example, NOT true evaluates to false and NOT false evaluates to true.

offset offset

Both FASTQ and file offset (in context of seek).

OR

A logical operator where x OR y is true if either x or y are true, but false if neither x nor y are true.

overplotting

Overplotting occurs when a plot contains a large amount of close data points that crowd the visualization and make understanding and extracting information from the graphic difficult.

pager, terminal pager

(1) A command-line tool used to view and scroll through a text file or text stream of data that's too large to display in a single terminal screen (e.g., the Unix tools less and more). (2) The terminal pager configured with the environmental variable $PAGER.

patch file

See *diff*.

plain text

A text format that does not contain special formatting or file encoding and is human readable.

POSIX Extended Regular Expressions or POSIX ERE

A variety of regular expression used by Unix programs that have support for additional features; contrast to *POSIX Basic Regular Expressions*. These extended feature regular expressions can often be enabled in Unix programs.

POSIX Basic Regular Expressions or POSIX BRE

See *POSIX Extended Regular Expressions*.

primary keys

A unique key used in a database table to identify a particular entry.

public key or private key

In SSH, an SSH public key is a key that's shared across systems to grant login access to the user with the associated private key. Public keys are shareable, but private keys must be protected by the user (see Chapter 2).

relational databases

A type of database used to store relationships among data contained across different tables.

relational database management system

The software system that implements the tools to work with a relational database.

repository

In Git, a repository is a directory containing code or other files that are being managed by Git.

right outer join

See *left outer join*.

R markdown

A variant of the *markdown* format that interweaves R code with markdown text. The R packages `rmarkdown` and `knitr` can be used to render R markdown files.

robust

In software, robust means protected against common data problems and software bugs; the opposite of fragile software, which may break silently and unexpectedly when it encounters certain data or parameters.

S3 object orientation

R's default object-orientation system; contrast to *S4 object orientation*, which is used by Bioconductor and some other R packages.

Sanger sequencing

A method of DNA sequencing commonly used before the advent high-throughput next-generation DNA sequencing methods. Usually Sanger sequencing can achieve read lengths of 700–1000bp.

schema

The organizational specification of a database, which describes how tables are structured, what columns they contain, and these columns' data types, etc.

serialization

In programming, serialization is the process of storing an object on disk so that it can be loaded in later.

shebang

The character combination #! used to indicate which program the shell should run a script with. Shebangs should occur on the first line of a script. For example, `#!/usr/bin/env python` is a commonly used Python shebang.

soft masked

See *hard masked*.

sorting keys

See *key*.

sorting stability

A characteristic of some sorting algorithms where tied entries are sorted such that their original ordering is maintained.

spreadsheet syndrome

A habit of spreadsheets users to utilize as few tables as possible, which can lead to unnecessary data redundancy and inefficient querying. In contrast, normalized databases split data into tables, which reduces redundancy.

SQL or Server Query Language

A declarative query language used to interact with a relational database management system.

SSH or Secure Shell

A protocol used to securely connect to a machine over a network connection and initiate an encrypted shell session through which one work with the remote machine.

standard error

In Unix, the standard error stream is used to relay error or other informational messages to the user.

standard input

In Unix, the standard input (also known as standard in) stream is used to stream data into a program, often from a file or another program's standard output (as with a Unix pipe).

standard output

In Unix, the standard output (also known as standard out) stream is a standard Unix stream used to output program results (and sometimes messages). Standard output can be redirected to a file through the Unix redirect operators > or >>.

unit testing

A method to test software where individual functions, methods, or subroutines are tested as separate units. Unit testing helps protect against bugs and software regression, which is when a newly introduced bug breaks previously functional code.

Variant Call Format or VCF

A plain-text tab-delimited format used to store variant positions and other variant information, and the genotypes of individuals.

version control system or VCS

A system for recording changes to files (usually software code) during their development over time. For example, Git is a version control system.

Bibliography

Raymond, Eric S. "The Cathedral & the Bazaar: Musings on linux and open source by an accidental revolutionary." O'Reilly Media, Inc., 2001.

Li, Heng, Jue Ruan, and Richard Durbin. "Mapping short DNA sequencing reads and calling variants using mapping quality scores." *Genome Research* 18.11 (2008): 1851-1858.

Cock, Peter JA, et al. "The Sanger FASTQ file format for sequences with quality scores, and the Solexa/Illumina FASTQ variants." *Nucleic Acids Research* 38.6 (2010): 1767-1771.

McIlroy, M. Douglas. "A Research UNIX Reader: Annotated Excerpts from the Programmer's Manual" (1971-1986). *http://doc.cat-v.org/unix/unix-reader/reader.pdf*

Quinlan, Aaron R., and Ira M. Hall. "BEDTools: a flexible suite of utilities for comparing genomic features." *Bioinformatics* 26.6 (2010): 841-842.

Quinlan, Aaron R. "BEDTools: The Swiss-Army Tool for Genome Feature Analysis." *Current Protocols in Bioinformatics* (2014): 11-12.

Dale, Ryan K., Brent S. Pedersen, and Aaron R. Quinlan. "Pybedtools: a flexible Python library for manipulating genomic datasets and annotations." *Bioinformatics* 27.24 (2011): 3423-3424.

Garrison E, Marth G. Haplotype-based variant detection from short-read sequencing. arXiv preprint arXiv:1207.3907 [q-bio.GN] 2012

Wickham, Hadley. "ggplot2." *Wiley Interdisciplinary Reviews: Computational Statistics* 3.2 (2011): 180-185.

Wickham, Hadley. "Reshaping data with the reshape package." *Journal of Statistical Software* 21.12 (2007): 1-20.

Trapnell, Cole, Lior Pachter, and Steven L. Salzberg. "TopHat: discovering splice junctions with RNA-Seq." *Bioinformatics* 25.9 (2009): 1105-1111.

Anders, Simon, Paul Theodor Pyl, and Wolfgang Huber. "HTSeq–A Python framework to work with high-throughput sequencing data." *Bioinformatics* (2014): btu638.

Lawrence, Michael, et al. "Software for computing and annotating genomic ranges." *PLoS Computational Biology* 9.8 (2013): e1003118.

Trapnell, Cole, et al. "Differential gene and transcript expression analysis of RNA-seq experiments with TopHat and Cufflinks." *Nature Protocols* 7.3 (2012): 562-578.

Li, Bo, and Colin N. Dewey. "RSEM: accurate transcript quantification from RNA-Seq data with or without a reference genome." BMC Bioinformatics 12.1 (2011): 323.

Riley, Tim, and Adam Goucher. *Beautiful Testing*. O'Reilly Media, Inc., 2009.

Noble, William Stafford. "A quick guide to organizing computational biology projects." *PLoS Computational Biology* 5.7 (2009).

Fonseca, Nuno A., et al. "Tools for mapping high-throughput sequencing data." *Bioinformatics* (2012).

Leek, Jeffrey T., et al. "Tackling the widespread and critical impact of batch effects in high-throughput data." *Nature Reviews Genetics* 11.10 (2010): 733-739.

Wilson, Greg, et al. "Best practices for scientific computing." *PLoS Biology* 12.1 (2014): e1001745.

Boslaugh, Sarah. *Statistics in a Nutshell*. O'Reilly Media, Inc., 2012.

Quinn, Gerald Peter, and Michael J. Keough. *Experimental Design and Data Analysis for Biologists*. Cambridge University Press, 2002.

Kreitman, Martin E., and Montserrat Aguadé. "Excess polymorphism at the Adh locus in Drosophila melanogaster." *Genetics* 114.1 (1986): 93-110.

Baggerly, Keith A., and Kevin R. Coombes. "Deriving chemosensitivity from cell lines: Forensic bioinformatics and reproducible research in high-throughput biology." *The Annals of Applied Statistics* (2009): 1309-1334.

Li, Heng, and Richard Durbin. "Fast and accurate short read alignment with Burrows–Wheeler transform." *Bioinformatics* 25.14 (2009): 1754-1760.

Li, Heng. "Towards Better Understanding of Artifacts in Variant Calling from High-Coverage Samples." arXiv preprint arXiv:1404.0929 (2014).

Nakamura, Kensuke, et al. "Sequence-specific error profile of Illumina sequencers." *Nucleic Acids Research* (2011): gkr344.

Fritz, Markus Hsi-Yang, et al. "Efficient storage of high throughput DNA sequencing data using reference-based compression." *Genome Research* 21.5 (2011): 734-740.
APA

Pages, H. pasillaBamSubset: Subset of BAM files from "Pasilla" experiment. R package version 0.2.0.

Kent, W. James, et al. "The human genome browser at UCSC." *Genome Research* 12.6 (2002): 996-1006.

Welter, Danielle, et al. "The NHGRI GWAS Catalog, a curated resource of SNP-trait associations." *Nucleic Acids Research* 42.D1 (2014): D1001-D1006.

Hernandez, Ryan D., et al. "Classic selective sweeps were rare in recent human evolution." *Science* 331.6019 (2011): 920-924.

Perry, George H., et al. "Comparative RNA sequencing reveals substantial genetic variation in endangered primates." *Genome Research* 22.4 (2012): 602-610.

Aird, Daniel, et al. "Analyzing and minimizing PCR amplification bias in Illumina sequencing libraries." *Genome Biology* 12.2 (2011): R18.

Myers, Simon, et al. "A common sequence motif associated with recombination hot spots and genome instability in humans." *Nature Genetics* 40.9 (2008): 1124-1129.

Myers, Simon, et al. "A fine-scale map of recombination rates and hotspots across the human genome." *Science* 310.5746 (2005): 321-324.

Kent, W. James, et al. "The human genome browser at UCSC." *Genome Research* 12.6 (2002): 996-1006.

Wickham, Hadley. "The split-apply-combine strategy for data analysis." *Journal of Statistical Software* 40.1 (2011): 1-29.

Wickham, Hadley. *Advanced R*. CRC Press, 2014.

Smit, Arian FA, Robert Hubley, and Phil Green. "RepeatMasker Open-3.0." (1996): 1996.

Spencer, Chris CA, et al. "The influence of recombination on human genetic diversity." *PLoS Genetics* 2.9 (2006): e148.

Cleveland, William S. *Visualizing Data*. Hobart Press, 1993.

Popper, Karl. *The Logic of Scientific Discovery*. Routledge, 2005.

Peng, Roger D. "Reproducible research in computational science." *Science*, 334.6060 (2011): 1226.

Siepel, Adam, et al. "Evolutionarily conserved elements in vertebrate, insect, worm, and yeast genomes." *Genome Research* 15.8 (2005): 1034-1050.

1000 Genomes Project Consortium. "An integrated map of genetic variation from 1,092 human genomes." *Nature* 491.7422 (2012): 56-65.

Kreibich, Jay. *Using SQLite*. O'Reilly Media, Inc., 2010.

Molinaro, Anthony. *SQL Cookbook*. O'Reilly Media, Inc., 2005.

Welter, Danielle, et al. "The NHGRI GWAS Catalog, a curated resource of SNP-trait associations." *Nucleic Acids Research* 42.D1 (2014): D1001-D1006.

Index

Symbols

About the Author

Vince Buffalo is currently a first-year graduate student in Graham Coop's lab at UC Davis in the Population Biology Graduate Group. Before starting his PhD in population genetics, Vince worked professionally as a bioinformatician in the Bioinformatics Core at the UC Davis Genome Center and in the Department of Plant Sciences.

An obsessive programmer since he was a young teenager, Vince was drawn to the statistical and computational problems of genomics. He works on open source bioinformatics tools in his work and free time, and enjoys fly fishing and cooking when away from the computer.

Colophon

The animal on the cover of *Bioinformatics Data Skills* is a European hamster (*Cricetus cricetus*). Also known as the Eurasian or black-bellied hamster, this species is much larger and than the dwarf hamsters that are commonly kept as pets. The European hamster is native to the low-lying farmland of Belgium, Russia, Romania, and the countries in between.

As a nocturnal species, European hamsters prefer to spend the day sleeping in one of the rooms of their complex burrow systems. At night, they emerge to forage for seeds, legumes, root vegetables, and insects. Using their elastic cheek pouches, hamsters can carry food back to their burrows and deposit it in special storage chambers. These chambers can grow very large and store up to 65 grams of food.

From October through March, European hamsters hibernate, waking every five to seven days to feed from the storage chambers. Once hibernation is over, the hamsters emerge to breed and raise their young. After a gestation period of only 20 days, litters of three to fifteen babies are born, and will stay with the mother for three weeks.

Although the overall population of European hamsters is healthy, it is considered critically endangered in some countries. In 2011, the Court of Justice in Luxembourg ruled that France needed to adjust its agricultural and urbanization policies to better protect the European hamster, or else face fines of up to $24.6 million.

Many of the animals on O'Reilly covers are endangered; all of them are important to the world. To learn more about how you can help, go to *animals.oreilly.com*.

The cover image is from Lydekker's *Royal Natural History*. The cover fonts are URW Typewriter and Guardian Sans. The text font is Adobe Minion Pro; the heading font is Adobe Myriad Condensed; and the code font is Dalton Maag's Ubuntu Mono.

Get even more for your money.

Join the O'Reilly Community, and register the O'Reilly books you own. It's free, and you'll get:

- $4.99 ebook upgrade offer
- 40% upgrade offer on O'Reilly print books
- Membership discounts on books and events
- Free lifetime updates to ebooks and videos
- Multiple ebook formats, DRM FREE
- Participation in the O'Reilly community
- Newsletters
- Account management
- 100% Satisfaction Guarantee

Signing up is easy:

1. Go to: oreilly.com/go/register
2. Create an O'Reilly login.
3. Provide your address.
4. Register your books.

Note: English-language books only

To order books online:
oreilly.com/store

For questions about products or an order:
orders@oreilly.com

To sign up to get topic-specific email announcements and/or news about upcoming books, conferences, special offers, and new technologies:
elists@oreilly.com

For technical questions about book content:
booktech@oreilly.com

To submit new book proposals to our editors:
proposals@oreilly.com

O'Reilly books are available in multiple DRM-free ebook formats. For more information:
oreilly.com/ebooks

O'REILLY®

CPSIA information can be obtained
at www.ICGtesting.com
Printed in the USA
LVOW03s0808290716

498228LV00007B/22/P